Sustainable Environmental Science and Water/Wastewater Treatment

Sustainable Environmental Science and Water/Wastewater Treatment

Editor

Kaan Yetilmezsoy

MDPI

Basel • Beijing • Wuhan • Barcelona • Belgrade • Novi Sad • Cluj • Manchester

Kaan Yetilmezsoy
Department of
Environmental Engineering
Yildiz Technical University
Istanbul
Turkey

Editorial Office
MDPI AG
Grosspeteranlage 5
4052 Basel, Switzerland

This is a reprint of the Special Issue, published open access by the journal *Sustainability* (ISSN 2071-1050), freely accessible at: www.mdpi.com/journal/sustainability/special_issues/B00C97X97O.

For citation purposes, cite each article independently as indicated on the article page online and using the guide below:

Lastname, A.A.; Lastname, B.B. Article Title. *Journal Name* **Year**, *Volume Number*, Page Range.

ISBN 978-3-7258-1542-5 (Hbk)
ISBN 978-3-7258-1541-8 (PDF)
https://doi.org/10.3390/books978-3-7258-1541-8

Contents

About the Editor

Kaan Yetilmezsoy

Kaan Yetilmezsoy is a professor of the Environmental Engineering Department, Faculty of Civil Engineering, at Yildiz Technical University, Davutpasa Campus, Esenler, Istanbul, Turkey. He received his B.Sc. degrees in both Environmental Engineering (1st rank graduate) and Civil Engineering (graduated as an honor student) from Yildiz Technical University (2nd rank graduate), Istanbul/Turkey, in 2002 and 2004, respectively. He also obtained his M.Sc. and Ph.D. degrees in Environmental Engineering from Yildiz Technical University, Istanbul/Turkey, in 2004 and 2008, respectively. His research interests are mathematical modeling, linear/non-linear regression analysis-based modeling, artificial neural networks, fuzzy logic, adaptive neuro-fuzzy inference systems, soft-computing techniques, experimental design, multi-objective optimization techniques, MATLAB®-based programming, scientific computation for engineers, the design of water and wastewater treatment plants, fluid mechanics and hydraulics, and post-treatment technologies. He has authored more than 200 publications. He also received "Outstanding Young Scientist Award (TÜBA-GEBİP)" in 2018 from the Turkish Academy of Sciences (TÜBA) for his academic achievements.

Preface

This reprint, entitled "Sustainable Environmental Science and Water/Wastewater Treatment", covers ground-breaking approaches to innovative water and wastewater treatment and environmental research that promote sustainability. It aims to highlight and disseminate sustainable, ecologically responsible and solution-oriented economic strategies and cutting-edge technologies to reduce anthropogenic/industrial environmental risks.

Kaan Yetilmezsoy
Editor

sustainability

MDPI

Article

Enhancing Sustainability in Italian Water Supply Pipes through Life Cycle Analysis

Isabela Maria Simion [1,*], Sara Pennellini [2,*], Eric Awere [3], Alessandro Rosatti [2] and Alessandra Bonoli [2]

1 Research Department, Iasi University of Life Sciences "Ion Ionescu de la Brad", Mihail Sadoveanu Alley 3, 700490 Iasi, Romania
2 Department of Civil, Chemical, Environmental and Material Engineering, University of Bologna, Via Terracini 28, 40131 Bologna, Italy; alessandro.rosatti2@unibo.it (A.R.)
3 Department of Civil Engineering, Cape Coast Technical University, Cape Coast P.O. Box DL 50, Ghana
* Correspondence: isabela.simion@iuls.ro (I.M.S.); sara.pennellini2@unibo.it (S.P.)

Abstract: The primary concern regarding the sustainability of the urban water cycle remains the performance of water supply systems. This, in turn, is determined by the functionality and sustainability of the system components, such as the pipe networks, pumps, and other appurtenances, which must be analyzed from an environmental perspective. The aim of the present study is to analyze the sustainability of two different types of water supply pipe materials that are commonly used, polyvinyl chloride and high-density polyethylene, using a comparative Life Cycle Analysis methodology. The functional unit was established in accordance with the water supply system that serves an Italian metropolitan city with a dimension of 9240 km, as one meter of water supply infrastructure, with 40 years as a life span. A cradle-to-gate analysis was conducted, starting from the production phase of the water pipelines to the maintenance phase, excluding the end of life and disposal phases. The chosen methodology was CML, justified by the fact that the results are more understandable and reproducible. Results comparison revealed a higher environmental impact during the production phase, while the maintenance phase had a very low impact. Notably, PVC pipe in comparison with HDPE material had a higher impact, except in two categories of impact: abiotic depletion and photochemical oxidation. The study contributes to the future development of alternative approaches for sustainable and eco-efficient water supply infrastructure designs and materials.

Keywords: environmental impacts; eco-friendly; life cycle analysis (LCA); pipe material; sustainability; water supply system

check for updates

Citation: Simion, I.M.; Pennellini, S.; Awere, E.; Rosatti, A.; Bonoli, A. Enhancing Sustainability in Italian Water Supply Pipes through Life Cycle Analysis. *Sustainability* **2024**, *16*, 2685. https://doi.org/10.3390/su16072685

Academic Editor: Kaan Yetilmezsoy

Received: 16 February 2024
Revised: 19 March 2024
Accepted: 21 March 2024
Published: 25 March 2024

1. Introduction

As the Brundtland Commission concluded that the fundamental concepts in policy generation are "sustainability" and "sustainable development", the new common goal has become "development that meets the needs of the present without compromising the ability of future generations to meet their own needs" [1,2].

In this context, water sustainability has become one of the most discussed topics worldwide in the last decade. Sectors involving water extraction from aquifers and water supply and sanitation (WSS) are facing increasing pressures, especially due to the impacts generated by the materials involved in terms of climate change, environmental pollution, ozone depletion, marine ecotoxicity, and waste accumulation [3–6]. When discussing water supply systems, the risks related to managing the system are primarily examined from the perspective of water consumers [7]. Water network operators are obligated to uphold services at an elevated operational standard, ensuring full adherence to safety and availability regulations concerning the cleanliness, purity, and taste of water. Additionally, they strive to promote environmental sustainability within waterworks infrastructure [8]. Specifically, there is a growing emphasis on the utilization of advanced technologies and analytical tools as indispensable tools for sustainability. These tools enhance our comprehension, facilitate efficient management, and promote judicious use of water resources [9].

After consulting studies on pipe utility, Fagan et al. [10] concluded that pipes have multiple uses, not limited to the construction industry, for erecting water supply and drainage systems for modern cities and housing, which are designed in accordance with general principles that are widely recognized and clearly defined in related legislation [11]. The most used are metal pipes; nevertheless, they are susceptible to damage due to the electrochemical reaction known as corrosion and present a potential risk of causing massive environmental and economic problems [4,12]. According to the EPA, corrosion is a part of the 15 mandatory water quality standards for drinking water contaminants [13]. The effects of the presence of corrosion are given by the metallic taste and the visible effects on the corroded pipes/bodies. In order to reduce these risks, non-metallic pipes, such as high-density polyethylene (HDPE) pipes and concrete pipes, have been used in recent years.

According to Hajibabaei et al. [14], the selection of the pipe material and the pipeline installation methodology are the main concerns in terms of impact assessment [15]. The most applied methodology to determine the environmental impact through an entire life cycle is Life Cycle Assessment (LCA). LCA is normally used to verify and ascertain the environmental impact of any activity and aims to examine all the effects produced on the environment [12,14,16].

Currently, the LCA methodology is used in many areas, such as eco-sustainable design, eco-labelling, waste management and other services, and the choice of production flows. This allow companies that use it to obtain the environmental impact of a product or service and be able to count on a competitive advantage derived from a market that is increasingly attentive and sensitive to eco-sustainability issues.

To evaluate the environmental impacts of a product, a service, or a process, the ISO standards of the 14040–14044:2006 series are available, which regulate the international Life Cycle Assessment (LCA) methodology [2,17–20]. This methodology allows the evaluation of energy and environmental loads along the entire life cycle. When speaking of "life cycle" analysis, we are referring to two aspects: the first is the verification of how much an entire system affects the environment, and the second lies in the analysis of the individual items of the cycle to improve the relative impact and identify critical issues in which to intervene.

In this study, life cycle analysis was developed for pipelines in the water supply sector, comparing various types of pipelines used within the Metropolitan City of Bologna aqueduct network, managed by an Italian multi-utility company. Studies conducted by the Italian multi-utility show that the pipeline materials present throughout the analyzed aqueduct network include polyvinyl chloride (PVC), high-density polyethylene (HDPE), cast-iron, steel, and cement/asbestos. Notably, PVC and cast-iron exhibit superior technical and mechanical performance. HDPE demonstrates the highest consistency within the networks, while steel boasts of a significant percentage of usage within the specific aqueduct network under analysis. Conversely, cement/asbestos pipelines are no longer considered, since the usage of such pipelines in recent years has been negligible. The use of polymeric pipes in water supply is gradually replacing cast iron and steel pipes. However, data on the environmental impacts of the polymeric pipes for water supply pipes in the Italian Metropolitan City of Bologna are scarce.

In specific, the study has as its objective assessing the environmental impact of two commonly used plastic pipes (PVC and HDPE) for water supply in the Metropolitan City of Bologna, using an LCA tool.

2. Materials and Methods

This study presents a Life Cycle Assessment of polyvinyl chloride (PVC) and high-density polyethylene (HDPE) pipes in the water supply system of an Italian metropolitan city. The environmental impact was assessed utilizing LCA methodology [18], a widely used approach for evaluating water supply systems [17,21,22]. LCA comprises four steps, presented in Figure 1.

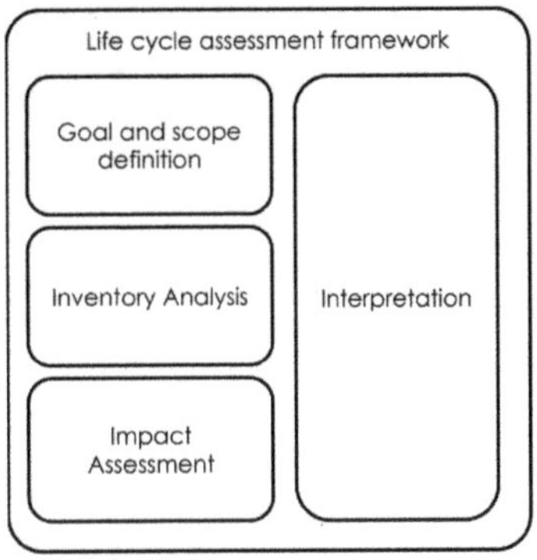

Figure 1. ISO framework for LCA ISO 14040.

This study employed SimaPro 9.2 software alongside the Ecoinvent v.3.8 [23,24] and Industrial Data 2.0 databases. This approach facilitated systematic and transparent modeling and analysis of diverse life cycles to assess the environmental impact of processes throughout selected life cycle stages, pinpointing critical areas (hotspots) across the entire chain [22,25]. SimaPro and Ecoinvent are commonly used for LCA studies due to their extensive availability of reliable data, user-friendly interfaces, and compatibility with each other. Ecoinvent provides a comprehensive and transparent database of environmental data, rigorously validated by experts. SimaPro, on the other hand, offers an intuitive platform for conducting LCA analysis, making the process more accessible and streamlined. Together, they ensure consistency, comparability, and accuracy in LCA results, making them preferred tools for researchers and practitioners in the field of environmental impact assessment. In addition, the CML-IA calculation method used to analyze various impact categories is a validated method employed by all LCA software.

2.1. Goal and Scope Definition

The objective of this LCA study was to assess the sustainability of two types of water pipes of different materials used for the water supply system in the Metropolitan City of Bologna to determine the type of plastic pipeline with the lowest environmental impact. The materials investigated were PVC pipe and HDPE pipe. Two different pipe diameters (90 mm and 125 mm) for water supply infrastructure were assessed. The specific goal was established to assess the environmental impacts resulting from the production, transportation, installation, and maintenance phases that constitute the life cycle of a water supply pipeline.

A functional unit (FU) serves as the criterion for comparing distinct phases by establishing relationships between inputs and outputs [18]. In an LCA for water supply infrastructure, inputs are used as the FU [21]. For this research, the choice of FU is one meter of water supply infrastructure with a life span of 40 years. The life span of the water pipelines at 40 years was selected in accordance with the minimum Italian legislative criteria and the guidelines of the multi-utility water company. This value ensures a safe pipe life span, although in practical cases, the pipes may last longer than the selected life span. A cradle-to-gate comparative analysis drove the assessment, starting from the pipelines' production to their maintenance phase, excluding the end-of-life and disposal stages.

2.2. Boundaries

The inputs are the materials, energy, and resources that enter the unit process, while the outputs are the products, waste, and emissions generated due to the process [23].

The LCA was conducted using a cradle-to-gate method and compared the environmental impacts of pipes made of different plastic materials (PVC and HDPE) and diameters (90 mm and 125 mm). The LCA considered the production, the transport, the installation, and the maintenance of the pipes in the assessment (Figure 2).

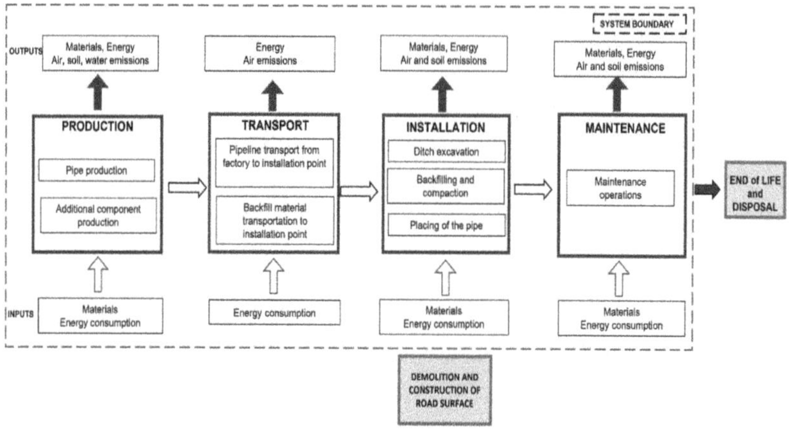

Figure 2. System boundary.

The life cycle typically commences with the extraction of raw materials, followed by their transportation to the pipe production factory. Subsequently, the manufactured pipelines are conveyed to the construction site for installation. This process encompasses the initial stages of material extraction and production, through to the final deployment of the pipelines. During the installation, a trench is excavated to accommodate the pipe, and the trench is backfilled after the pipe is installed [26,27]. Then, the maintenance operations for the pipeline are considered.

2.3. *Life Cycle Inventory (LCI) Analysis*

The primary inventory analysis was conducted utilizing the data provided by the waterworks company. To obtain product-specific data, the material quantities were estimated using precise measurements of dimensions or weights, ensuring accuracy in the assessment. Furthermore, process-specific data were meticulously gathered by performing a comprehensive time study analysis of each phase involved. Secondly, the collected data underwent rigorous scrutiny and evaluation to build the LCA model, ensuring its reliability and robustness. Finally, both primary data gathered directly from observations and measurements and secondary data obtained from reliable sources were synergically integrated to refine and enrich the study's modeling process. Background data was taken from the Ecoinvent v 3.08 and Industrial Data 2.0 databases, as well from the relevant literature, such as the guidelines followed by the multi-utility water company and [14,28].

The inventory of phases examined in this LCA study, as defined in SimaPro software, is described in the following subsections. Table 1 summarizes the list of items included in or excluded from the LCA inventory based on the LCA phases considered.

Table 1. Overview of items included and excluded in the analyzed phases of LCA.

LCA Phase	Included in the Analysis	Excluded from the Analysis
Production	Raw materials, pipe manufacturing, additional components for pipes	Production, maintenance, and end-of-life of production equipment
Transportation	Transportation distance, vehicle type, fuel consumption	Production, maintenance, and end-of-life of vehicles, transportation of raw material to the factory
Installation	Equipment for excavation and place of the pipes, materials required in trenches	Production and maintenance of installation machinery, pipeline dewatering, hydrostatic testing of the pipeline
Maintenance	Maintenance operations	Production and maintenance of equipment used in this phase

2.3.1. Production

The production of a typical HDPE pipe involves the use of HDPE material derived from the processing of granular high-density polyethylene, which is processed through the following two industrial processes [28]:

- Extrusion: industrial process for the creation of so-called 2D elements (duct);
- Injection molding: industrial process for forming elements defined as 3D (bends, fitting elements).

Similar production processes are followed for the manufacture of PVC pipe, starting from the processing of granular polyvinyl chloride.

The data on the normal pressure and diameter of pipes to be used were provided in the guidelines followed by the company that operates the waterworks.

2.3.2. Transport

The amount of environmental impact incurred during transportation is intricately tied to several factors, including the distance traveled, the specific type of vehicle employed, and the vehicle fuel consumption. This relationship underscores the importance of considering various variables when evaluating the environmental implications of transportation processes. The data about the transportation was provided by the waterworks company. The transportation distance from the pipe production factory to the installation site was considered equal to 215 km for HDPE pipelines and 175 km for the PVC pipeline.

On the other hand, the transport distance of backfill materials (sand and gravel) was assumed to be constant at 20 km.

2.3.3. Installation

The installation phase consists of trench excavation, sand bedding, laying of the pipe, backfilling, and compaction of the surface course. The dimensions of the trench are presented in Figure 3. In all cases, the trench was considered to be vertical. The specifications of the trench and thickness of the bed were indicated by the waterworks company guidelines and are shown in Figure 3. The depth of pipe was taken as 1 m from the ground surface to the top of the pipe. The pipe depth is based on the guidelines provided by the multi-utility company, and it is applicable throughout the Emilia-Romagna region in Italy. The pipes were assumed to be laid under unpaved road. Gravel was used for bedding, whereas sand was used for backfilling.

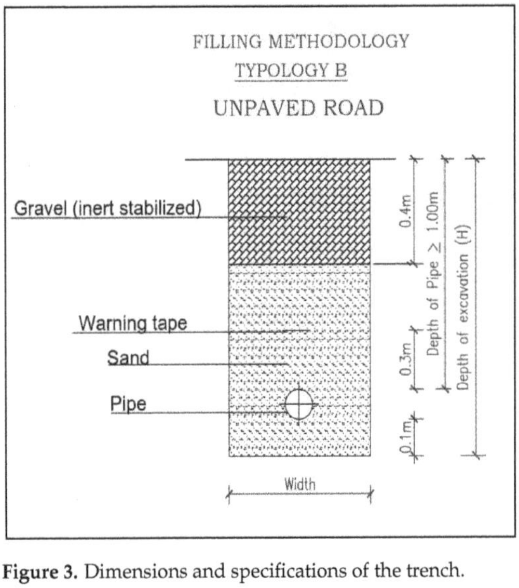

Figure 3. Dimensions and specifications of the trench.

A hydraulic excavator and compactor were considered for excavation, backfilling placement, and compaction of the surface course. The operating hours of the machines were taken from the literature [13,28].

2.3.4. Maintenance

The last phase of the LCA is represented by extraordinary maintenance of the water supply pipeline in the event of a burst. The maintenance activities are designed to restore to optimal operating conditions a water supply pipeline affected by a point break. The guidelines supplied by the company that operates the waterworks provide pipe material rupture rates over the 40-year life span and maintenance activities.

For a 1 km long pipeline that has a burst, the maintenance activities to be performed consist of replacing 1.5 m length of pipeline near the breakpoint. Therefore, the maintenance phase involves production of a new part of the pipeline, transportation to the site, excavation near the breaking point, replacement and installation of the deteriorated part, backfilling with new material, and disposal of the excavated material. However, in this study, the disposal of excavated material is neglected because the environmental impact is assumed to be insignificant.

2.4. Life Cycle Impact Assessment

The significance of the potential environmental impacts is evaluated by utilizing the comprehensive data obtained from Life Cycle Inventory (LCI). The analysis involves relating specific impacts with the inventory data, adhering to the guidelines outlined in ISO 14040:2006 [29]. Through this methodical approach, the LCI results are effectively translated into the environmental impacts from each scenario considered, thus providing valuable insights into the environmental ramifications of the assessed activities.

In this study, the CML-IA 2 baseline v 3.08/EU25 method was used, and mid-point impact categories investigated [30], focusing on the following seven (7) impact categories: Abiotic Depletion (ADP), Global Warming Potential (GWP100a), Ozone Layer Depletion Potential (ODP), Human Toxicity Potential (HTP), Photochemical Oxidation (PO), Acidification Potential (AP), and Eutrophication Potential (EP). These impact categories were selected after an extensive literature review and based on their application in environmental impact assessments of water supply pipe systems [12,14,25,28,30]. In addition, global warming potential, abiotic depletion, and human toxicity are used to highlight impact categories that relate to climate change.

3. Results

The findings of the study are presented in two parts. The first part focuses on the mid-point impact assessment, concentrating on the environmental impacts of comparing two common water pipe materials (namely PVC and HDPE with 90 and 125 mm diameters) to identify sustainable solutions among these alternatives. The second part shows the environmental impact associated with water pipe materials for the analyzed stages of the life cycle assessment.

Table 2 displays the impact assessment results for each impact category as well as for the CML-IA 2 mid-point level. Tables 3 and 4 provide insight into the relative contribution of each stage to the overall impacts.

Table 2. Impact assessment results for the various water pipe materials.

Impact Category	Units	Water Pipe Material			
		HDPE		PVC	
		Ø 90	Ø 125	Ø 90	Ø 125
Abiotic depletion	kg Sb eq	1.38×10^{-5}	1.47×10^{-5}	1.15×10^{-4}	1.74×10^{-4}
Global warming (GWP100a)	kg CO_2 eq	10.335	14.485	13.185	17.790
Ozone layer depletion (ODP)	kg CFC-11 eq	1.15×10^{-6}	1.24×10^{-6}	4.06×10^{-6}	5.85×10^{-6}

Table 2. *Cont.*

Impact Category	Units	Water Pipe Material			
		HDPE		PVC	
		Ø 90	Ø 125	Ø 90	Ø 125
Human toxicity	kg 1.4-DB eq	2.482	3.254	7.752	11.377
Photochemical oxidation	kg C_2H_4 eq	2.55×10^{-3}	3.94×10^{-3}	2.71×10^{-3}	3.71×10^{-3}
Acidification	kg SO_2 eq	0.034	0.046	0.053	0.072
Eutrophication	kg PO_4- eq	9.19×10^{-3}	0.013	0.016	0.022

Note: Values refer to functional unit of 1 m of water supply infrastructure.

Table 3. SimaPro results—normalized (dimensionless) values of impact of HDPE 90 mm diameter water pipe.

Impact Category	LCA Phase				
	Transportation	Production	Installation	Maintenance	Total
Abiotic depletion	1.18×10^{-13}	2.66×10^{-14}	1.79×10^{-14}	1.12×10^{-17}	1.62×10^{-13}
Global warming (GWP100a)	5.58×10^{-13}	9.56×10^{-13}	5.43×10^{-13}	1.28×10^{-16}	2.06×10^{-12}
Ozone layer depletion	5.83×10^{-15}	1.83×10^{-15}	5.17×10^{-15}	8.81×10^{-19}	1.28×10^{-14}
Human toxicity	1.42×10^{-13}	1.24×10^{-13}	5.36×10^{-14}	2.04×10^{-17}	3.2×10^{-13}
Photochemical oxidation	3.93×10^{-14}	1.82×10^{-13}	7.9×10^{-14}	1.76×10^{-17}	3×10^{-13}
Acidification	2.35×10^{-13}	4.7×10^{-13}	5.12×10^{-13}	7.79×10^{-17}	1.22×10^{-12}
Eutrophication	1.05×10^{-13}	3.36×10^{-13}	2.55×10^{-13}	4.32×10^{-17}	6.97×10^{-13}

Note: Values refer to functional unit of 1 m of water supply infrastructure.

Table 4. SimaPro results—normalized (dimensionless) values of impact of PVC 90 mm diameter water pipe.

Impact Category	LCA Phase				
	Transportation	Production	Installation	Maintenance	Total
Abiotic depletion	1.35×10^{-12}	1.01×10^{-13}	1.23×10^{-12}	1.79×10^{-14}	1.35×10^{-12}
Global warming (GWP100a)	2.62×10^{-12}	4.79×10^{-13}	1.6×10^{-12}	5.43×10^{-13}	2.62×10^{-12}
Ozone layer depletion	4.55×10^{-14}	5.01×10^{-15}	3.53×10^{-14}	5.17×10^{-15}	4.55×10^{-14}
Human toxicity	1×10^{-12}	1.22×10^{-13}	8.24×10^{-13}	5.36×10^{-14}	1×10^{-12}
Photochemical oxidation	3.2×10^{-13}	3.38×10^{-14}	2.07×10^{-13}	7.9×10^{-14}	3.2×10^{-13}
Acidification	1.89×10^{-12}	2.01×10^{-13}	1.18×10^{-12}	5.12×10^{-13}	1.89×10^{-12}
Eutrophication	1.19×10^{-12}	9.02×10^{-14}	8.47×10^{-13}	2.55×10^{-13}	1.19×10^{-12}

Note: Values refer to functional unit of 1 m of water supply infrastructure.

Characterization

The potential environmental impacts associated with each plastic water pipeline considered are shown in Figures 4 and 5. The results show that the HDPE pipe was the most environmentally friendly alternative for all the seven impact categories for the 90 mm diameter pipe, and in six out of seven impact categories for the 125 mm diameter pipe (except photochemical oxidation).

A comparative analysis reveals that PVC pipes exhibit the highest level of environmental impact among the considered materials. Table 2 shows that 17.8 kg CO_2 eq is released in the GWP categories of 125 mm diameter PVC pipe compared to 14.5 kg CO_2 eq of 125 mm diameter HDPE pipe. The greatest evidence of impacts can be seen between HDPE and PVC pipes in the categories of ADP (impact of PVC is 91% greater than that the HDPE pipe) and ozone layer depletion (impact of PVC pipe is 78% greater than that of HDPE).

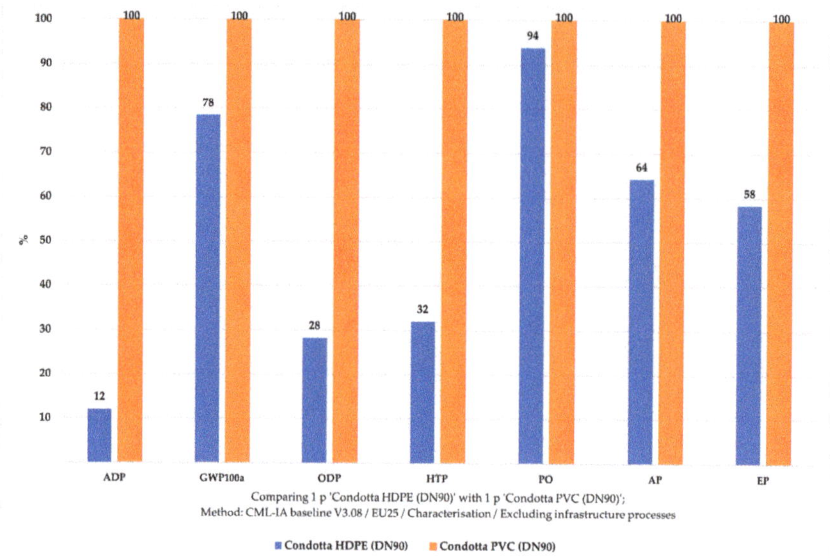

Figure 4. Comparison of impact results for HDPE and PVC with 90 mm diameter.

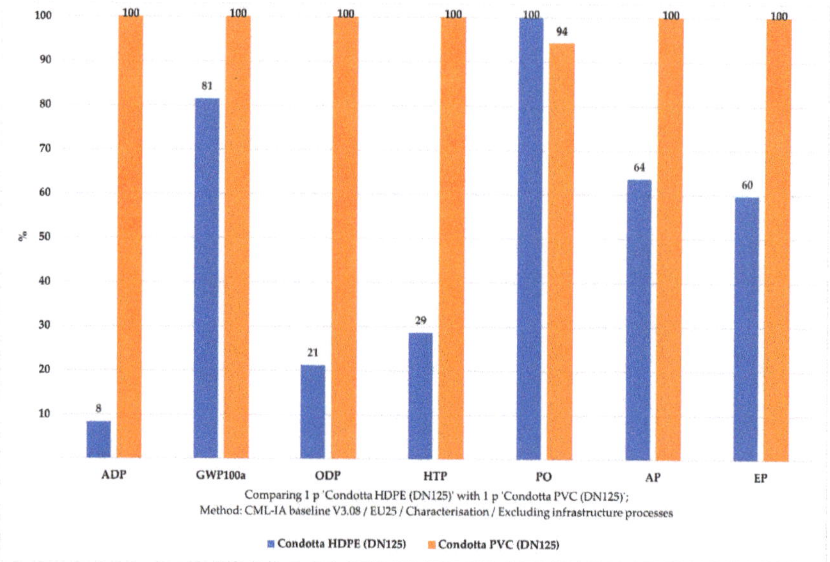

Figure 5. Comparison of impact results for HDPE and PVC with 125 mm diameter.

The 125 mm diameter pipes have higher environmental impacts than the 90 mm diameter pipes for both HDPE and PVC. The impact of the 125 mm diameter HDPE pipes was 6–35% higher than the 90 mm diameter pipe. Similarly, the 125 mm diameter PVC had an impact 26–34% higher than the 90 mm diameter.

Figure 6 illustrates the contribution of each life cycle stage to the environmental impact categories for the 90 mm diameter HDPE pipe. The same trend of results was obtained for the 125 mm diameter HDPE pipelines.

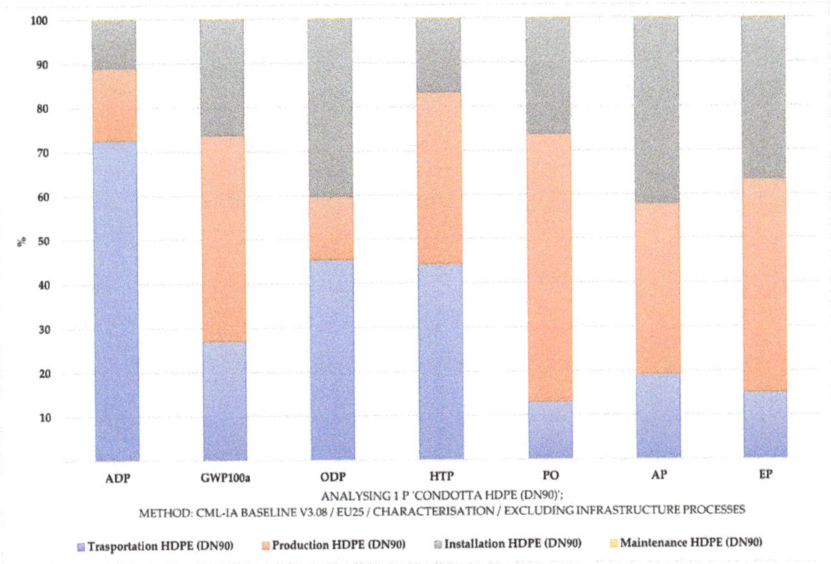

Figure 6. LCA phase contribution for HDPE 90 mm diameter pipe.

Figure 6 also indicates that the production phase is the largest life cycle impact stage in all categories except for abiotic depletion and ozone layer depletion. The production phase of the 90 mm diameter HDPE pipe contributes 16–61% for all impact categories. The installation phase of the 90 mm diameter HDPE pipe contributes 11–42% of all the impact categories. The transportation phase contributes 13–72% of all the impact categories. The contribution from the maintenance phase is insignificant, with less than 1%.

The phases of production and transportation are the main contributors, respectively responsible for over 73% of the impact in four out of seven impact categories (i.e., ADP, GWP100a, HTP, PO) and more than 57% in three out of seven impact categories (i.e., ODP, AP, EP).

Moreover, the LCA results of each impact category for the 90 mm diameter HDPE pipe are presented in Table 3. To provide clarity, the results for each category are normalized based on the scenario with the highest impact in that category. This normalization process ensures a clear and comparable understanding of the relative environmental implications across different scenarios [31].

Life cycle analysis of PVC pipelines shows a common trend for HDPE pipelines in the distribution of impacts with respect to the various stages considered.

Figure 7 indicates that the production phase is the largest life cycle impact stage in all categories. The production phase of PVC 90 mm diameter pipe contributes between 61 and 91% in all impact categories. The installation phase contributes 1–27% of all the impact categories, whereas the transportation phase contributes 7–18% of all the impact categories. The maintenance phase has an insignificant contribution of less than 1%.

Moreover, the LCA results of 90 mm diameter PVC water pipe for each impact category are presented in Table 4. To provide clarity, the results for each impact category are normalized based on the scenario with the highest impact in that category.

An uncertainty analysis was performed utilizing the Monte Carlo simulation technique, incorporating a 95% confidence interval.

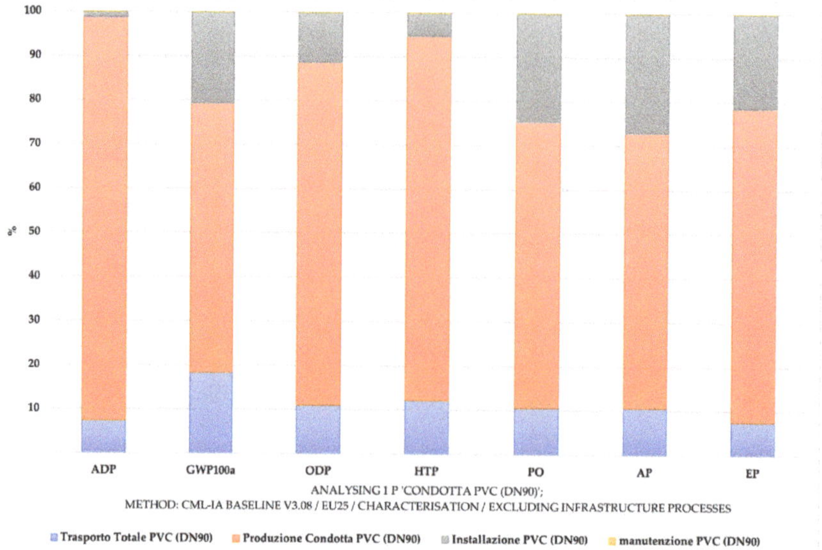

Figure 7. LCA phase contribution for PVC 90 mm diameter pipe.

4. Discussion

The high-density polyethylene pipes were the least impactful throughout their life cycle. PVC pipes have higher environmental impacts in all categories. Recio et al. [32] reported that the energy consumption (1055 kWh) and CO_2 emissions (454 kg) for producing 125 mm diameter HDPE pipe were 1.3% and 0.4% more than for PVC pipe. The production of PVC pipe has been noted by Alsabri and Al-Ghamdi [33] to consume more energy and contribute more to global warming than polyethylene pipe. This may be attributed to the use of chlorine in the manufacture of PVC. High energy consumption during the manufacturing of chlorine results in CO_2 and SO_2 emissions, which contributes to the high GWP100a and AP [34]. Similarly, in the production of PVC in China, Ye et al. [35] cited chlorine as the greatest contributor to human, terrestrial, and marine ecotoxicity.

Sanjuan-Delmas et al. [28] compared the environmental impacts of 90 mm diameter HDPE, PVC, ductile iron, and glass fiber-reinforced polyester pipes. They found similar environmental impacts for PVC and HDPE pipes, but PVC pipes were up to 10% less impactful in terms of ADP, AP and PO. Hajibabaei et al. [14] found PVC pipe to have a lower environmental impact on resources than HDPE pipe of the same size.

As in the previous studies regarding the life cycle impact assessment of pipelines used in water supplies systems [14,36], we observed that the impacts are generated by the primary material. The potential environmental impact of abiotic depletion, global warming, ozone layer depletion, photochemical oxidation, acidification, and eutrophication are more significant in the case of PVC pipe in comparison with HDPE. This finding underscores the significant role of PVC in contributing to environmental effects across various parameters and emphasizes the importance of exploring alternative materials to mitigate such impacts.

The current study revealed that the 125 mm diameter pipes had up to 35% higher environmental impact than the 90 mm diameter pipes for both HDPE and PVC. The production, transport, and installation of the pipes could be responsible for this observation. The weight of the pipe influences the amount of energy required for transport. In addition, the environmental impact of installing 90 mm diameter pipes (irrespective of material) is expected to be lower than the 125 mm diameter pipes. This is because the width and depth of the trenches for the 125 mm diameter pipes are more than that for the 90 mm diameter pipes. This increases the amount of energy and materials required for the production, transportation, and installation of the bigger diameter pipes [28]. In terms of the phases that contribute to the environmental impact, the results revealed that the production phase contributes to all impact categories except for photochemical oxidation, followed by the installation and transportation phase. This trend applies to both types of materials analyzed in the study.

The greatest contributor to the installation phase is the energy consumption for excavation, transport of backfilling material, backfilling, and compaction. To reduce the environmental impact, efficient trenching methods should be explored. For instance, the use of trenchless methods of pipe installation has been found to be less impactful than other traditional methods [14], such as open cut trenching, as was used in the current study area.

To address the environmental impact of end-of-life pipes, material recycling is gaining global attention. Ait-Touchente et al. [37] reviewed the literature on recent advances in PVC recycling. Their review shows that PVC recycling has greater environmental benefits and could significantly reduce greenhouse gas emissions. Similarly, the results of a study by Ye et al. [35] show that the production of recycled PVC has less impact than virgin PVC in all the impact categories. According to data from the Ecoinvent database, recycled HDPE granules are indeed less impactful compared to virgin ones in the categories AD, GWP100a, ODP, PO, and AP. However current Italian regulations do not permit the full utilization of recycled plastic materials for water supply pipelines, only as an addition to virgin material (see BS EN ISO 1452-1:2009 and BS EN 12201-1:2011) [38,39].

5. Conclusions

The environmental impacts of water supply infrastructure materials and their assessment through Life Cycle Analysis (LCA) are crucial topics for advancing sustainability in urban planning and construction practices. Addressing these issues contributes to the broader goals of reducing environmental footprints and promoting sustainable development in the context of global challenges. In this study, LCA methodology was used to assess the environmental impacts of 90 mm and 125 mm diameter PVC and HDPE pipes for the water supply system in an Italian metropolitan city.

A cradle-to-gate method was used to identify the most sustainable option from an environmental perspective. A functional unit of 1 m of water supply pipe was considered for a life span of 40 years. The LCA phases considered were production, transportation, installation, and maintenance of the pipes. The end-of-life and disposal stages of the pipe were excluded. The seven (7) impact categories assessed were Abiotic Depletion (ADP), Global Warming Potential (GWP100a), Ozone Layer Depletion Potential (ODP), Human Toxicity Potential (HTP), Photochemical Oxidation (PO), Acidification Potential (AP), and Eutrophication Potential (EP).

Considering both pipe diameters, the high-density polyethylene pipes are the least impactful throughout their life cycle. The PVC pipes had higher environmental impacts. In terms of the LCA phases, the production stage of the pipes had the largest environmental impact (61% of all seven impact categories). The transportation and installation phases showed similar impact contributions. Maintenance of the pipes had insignificant environmental impact. When the diameter of the pipe increased from 90 mm diameter to 125 mm diameter, the environmental impacts increased by about one-third for both materials.

The contribution of the 125 mm diameter PVC pipe to GWP was 17.8 kg CO_2 eq compared to 14.5 kg CO_2 eq for HDPE pipe. Considering the 90 mm diameter pipes, the values for the 125 mm diameter pipes were reduced by 26% for the PVC pipe and 29% for the HDPE pipe. HDPE pipes are environmentally sustainable options to replace PVC pipes for the water supply in this Italian metropolitan city.

The development of a tool based on LCA methodology for calculating the environmental impact assessment of water pipes, along with the incorporation of sustainable indices in public procurement processes for purchasing water supply pipes, derived from global warming potential results, are useful in developing alternative approaches for sustainable and eco-efficient water supply infrastructure design and materials.

The environmental impact of the use, end-of-life, and disposal phases of the pipes was not assessed in this study, even though it could be significant. This could change the impacts related to the various phases of the life cycle. However, the exclusion of end-of-life and disposal phases was made because the multi-utility company does not manage the end of life of the pipelines. Future research should extend this study to include the use, end-of-life, and disposal phases of the pipes.

Moreover, further research should be conducted to assess the sustainability of PVC and HDPE manufacturing processes as well as the lifetime of the pipes. The lifespan of the pipe affects the environmental impact of its application. From an economic perspective, an analysis of the manufacturing process chain could be performed, using suitable economic indicators. In terms of social sustainability, it is imperative to consider the multifaced impact of industry on the local community. This encompasses not only the direct effects on industry operators but also the broader implications of organization policies, community engagement initiatives, customer satisfaction metrics, and options for product take-back programs. By employing a range of social indicators, stakeholders can comprehensively evaluate the social footprint of industry activities, identify areas for improvement, and implement strategies to foster positive social outcomes and community well-being. Finally, the use of organic bioplastic pipes could be explored.

Author Contributions: Conceptualization, I.M.S. and S.P.; methodology, S.P., A.R. and A.B.; software, S.P.; validation, S.P., A.R. and A.B.; writing—original draft preparation, I.M.S., E.A. and S.P.; writing—review and editing, I.M.S., S.P., A.B., A.R. and E.A.; visualization, I.M.S. and S.P.; supervision, I.M.S. and S.P.; project administration, A.B.; funding acquisition, A.B. All authors have read and agreed to the published version of the manuscript.

Funding: This research received no external funding.

Institutional Review Board Statement: Not applicable.

Informed Consent Statement: Not applicable.

Data Availability Statement: The data presented in this study are available on request from the corresponding author.

Acknowledgments: The authors would like to acknowledge the Hera s.p.a. multi-utility for providing the case study and for their valuable collaboration throughout the project.

Conflicts of Interest: The authors declare no conflicts of interest.

References

1. Olosutean, H.; Cerciu, M. Water Sustainability in the Context of Global Warming: A Bibliometric Analysis. *Sustainability* **2022**, *14*, 8349. [CrossRef]
2. World Commission on Environment and Development. Report of the World Commission on Environment and Development: Our Common Future towards Sustainable Development 2. Part II. Common Challenges Population and Human Resources 4. Available online: https://sustainabledevelopment.un.org/content/documents/5987our-common-future.pdf (accessed on 15 March 2023).
3. Sustainable Corporation. Life Cycle Assessment of Pvc Water and Sewer Pipe and Comparative Sustainability Analysis of Pipe Materials. 2017. Available online: https://www.uni-bell.org/files/Reports/Life_Cycle_Assessment_of_PVC_Water_and_Sewer_Pipe_and_Comparative_Sustainability_Analysis_of_Pipe_Materials.pdf (accessed on 2 January 2023).
4. Chohan, I.M.; Ahmad, A.; Sallih, N.; Bheel, N.; Ali, M.; Deifalla, A.F. A Review on Life Cycle Assessment of Different Pipeline Materials. *Results Eng.* **2023**, *19*, 101325. [CrossRef]
5. Haidery, J.A.; Baş, B. Life Cycle Assessment of Construction of Water Supply Pipelines: A Case Study from Van, Turkey. *Int. J. Environ. Geoinformatics* **2020**, *7*, 23–32. [CrossRef]
6. Bilalova, S.; Newig, J.; Tremblay-Lévesque, L.C.; Roux, J.; Herron, C.; Crane, S. Pathways to Water Sustainability? A Global Study Assessing the Benefits of Integrated Water Resources Management. *J. Environ. Manag.* **2023**, *343*, 118179. [CrossRef] [PubMed]
7. Pietrucha-Urbanik, K.; Tchórzewska-Cieślak, B. Approaches to Failure Risk Analysis of the Water Distribution Network with Regard to the Safety of Consumers. *Water* **2018**, *10*, 1679. [CrossRef]
8. Rak, J.R.; Pietrucha-Urbanik, K. An Approach to Determine Risk Indices for Drinking Water-Study Investigation. *Sustainability* **2019**, *11*, 3189. [CrossRef]
9. Pietrucha-Urbanik, K.; Rak, J. Water, Resources, and Resilience: Insights from Diverse Environmental Studies. *Water* **2023**, *15*, 3965. [CrossRef]

10. Fagan, J.E.; Reuter, M.A.; Langford, K.J. Dynamic Performance Metrics to Assess Sustainability and Cost Effectiveness of Integrated Urban Water Systems. *Resour. Conserv. Recycl.* **2010**, *54*, 719–736. [CrossRef]
11. Vahidi, E.; Jin, E.; Das, M.; Singh, M.; Zhao, F. Comparative Life Cycle Analysis of Materials in Wastewater Piping Systems. *Procedia Eng.* **2015**, *118*, 1177–1188. [CrossRef]
12. Amores, M.J.; Meneses, M.; Pasqualino, J.; Antón, A.; Castells, F. Environmental Assessment of Urban Water Cycle on Mediterranean Conditions by LCA Approach. *J. Clean. Prod.* **2013**, *43*, 84–92. [CrossRef]
13. U.S. Environmental Protection Agency. Secondary Drinking Water Standards: Guidance for Nuisance Chemicals. Available online: https://www.epa.gov/sdwa/secondary-drinking-water-standards-guidance-nuisance-chemicals (accessed on 24 September 2023).
14. Hajibabaei, M.; Nazif, S.; Tavanaei Sereshgi, F. Life Cycle Assessment of Pipes and Piping Process in Drinking Water Distribution Networks to Reduce Environmental Impact. *Sustain. Cities Soc.* **2018**, *43*, 538–549. [CrossRef]
15. Simion, I.M.; Hlihor, R.M.; Rosca, M.; Filote, C.; Cozma, P. Sustainable Cost Indicators Used in Biosorption Process Applied for Heavy Metals Removal. In Proceedings of the 2021 International Conference on e-Health and Bioengineering, Iasi, Romania, 18–19 November 2021; pp. 4–7. [CrossRef]
16. Zanni, S.; Awere, E.; Bonoli, A. *Life Cycle Sustainability Assessment: An Ongoing Journey;* Elsevier Inc.: Amsterdam, The Netherlands, 2020; ISBN 9780128183557.
17. Zanni, S.; Cipolla, S.S.; di Fusco, E.; Lenci, A.; Altobelli, M.; Currado, A.; Maglionico, M.; Bonoli, A. Modeling for Sustainability: Life Cycle Assessment Application to Evaluate Environmental Performance of Water Recycling Solutions at the Dwelling Level. *Sustain. Prod. Consum.* **2019**, *17*, 47–61. [CrossRef]
18. European Commission. *ILCD Handbook—General Guide on LCA—Detailed Guidance;* European Commission: Luxembourg, 2010; ISBN 9789279190926. Available online: https://eplca.jrc.ec.europa.eu/uploads/ILCD-Handbook-General-guide-for-LCA-DETAILED-GUIDANCE-12March2010-ISBN-fin-v1.0-EN.pdf (accessed on 10 March 2022).
19. Muralikrishna, I.V.; Manickam, V. *Chapter Five—Life Cycle Assessment*; Muralikrishna, I.V., Manickam, V.B.T.-E.M., Eds.; Butterworth-Heinemann: Oxford, UK, 2017; pp. 57–75. ISBN 978-0-12-811989-1.
20. Guinee, J.B. Handbook on Life Cycle Assessment Operational Guide to the ISO Standards. *Int. J. Life Cycle Assess.* **2002**, *7*, 311–313. [CrossRef]
21. Cherubini, F.; Bargigli, S.; Ulgiati, S. Life Cycle Assessment (LCA) of Waste Management Strategies: Landfilling, Sorting Plant and Incineration. *Energy* **2009**, *34*, 2116–2123. [CrossRef]
22. Arfelli, F.; Ciacci, L.; Vassura, I.; Passarini, F. Nexus Analysis and Life Cycle Assessment of Regional Water Supply Systems: A Case Study from Italy. *Resour. Conserv. Recycl.* **2022**, *185*, 106446. [CrossRef]
23. Wernet, G.; Bauer, C.; Steubing, B.; Reinhard, J.; Moreno-Ruiz, E.; Weidema, B. The Ecoinvent Database Version 3 (Part I): Overview and Methodology. *Int. J. Life Cycle Assess.* **2016**, *21*, 1218–1230. [CrossRef]
24. Pré Simapro Database Manual. 2023, pp. 3–58. Available online: https://simapro.com/wp-content/uploads/2023/04/DatabaseManualMethods.pdf (accessed on 1 April 2023).
25. Asadollahfardi, G.; Panahandeh, A.; Moghadam, E.I.; Masoumi, S.; Tayebi Jebeli, M. Environmental Life Cycle Assessment of Different Types of the Municipal Wastewater Pipeline Network. *Environ. Qual. Manag.* **2023**, *32*, 97–110. [CrossRef]
26. Blanco, A.; Pujadas, P.; Cavalaro, S.H.P.; Aguado, A. Methodology for the Design of Controlled Low-Strength Materials. Application to the Backfill of Narrow Trenches. *Constr. Build. Mater.* **2014**, *72*, 23–30. [CrossRef]
27. Asgarihajifirouz, M.; Dong, X.; Shiri, H. Assessment of the Response of Trenched–Backfilled Pipelines to Strike-Slip Faults: An Analytical Approach. *Geosciences* **2023**, *13*, 47. [CrossRef]
28. Sanjuan-Delmás, D.; Petit-Boix, A.; Gasol, C.M.; Villalba, G.; Suárez-Ojeda, M.E.; Gabarrell, X.; Josa, A.; Rieradevall, J. Environmental Assessment of Different Pipelines for Drinking Water Transport and Distribution Network in Small to Medium Cities: A Case from Betanzos, Spain. *J. Clean. Prod.* **2014**, *66*, 588–598. [CrossRef]
29. *ISO 14040:2006;* Environmental Management—Life Cycle Assessment—Principles and Framework. ISO: Geneva, Switzerland, 2006.
30. Garfí, M.; Cadena, E.; Sanchez-Ramos, D.; Ferrer, I. Life Cycle Assessment of Drinking Water: Comparing Conventional Water Treatment, Reverse Osmosis and Mineral Water in Glass and Plastic Bottles. *J. Clean. Prod.* **2016**, *137*, 997–1003. [CrossRef]
31. Hélias, A.; Esnouf, A.; Finkbeiner, M. Consistent Normalization Approach for Life Cycle Assessment Based on Inventory Databases. *Sci. Total Environ.* **2020**, *703*, 134583. [CrossRef] [PubMed]
32. Recio, J.M.B.; Narváez, R.P.; Guerrero, P.J. Estimate of Energy Consumption and CO_2 Emission Associated with the Production, Use and Final Disposal of PVC, Aluminium and Wooden Windows. Département Proj. d'Engineyeria, Universitai Politecnica de Catalunya Environmental Modelling Laboratory Report: PVC-Tub-200512-2 Barcelona. December 2005. Available online: http://www.tosatti.net/images/PDF/reportlca.pdf (accessed on 5 October 2023).
33. Alsabri, A.; Al-Ghamdi, S.G. Carbon Footprint and Embodied Energy of PVC, PE, and PP Piping: Perspective on Environmental Performance. *Energy Rep.* **2020**, *6*, 364–370. [CrossRef]
34. Comaniţă, E.-D.; Ghinea, C.; Roşca, M.; Simion, I.M.; Petraru, M.; Gavrilescu, M. Environmental Impacts of Polyvinyl Chloride (PVC) Production Process. In Proceedings of the 2015 E-Health and Bioengineering Conference (EHB), Iaşi, Romania, 19–21 November 2015; pp. 1–4.
35. Ye, L.; Qi, C.; Hong, J.; Ma, X. Life Cycle Assessment of Polyvinyl Chloride Production and Its Recyclability in China. *J. Clean. Prod.* **2017**, *142*, 2965–2972. [CrossRef]

36. Venkatesh, G.; Brattebø, H. Assessment of Environmental Impacts of an Aging and Stagnating Water Supply Pipeline Network: City of Oslo, 1991–2006. *J. Ind. Ecol.* **2012**, *16*, 722–734. [CrossRef]
37. Ait-Touchente, Z.; Khellaf, M.; Raffin, G.; Lebaz, N.; Elaissari, A. Recent Advances in Polyvinyl Chloride (PVC) Recycling. *Polym. Adv. Technol.* **2024**, *35*, e6228. [CrossRef]
38. *ISO 1452-1:2009*; Plastics Piping Systems for Water Supply and for Buried and Above-Ground Drainage and Sewerage under Pressure. ISO: Geneva, Switzerland, 2009.
39. *BS EN 12201-1:2011*; Plastics Piping Systems for Water Supply, and for Drainage and Sewerage under Pressure. Polyethylene (PE)–General. British Standards Institution: London, UK, 2011.

sustainability

MDPI

Article

From Scarcity to Abundance: Nature-Based Strategies for Small Communities Experiencing Water Scarcity in West Texas/USA

Luis Carlos Soares da Silva Junior [1,*], David de Andrade Costa [2,3] and Clifford B. Fedler [1]

[1] Civil, Environmental, and Construction Engineering Department, Texas Tech University, Lubbock, TX 79409, USA; clifford.fedler@ttu.edu

[2] Department of Ecology and Conservation Biology, Texas A&M University, College Station, TX 77843, USA; david.costa@iff.edu.br

[3] Doctoral Program in Modeling and Technology for the Environment Applied to Water Resources, Instituto Federal Fluminense, São João da Barra 28200-000, RJ, Brazil

* Correspondence: lsilvaju@ttu.edu

Abstract: Water scarcity is one of the global challenges that threatens economic development and imposes constraints on societal growth. In the semi-arid expanse of West Texas, small communities are struggling with both growing populations and decreasing water resources in the regional aquifer. This study compares two nature-based methods that could solve this problem. The first approach uses ponds and wetlands to make natural processes work together to treat the wastewater that the community receives. We applied a novel Pond-in-Pond system, which offers advantages compared to conventional pond system configurations. This system unlocks strategic hydrodynamic advantages by introducing a deeper anaerobic pit surrounded by berms, which then outflows into a larger pond. The second approach consists of an alternative strategy which integrates waste stabilization ponds, a storage basin, and the reuse of wastewater for crop irrigation—a feat that not only treats water but also enriches soil fertility. Both approaches were analyzed in terms of economic potential and pollution control. The land application had a better return on investment and emphasized the importance of innovative solutions for sustainable water management in arid regions, offering economic and community benefits. The application conveys a clear message: where water is scarce, innovation can grow; where problems are big, solutions are available; and where nature's processes are understood, they can be used.

Keywords: Pond-in-Pond system; wetlands; nature-based solutions; water reuse; wastewater treatment

Citation: Silva Junior, L.C.S.d.; Costa, D.d.A.; Fedler, C.B. From Scarcity to Abundance: Nature-Based Strategies for Small Communities Experiencing Water Scarcity in West Texas/USA. *Sustainability* **2024**, *16*, 1959. https://doi.org/10.3390/su16051959

Academic Editor: Kaan Yetilmezsoy

Received: 16 January 2024
Revised: 22 February 2024
Accepted: 24 February 2024
Published: 27 February 2024

1. Introduction

When considering water scarcity and growing concerns about sustainability, water resource management is crucial for achieving Sustainable Development Goal 6, which targets access to water and sanitation for all by 2030 [1]. Nevertheless, supplying an adequate quantity of water with suitable quality has proven to be a challenging task for many governments. Water scarcity occurs when the demand for water exceeds its availability [2,3]. However, the discussion on water demand remains an open topic, primarily approached in the context of real needs and water efficiency [4]. Beyond its tangible aspects, the social construction of water scarcity significantly influences how societies perceive, discuss, and respond to this challenge [5]. The discourse surrounding water scarcity encapsulates the ways in which societies conceptualize and communicate about the availability, distribution, and management of water resources [6]. This approach is not merely a reflection of objective water availability; it is intertwined with societal values, power dynamics, and cultural narratives [7]. These narratives often surpass scientific assessments, incorporating cultural, political, and economic factors [8].

For example, in India, water scarcity is not solely a result of physical shortages but is also linked to issues of governance, unequal access, and power dynamics [9]. Similarly, in Jordan, the world's second most water-scarce country, the narrative of water scarcity is shaped by geopolitical tensions, regional conflicts, and historical water use patterns [10]. Addressing disparities in water distribution and equity across societies requires an in-depth analysis of water management practices and current allocations to identify beneficiaries, excluded groups, and marginalized communities [11], extending beyond the singular focus on water insufficiency [12]. These concepts extend to water quality [13], since access to water is limited if the available water does not meet quality standards for societal uses. In some instances, researchers have pointed out how the perception of water can be used to justify the monetization of water services [14,15]. Regardless, addressing water-related challenges should be approached with a sustainable perspective.

The increasing demand for water for human consumption, industrial, agricultural, energy, and ecological activities has been driving innovative and sustainable practices. Among the various water uses, agriculture accounts for a substantial portion of the demand, responsible for approximately 70% of global consumption, reaching 80% in South America and 90% in South Asia [16–18].

A forward-looking strategy to address agricultural water use efficiency is imperative for sustainable practices. Future trends underscore precision agriculture technologies [19], smart irrigation systems [20], and data-driven decision making processes in optimizing water application, minimizing waste, and maximizing crop yields [21]. Complementing these technological advancements, strategic research and development investments in water-efficient technologies [22], the promotion of drought-resistant crops [23], and the adoption of advanced irrigation practices [24] are critical components. Especially in arid regions, where rainfall and surface water resources are scarce, underground aquifers emerge as a predominant water supply alternative, particularly important for agricultural activities [25]. However, the overexploitation of these aquifers frequently exceeds their natural replenishment capacity, resulting in water level depletion and subsidence [26,27]. In the United States of America, the Ogallala Aquifer provides water for agriculture in eight states of the country [28], with important effects on the economy [29]; but the aquifer has experienced water depletion over the years, gaining attention from scientists [30,31] and demanding management actions and new solutions [32].

A promising approach to address water scarcity can involve the implementation of sustainable practices in wastewater treatment [33]. Rather than discarding the treated wastewater, the growing practice of agricultural reuse highlights the potential to effectively treat wastewater and promote a sustainable approach to the effluent by irrigating crops [34].

Waste Stabilization Ponds (WSP) are nature-based solutions that treat wastewater and are still largely used, especially in small and rural communities. This technology relies on large residence times to settle particles and to biologically degrade organic material present in the municipal wastewater with microorganisms. Several authors have investigated possible configurations of ponds in terms of area loading, volume, residence time, biokinetics, etc., [35] as well as the hydrodynamic conditions of these systems and the influence of effluent input on recirculation patterns and treatment efficiency [36]. An innovative configuration identified in the literature involves a pond with a deeper sub-basin near the inlet [37], subsequently enhanced through the incorporation of berms [38] and increased depth at the center of the pond, giving rise to the Pond-in-Pond (PIP) concept [39].

The PIP concept offers advantages such as low operational cost and higher long-term effluent quality reliability with significant organic matter reduction. The treated water can be utilized for agricultural irrigation or further treated and discharged into water bodies. Drawbacks of unplanned water reuse for irrigation include potential chemical and biological contaminations [40], and soil salinity increases due to accumulation of the salts present in the recycled water in the soil, which adversely affects plant growth [41]. Thus, it is crucial to develop appropriate engineering systems and management practices. Adequate monitoring and maintenance are also essential to ensure the efficiency of the process and the compliance with environmental standards.

There are three main land application processes [35]: slow rate, overland flow, and rapid infiltration. Slow rate involves applying effluent to vegetated soil for filtration, achieving high removal rates for organic matter, nutrients, and pathogens. However, it requires better pre-treatment, moderate soil permeability, and a large area. Overland flow systems apply effluent over plants, resulting in lower removal efficiencies. This method has a higher loading rate, requiring less area and pre-treatment. Rapid infiltration applies wastewater to a basin for percolation, working well with highly permeable soils and requiring no vegetation.

Besides appropriate engineering design, financial viability in wastewater treatment is crucial to ensure long-term operations. When comparing different solutions, it is essential to consider not only the initial installation costs but also the ongoing operational and maintenance costs over time. Nature-based treatment systems offer long-term economic benefits by reducing operational costs such as energy, personnel and chemical costs. Additionally, reusing treated water for non-potable purposes, such as crop irrigation, can generate income for the municipality [42]. When assessing financial viability, it is also important to consider the environmental and social benefits provided by alternative wastewater treatment solutions.

A gap in the literature has been identified, where few studies comprehensively address both engineering and financial feasibility aspects related to sustainability in wastewater reuse for agricultural irrigation. An interdisciplinary approach is crucial for developing effective and realistic solutions that can be applied in the real world. Integrating technical expertise with economic analyses collaborates in developing sustainable strategies that are technically efficient, environmentally sound, and economically viable, thereby generating social benefits through support for small rural families.

In this context, the aim of this study is to evaluate the engineering and financial feasibility aspects of two nature-based methods for wastewater treatment and reuse. The first method focuses on treating the wastewater in a PIP system, followed by constructed wetlands to polish the effluent for discharge. The second also uses the PIP, to reduce the organic matter, followed by a storage basin, to provide year-round water storage, and an engineered wastewater reuse system for alfalfa irrigation using the treated water that contains nutrients to enhance soil fertility. Economic potential and pollution control were evaluated for both methods.

2. Materials and Methods

The methodology involves three key steps for effective project planning: first, define design parameters to establish project specifications; next, estimate costs for a realistic financial overview; finally, perform a financial analysis to evaluate economic viability. The project incorporates sustainability considerations across social, environmental, and economic dimensions to ensure a comprehensive and responsible design.

2.1. Study Area

The northwest part of Texas is a vast region characterized by a flat and elevated landscape that stretches for hundreds of kilometers. Rainfall is generally limited and irregular, making water management an important aspect for agriculture and daily life. The state of Texas, in general, exhibits a significant variation in precipitation, with annual average precipitation (Figure 1) ranging from 61 inches in the coastal region, gradually decreasing to 18 inches in the northwest region, and 9 inches in the extreme west.

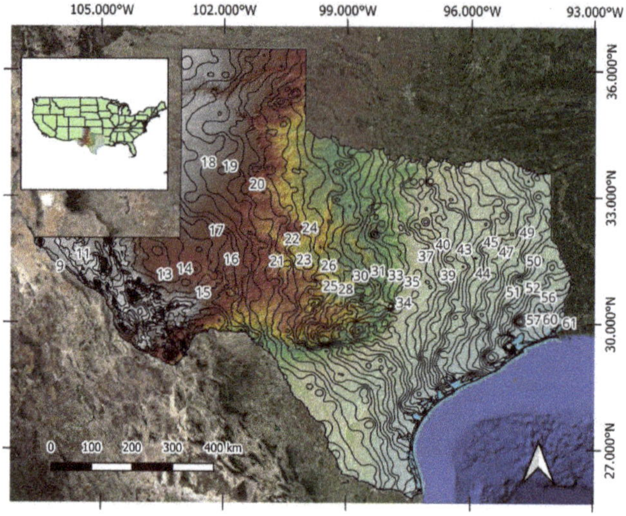

Figure 1. Annual average precipitation isolines, in inches per year, from 1981 to 2010. Public data obtained from the Natural Resources Conservation Service (NRCS).

Agriculture in the northwest region plays a significant role in the regional economy, with farms dedicated to the production of cotton, livestock, and grains. The population is generally dispersed, with smaller towns and rural communities distributed across the territory. Prominent cities in the region include Lubbock, Amarillo, and Abilene. The area is also known for the Ogallala Aquifer (Figure 2), an underground source of freshwater. However, overexploitation and the growing demand for water pose challenges to the sustainability of the aquifer. Figure 2 highlights water extraction wells in the region, obtained from The Texas Water Development Board (TWDB), Drillers Report (SDR) Database, with records of approximately 117,000 wells in the Texas portion of the Ogallala Aquifer.

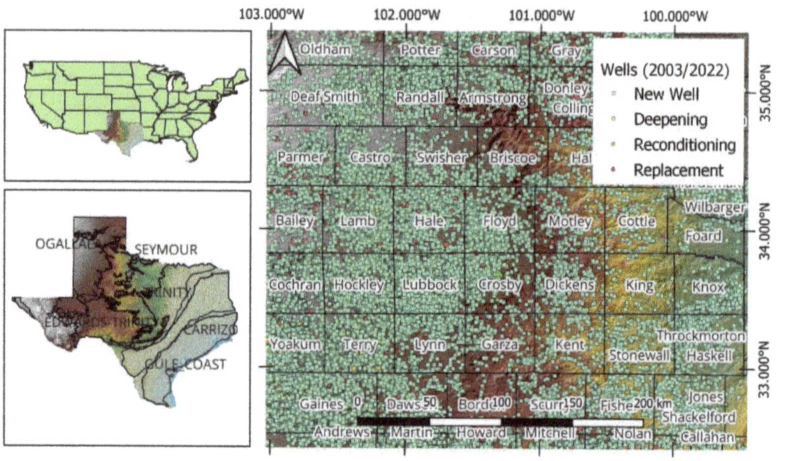

Figure 2. Location map highlighting northwest Texas, cities, and water abstraction wells relying on the Ogalala aquifer.

2.2. Population Estimative

Based on census data from 2020, the West Texas region, limited to counties with a population ranging from 1000 and 100,000 inhabitants, presented an average population of 7791 for the reference year. The region has been experiencing an annual population growth rate of 4.1% [43]. Projecting over 20 years, the average population will reach 17,108 inhabitants.

According to Texas Water Development Board [44], water consumption in Texas averages 86 gallons (325 L) per capita per day. Considering an 80% return rate, the daily production of wastewater for the average population would be 0.53 MGD (23.2 L/s) and would reach 1.18 MGD (51.7 L/s) in year 20.

2.3. Nature-Based Solutions for Wastewater Treatment

The nature-based solutions used for municipal wastewater treatment explored were the Pond-in-Pond systems, free water surface constructed wetlands (FWS-CW) for stream discharge, and land application systems for effluent reuse. Pond systems provide extensive suspended solids and Biochemical Oxygen Demand (BOD) removal and are primarily used as the main treatment unit for small municipalities. BOD and total suspended solids are the only effluent limitation parameters for stabilization ponds, according to Texas Commission on Environmental Quality (TCEQ) standards [45]. Moreover, BOD is the only recommended parameter, according to the United States Environmental Protection Agency (EPA) [46], for land application.

Although ponds are largely used as the main treatment unit, it can be challenging to achieve year-round consistent BOD removal in ponds. Therefore, an additional unit is usually added downstream to improve wastewater treatment. The most suited nature-based solution for a wastewater treatment unit to polish pond's effluents is constructed wetlands.

Pond and wetland operations are simple and are based on physical processes. In the anaerobic pit of the PIP, solids settling is the predominant process, whereas in the outer pond, the wind-driven aeration promotes oxygen transfer, and, therefore, the oxidation of organic matter. In wetlands, the adsorption of contaminants in macrophytes is the major process.

2.4. Pond and Constructed Wetland Designs

The PIP designed for both solutions has a design similar to the system studied by Adhikari and Fedler [39], also located in Colorado City, a small city in West Texas. The geometrical configuration for the PIP was a squared pit, with a 1:1 Length-to-Width Ratio (LWR) and 6 m depth. The outer pond was designed with a 3:1 LWR and 3 m depth, and the walls had a 3:1 slope factor. The drawings were presented in previous publications [35,38], and a picture of this PIP system was presented by Silva Junior and Fedler [36].

Adhikari and Fedler [39] assumed the incoming influent had 200 mg/L of BOD concentration. The Pond-in-Pond was designed according to Adhikari and Fedler [39] with an anaerobic pit and an outer pond. The anaerobic pit, responsible for settling most of the suspended solids, was designed with a 2-day Hydraulic Retention Time (HRT). Additionally, based on these authors, it was estimated that this pit would remove 45% of the organic matter, in the form of BOD. The outer pond was designed using an iterative model to estimate the necessary time to simulate the biodegradation of BOD to reach below 40 mg/L, using a first-order decay model $\left(C = C_{in} \times e^{-kt}\right)$. Based on empirical results, these authors analyzed the PIP's performance over the years, and accounting for seasonal variations, such as wind conditions, a biokinetic coefficient was estimated. When adjusted for the lowest regional temperature, 5 °C, this coefficient was 0.069 d^{-1}.

Although an effluent with 40 mg/L of BOD is below the recommended BOD concentration of 100 mg/L by EPA [46], it is not yet suitable for discharge, according to TCEQ standards, which recommend 30 mg/L of BOD [45]. Therefore, for stream discharge, an additional treatment unit would be required. The wetland design was designed according to Reed et al.'s (1995) [47] procedure to dimension FWS-CW. This wetland was designed with a 3:1 LWR and 0.67 m depth. The assumptions were an influent BOD of 40 mg/L, as previously calculated for the PIP, a desired effluent of 20 mg/L for discharge, a media porosity of 0.65, an initial influent temperature of 5 °C, and a biokinetic coefficient of 0.68 day^{-1}. Lastly, the design was checked for ice formation, summer condition, and hydraulics.

2.5. Land Application Design

The land application system used the final projection of inflow and was designed for 1.25 MGD (54.8 L/s). It assumed a salts concentration of 500 mg/L, electrical conductivity with a 10% decrease in production of 3.4 mmhos/cm [48], and climatic data for the city of Lubbock; the evapotranspiration calculation was based on the Penman–Monteith model, proposed by the FAO [49] and used in studies [50–52] around the world, and the crop evaluated was Alfalfa.

Although the TCEQ [46] requires its standard model for the design of land application, it is very conservative and does not account for storage in the vadose zone. This method promotes larger land areas and storage volumes, which increases the capital costs. On the other hand, the Texas Tech University (TTU) method, introduced by Fedler and Borrelli [39], was presented as an alternative water balance approach. It was later improved by Fedler [35] to include nitrogen and salt.

$$SM_i = SM_{i-1} + P_i + I_i - ET_i - L_i \pm S_i$$

where SM_i represents the soil moisture in month i; SM_{i-1} is soil moisture in the previous month, $i-1$; P_i is the precipitation in month i; I_i is the irrigation provided to the crop in month i; ET_i is the calculated evapotranspiration for month i; L_i is the amount of leaching to the groundwater for month i; and S_i represents the water going in and out of the storage unit. All units are expressed length of water per time, e.g., m/month.

As described by Fedler [35], this method offers a lesser storage requirement, decreasing the construction costs; yet it requires the largest land area, promoting more land to grow crops. Moreover, this method offers the advantage of considering nitrogen uptake, therefore, reducing groundwater contamination risks, and it accounts for salts balance, which can severely impact the crop yield. This is also a sustainable method because it uses the vadose zone to store water, reducing pond requirements and water losses to evaporation.

The crop selection was based on the regional perspective to achieve a crop that is usually grown in the area and yields a competitive selling price to maximize the revenue income to the land application system. Based on local observations, alfalfa crops are significantly grown in the region, can reach up to 4 cycles per year in the area, and have a high selling price, when compared to other crops.

Nitrogen was also simulated to prevent groundwater contamination by nitrate, which is the soluble fraction of nitrogen. The model assumes effluent nitrogen concentration from the pond system of 30 mg/L of N as total nitrogen, 400 lb/ac (44.8 g/m^2) of nitrogen uptake by alfalfa [49], the effluent flowrate applied for irrigation, and the land area for each phase.

Additionally, the salts balance was considered to prevent soil salinity from excessively deteriorating crop efficiency. The salts modeling was carried out by calculating the leaching rate according to FAO [49] standards and assuming 500 mg/L of total salts concentration in the effluent.

2.6. Financial Estimations

From the design calculations, estimating a median land price of USD 1880/acre (USD 4665/ha) [53] in the West Texas region and USD 4250/acre-ft (USD 3.4/m^3) for excavation costs [54]. An additional 50% was added for areas in both ponds and wetlands to allow for sloping walls, driveways, and setbacks.

There was a projected yield of 7.2 tons/acre-year (17.8 tons/ha-year) [55], USD 217/ton of alfalfa hay in Texas [55], and a historical difference in Consumer Price Index (CPI) and Producer Price Index (PPI) of 0.60% [56] to compare the cash flow between inflation and producing goods over time.

3. Results

Based on the design premises, both systems were modeled and designed to achieve their goals: (1) polish and discharge the effluent in a stream, and (2) store and irrigate an alfalfa crop. Thereafter, the land application design selection and financial evaluation were performed.

3.1. Stream Discharge System

As a result of the design calculations and assumptions, a total of two Pond-in-Ponds were implemented. The PIP dimensions (Figure 3) were 240 m in length, 80 m in width, and 3 m in depth, and the anaerobic pit was designed with a 6 m depth. The first was implemented in year 0 of operation and the second one after 5 years of operation to handle the growing influent flow until year 20. Based on those dimensions and the adjusted kinetic coefficient, a 40 mg/L effluent concentration and below would be achieved throughout the lifespan of the system, considering the flow variations and ponds' construction.

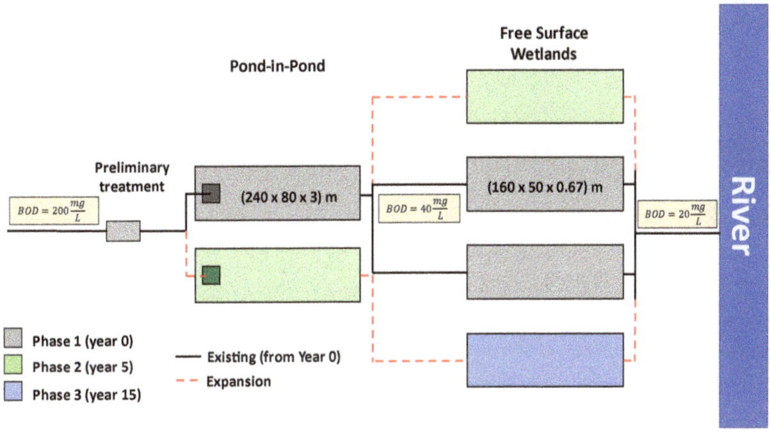

Figure 3. Stream discharge system conceptual design representing a preliminary treatment, comprising a screening process, Pond-in-Ponds reducing the wastewater BOD under 40 mg/L, and the free water surface wetlands polishing the effluent under 20 mg/L to discharge in a surface water.

For the second system, the constructed wetland with free water surface, the dimensions designed were 160 m in length, 50 m in width, and 0.67 m in depth. The filling material considered had 0.65 porosity, and the flows were calculated for summer and winter conditions. The first two cells were designed to be implemented in year 0 of operations, a third one in year 5, and a fourth one in year 15. The calculated dimensions should provide 20 mg/L of BOD consistently throughout the lifespan of the treatment system, below the 30 mg/L BOD concentration for effluent discharge in streams, according to the regulatory limit for nature-based systems in Texas [45].

3.2. Agricultural Reuse System

For the land application system, the PIP system was the same as the previous system designed for the stream discharge system. The land application design was decided upon the evaluation between the two proposed alternatives, TCEQ and TTU methods. The decision was based on reducing the storage volume, which has the biggest impact on the capital costs, and increasing the irrigation area, to increase the crop production and the operational revenue. The TTU method outperforms the TCEQ method in both aspects, with a 16% larger crop area and 42% less storage volume needed (Table 1).

Table 1. Summary on land application methods comparison.

Parameters	TCEQ	TTU	Units
Irrigation area	205 (83)	238 (96)	Acre (ha)
Storage volume	308 (378,000)	179 (221,000)	ac-ft (m^3)
HRT in storage tank	100	58	days

Therefore, for the land application design using the TTU method (Figure 4), a storage pond with 125 acre-ft or 1.5×10^5 m^3 was designed with a square configuration and a superficial area of 27,500 m^2 (165 m $\times$ 165 m; water dimensions), an 8 m depth, and a sloping wall of 1:3. The land application site for the crop irrigation was designed to be 238 acre or 9.6×10^5 m^2 (96 ha).

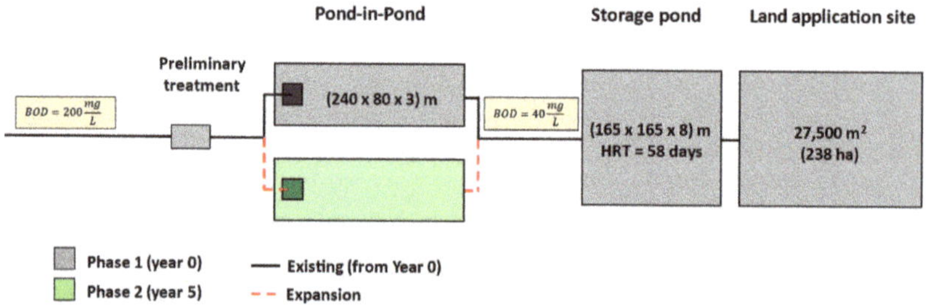

Figure 4. Agricultural reuse system conceptual design representing a preliminary treatment, comprising a screening process, Pond-in-Ponds reducing the wastewater BOD under 40 mg/L, a storage pond to provide year-long water for irrigation and the land application system used to produce alfalfa with reused water.

In addition to the BOD analysis, which is the major requirement from regulatory standards in Texas, the simulated nitrogen uptake was 82% of the potential update by alfalfa. This simulation provides evidence that there would be no nitrogen contamination to groundwater due to plants' nitrogen consumption. Lastly, the salts balance performed indicated that the expected crop yield decrement would be under 3%, which is within the recommended range [49].

3.3. *Financial Analysis and Cash Flow Projection*

The cost estimation was based on the areas and volumes and is summarized in Table 2. The PIPs have a higher area and volume than the constructed wetlands, which led to higher land and excavation costs. The stream discharge system had a total cost of USD 428,000, while the agricultural reuse system totaled USD 1,585,000, which is over three times the discharge system.

Based on the estimated 7.2 ton per acre-year (17.8 ton per ha-year) alfalfa yield, the land application system would produce 1715 tons of alfalfa per year, on average. At a price of USD 217 per ton, the median for the study area, the average revenue would be over USD 372,000 per year, which would be doubled after year 10, with the projected system expansion.

The financial analysis performed to compare both systems with the cashflow projection (Figure 5) shows the benefits of choosing the agricultural reuse system over the discharge system. With a payback period of 3.7 years after the beginning of operations, the crop irrigation needs a short period to fully offset the initial costs invested.

Table 2. Summary of design parameters and estimated costs for each component in both design approaches.

Parameter	Pond-in-Ponds	Constructed Wetlands	Land Application	Units
Area	9.5 (3.8)	2.4 (1.0)	238 (96)	acre (ha)
Storage volume	80 (98,700)	13 (16,000)	124 (153,000)	ac-ft (m^3)
Estimated land cost	27	7	448	$1000
Estimated excavation cost	340	55	527	$1000
Estimated total cost	366	62	1219	$1000

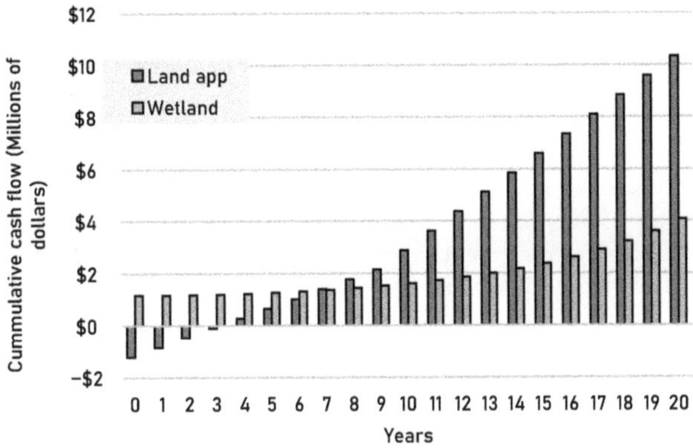

Figure 5. Cashflow projection for both projects over a 20-year period. With a high revenue, the land application system proves to be more profitable over the years.

4. Discussion

West Texas is a developing, very agriculture-oriented region with a semi-arid climate that relies on the Ogallala aquifer as its main fresh water source. With the depletion of this water source over the last few decades, this region needs sustainable solutions to approach water scarcity and sustain its growth [57]. Pond-in-Pond technology is proved to be a more efficient solution for wastewater treatment than conventional ponds. PIPs have been used in some municipalities in the region as the main treatment technology [39].

Constructed wetlands are often used as solutions to polish the effluent before stream discharge [58]. Although not evaluated, constructed wetlands also offer additional benefits such as the removal of nutrients like N, P, and emerging contaminants [59]. Although most treatment systems focus on treating the wastewater for stream discharge, the need to augment the water supplies enforced sustainable solutions to overcome water scarcity [60].

Land application systems represent a sustainable approach to reuse water [34]. The TTU method is a slow-rate system that was superior to its counterpart, the TCEQ's method. Since the TTU method uses the vadose zone for storage, it reduces the surface water storage and evapotranspiration, and allows for a larger crop area, which promotes higher crop production [35].

Moreover, the TTU method controls the nitrogen leaching to groundwater which makes it a safer method, as nitrate contamination is a serious public hazard [61], and is especially important in West Texas, characterized by substantial groundwater abstraction, as displayed in Figure 2. Another important aspect that this land application method manages is the salt balance [35]. Salt control is important to avoid an excessive increase in soil salinity, as it has a significant negative impact on crop production [62].

Although wastewater treatment and reuse are consistently presented as a sustainable pathway, often studies lack financial analysis on its impact, especially for agricultural reuse projects. The financial analysis demonstrates that agricultural reuse, often referred to as non-profitable, when compared to industrial reuse, has a high potential to be explored.

As both systems were designed and the costs were analyzed, the agricultural reuse system had a much higher capital cost when compared to the discharge system. But to analyze the projected costs over time, cash flow analysis highlighted the quick payback period, which is often more than 10 years in municipal projects. This reinforces that this agricultural reuse solution is a profitable approach to valuing water, a resource that is usually considered waste after wastewater treatment.

Small-scale irrigation through wastewater reuse offers substantial advantages for local communities and families, providing a sustainable alternative that positively impacts the local economy. Particularly in communities with limited water resources, it becomes a viable solution, addressing water scarcity and bolstering agricultural productivity [63]. To further enhance the relevance and effectiveness of such initiatives, it is crucial to consider local environmental demands, tailor solutions to the specific needs of each community, and align with regional policies [64]. For example, Israel has been a pioneer in wastewater reuse for agriculture, particularly in arid regions [65]. Similarly, California has introduced decentralized wastewater treatment systems to recover water for the irrigation of landscapes [66].

Agriculture requires effective soil nutrient management for plant development. Among the macronutrients, nitrogen and phosphorus are also found in wastewater [67]. Unlike nitrogen, the limited availability of phosphorus (P) underscores the importance of utilizing the nutrient content in treated wastewater. This approach presents a sustainable method for agriculture while addressing the environmental impact of excessive fertilizer use [68]. Also, the implementation of wastewater reuse systems creates economic opportunities within the community and stimulates new job positions [69], especially for businesses involved in wastewater treatment and irrigation infrastructure.

Despite the benefits of wastewater reuse, it is essential to acknowledge the oversight in assessing nutrient transport post land application of treated wastewater. This aspect is recommended for a comprehensive understanding of the environmental impact, as effective nutrient management can enhance soil fertility and contribute to sustainable agriculture. Policies and regulations present substantial importance to the widespread implementation of wastewater reuse systems on a global scale [70]. Addressing public health concerns and community resistance is also essential for the successful adoption of wastewater reuse systems [71].

The reuse of wastewater can have environmental, economic, and human health impacts [72]. The economic impact is generally positive, as it increases productivity and reduces costs associated with water and fertilizers. From a health perspective, there may be exposure to pathogens and heavy metals. Environmentally, impacts on the soil can occur due to the introduction of salts, increased microbiological activity, and other chemicals in the soil [73].

These factors must be managed in a reuse project. For example, conventional units of wastewater treatment plants often exhibit elevated levels of pathogens [74,75], even after treatment, requiring an additional disinfection step. Conversely, natural ponds contribute to decreased pathogen concentrations through the action of solar radiation [76]. Nonetheless, depending on the intended application, a disinfection step may still be necessary.

One of the problems that can arise from applying wastewater to land for crop irrigation is bad odor; thus, it is important to stress the importance of wastewater pre-treatment prior to its land application. An excess of suspended solids can cause problems in the irrigation system, i.e., blocking the sprinkler head, and organic matter could lead to an odor nuisance. The recommended limit for biological oxygen demand concentration in land-applied effluent is 100 mg/L [77]. On the other hand, even after treatment, some studies have focused on pharmaceuticals, endocrine disruptors, microplastics, and other chemicals which may be absorbed by crops irrigated with treated wastewater [72,78,79]. While wastewater reuse presents a sustainable water resource, addressing and mitigating potential risks associated with the presence of these chemicals is still an open topic in science frontiers.

In arid lands, where water scarcity and rising costs pose challenges to traditional agriculture, wastewater treatment and reuse gains even more significance. The nutrient-rich nature of treated wastewater becomes particularly valuable, potentially reducing expenses on external fertilizers and fostering agricultural sustainability.

5. Conclusions

The Pond-in-Pond (PIP) system is an established nature-based wastewater treatment solution focused on small and rural communities. The PIP is known for outperforming conventional ponds, given its superior treatment capacity and reliability over time. But most of the research developed over the years only focused on the technical aspect of the treatment, neglecting the sustainable potential this technology has to offer. This technology also offers resource recovery from the PIP, with the nutrient-rich effluent that can be reused in a more sustainable and safe way for crop irrigation.

This study compared two nature-based methods to treat wastewater and addresses water scarcity in small West Texas communities. The first system analyzed was composed of PIP and constructed wetlands to treat and discharge the effluent wastewater into a stream according to state regulation standards. Alternatively, the second strategy utilizes the potential of the effluent to reuse the water for alfalfa production. Although reuse is commonly referred to as a sustainable option, we analyzed its financial feasibility to prove that not only does it provide a sustainable resource, but it also generates revenue for the municipalities.

The land application water balance method selected was the TTU method, which provides lesser storage and increases the productive crop area, it was superior to the TCEQ's method. Cash flow analysis showed that agricultural reuse would be more profitable over the years by generating income that would pay the increased investments back in under four years. This result reinforces the need to promote this technology in small communities that have limited investment capacity and a shortage of specialized labor.

This study is particularly relevant for arid regions, especially those that struggle with water scarcity, such as the ones in West Texas. Sustainable water practices outweigh traditional methods, demonstrating how innovative approaches can tackle water scarcity worldwide.

Author Contributions: Conceptualization, L.C.S.d.S.J. and C.B.F.; methodology, L.C.S.d.S.J. and D.d.A.C.; writing—original draft preparation, L.C.S.d.S.J. and D.d.A.C.; writing—review and editing, L.C.S.d.S.J., D.d.A.C. and C.B.F. All authors have read and agreed to the published version of the manuscript.

Funding: The second author acknowledges the financial support provided by the Brazilian National Council for Scientific and Technological Development (CNPq).

Institutional Review Board Statement: Not applicable.

Informed Consent Statement: Not applicable.

Data Availability Statement: The raw data supporting the conclusions of this article will be made available by the authors on request.

Acknowledgments: The authors acknowledge their respective institutions.

Conflicts of Interest: The authors declare no conflicts of interest.

References

1. UN. *Transforming Our World: The 2030 Agenda for Sustainable Development*; United Nations (UN): New York, NY, USA, 2015.
2. He, C.; Liu, Z.; Wu, J.; Pan, X.; Fang, Z.; Li, J.; Bryan, B.A. Future Global Urban Water Scarcity and Potential Solutions. *Nat. Commun.* **2021**, *12*, 4667. [CrossRef] [PubMed]

3. Rijsberman, F.R. Water Scarcity: Fact or Fiction? *Agric. Water Manag.* **2006**, *80*, 5–22. [CrossRef]
4. Shahangian, S.A.; Tabesh, M.; Yazdanpanah, M. Psychosocial Determinants of Household Adoption of Water-Efficiency Behaviors in Tehran Capital, Iran: Application of the Social Cognitive Theory. *Urban Clim.* **2021**, *39*, 100935. [CrossRef]
5. Schmidt, J.J. Water as Global Social Policy—International Organizations, Resource Scarcity, and Environmental Security. In *International Organizations in Global Social Governance*; CRC: Boca Raton, FL, USA, 2021; pp. 275–296.
6. Bréthaut, C.; Ezbakhe, F.; McCracken, M.; Wolf, A.; Dalton, J. Exploring Discursive Hydropolitics: A Conceptual Framework and Research Agenda. *Int. J. Water Resour. Dev.* **2022**, *38*, 464–479. [CrossRef]
7. Turhan, Y. The Hydro-Political Dilemma in Africa Water Geopolitics: The Case of the Nile River Basin. *African Secur. Rev.* **2021**, *30*, 66–85. [CrossRef]
8. Al-Muqdadi, S.W.H. The Spiral of Escalating Water Conflict: The Theory of Hydro-Politics. *Water* **2022**, *14*, 3466. [CrossRef]
9. Vasani, H. International Water Law and Hydropolitics: An Enquiry into the Water Conflict between India and Nepal. *Water Int.* **2023**, *48*, 259–281. [CrossRef]
10. Hussein, H.; Natta, A.; Yehya, A.A.K.; Hamadna, B. Syrian Refugees, Water Scarcity, and Dynamic Policies: How Do the New Refugee Discourses Impact Water Governance Debates in Lebanon and Jordan? *Water* **2020**, *12*, 325. [CrossRef]
11. Hussein, H. An Analysis of the Framings of Water Scarcity in the Jordanian National Water Strategy. *Water Int.* **2019**, *44*, 6–13. [CrossRef]
12. Hussein, H. Lifting the Veil: Unpacking the Discourse of Water Scarcity in Jordan. *Environ. Sci. Policy* **2018**, *89*, 385–392. [CrossRef]
13. Naidoo, S. Understanding Social Constructionism of Water Quality. In *Social Constructions of Water Quality in South Africa*; Springer International Publishing: Cham, Switerland, 2022; pp. 13–29.
14. Hellberg, S. Scarcity as a Means of Governing: Challenging Neoliberal Hydromentality in the Context of the South African Drought. *Environ. Plan. E Nat. Sp.* **2020**, *3*, 186–206. [CrossRef]
15. Brisman, A.; McClanahan, B.; South, N.; Walters, R. The Politics of Water Rights: Scarcity, Sovereignty and Security. In *Water, Governance, and Crime Issues*; Springer International Publishing: Cham, Switerland, 2020; pp. 17–29.
16. Ungureanu, N.; Vlăduț, V.; Voicu, G. Water Scarcity and Wastewater Reuse in Crop Irrigation. *Sustainability* **2020**, *12*, 9055. [CrossRef]
17. Food and Agriculture Organization. *The State of the World's Land and Water Resources for Food and Agriculture 2021—Systems at Breaking Point*; FAO: Roma, Italy, 2022; ISBN 978-92-5-136127-6.
18. Rosa, L.; Chiarelli, D.D.; Rulli, M.C.; Dell'Angelo, J.; D'Odorico, P. Global Agricultural Economic Water Scarcity. *Sci. Adv.* **2020**, *6*, eaaz6031. [CrossRef] [PubMed]
19. Liu, W.; Shao, X.-F.; Wu, C.-H.; Qiao, P. A Systematic Literature Review on Applications of Information and Communication Technologies and Blockchain Technologies for Precision Agriculture Development. *J. Clean. Prod.* **2021**, *298*, 126763. [CrossRef]
20. Bwambale, E.; Abagale, F.K.; Anornu, G.K. Smart Irrigation Monitoring and Control Strategies for Improving Water Use Efficiency in Precision Agriculture: A Review. *Agric. Water Manag.* **2022**, *260*, 107324. [CrossRef]
21. Kamble, S.S.; Gunasekaran, A.; Gawankar, S.A. Achieving Sustainable Performance in a Data-Driven Agriculture Supply Chain: A Review for Research and Applications. *Int. J. Prod. Econ.* **2020**, *219*, 179–194. [CrossRef]
22. Geetha Varma, V. Water-Efficient Technologies for Sustainable Development. In *Current Directions in Water Scarcity Research*; Elsevier: Amsterdam, The Netherlands, 2022; pp. 101–128.
23. Fan, T.; Li, S.; Zhao, G.; Wang, S.; Zhang, J.; Wang, L.; Dang, Y.; Cheng, W. Response of Dryland Crops to Climate Change and Drought-Resistant and Water-Suitable Planting Technology: A Case of Spring Maize. *J. Integr. Agric.* **2023**, *22*, 2067–2079. [CrossRef]
24. Kumar, P.; Udayakumar, A.; Anbarasa Kumar, A.; Senthamarai Kannan, K.; Krishnan, N. Multiparameter Optimization System with DCNN in Precision Agriculture for Advanced Irrigation Planning and Scheduling Based on Soil Moisture Estimation. *Environ. Monit. Assess.* **2023**, *195*, 13. [CrossRef]
25. Madramootoo, C.A. Sustainable Ground Water Use in Agriculture. *Irrig. Drain.* **2012**, *61*, 26–33. [CrossRef]
26. Mayer, A.; Heyman, J.; Granados-Olivas, A.; Hargrove, W.; Sanderson, M.; Martinez, E.; Vazquez-Galvez, A.; Alatorre-Cejudo, L.C. Investigating Management of Transboundary Waters through Cooperation: A Serious Games Case Study of the Hueco Bolson Aquifer in Chihuahua, Mexico and Texas, United States. *Water* **2021**, *13*, 2001. [CrossRef]
27. Mittelstet, A.R.; Smolen, M.D.; Fox, G.A.; Adams, D.C. Comparison of Aquifer Sustainability Under Groundwater Administrations in Oklahoma and Texas1. *JAWRA J. Am. Water Resour. Assoc.* **2011**, *47*, 424–431. [CrossRef]
28. Lauer, S.; Sanderson, M.R.; Manning, D.T.; Suter, J.F.; Hrozencik, R.A.; Guerrero, B.; Golden, B. Values and Groundwater Management in the Ogallala Aquifer Region. *J. Soil Water Conserv.* **2018**, *73*, 593–600. [CrossRef]
29. Terrell, B. Ogallala Aquifer Depletion: Economic Impact on the Texas High Plains. *Water Policy* **2002**, *4*, 33–46. [CrossRef]
30. Basso, B.; Kendall, A.D.; Hyndman, D.W. The Future of Agriculture over the Ogallala Aquifer: Solutions to Grow Crops More Efficiently with Limited Water. *Earth's Futur.* **2013**, *1*, 39–41. [CrossRef]
31. Deines, J.M.; Schipanski, M.E.; Golden, B.; Zipper, S.C.; Nozari, S.; Rottler, C.; Guerrero, B.; Sharda, V. Transitions from Irrigated to Dryland Agriculture in the Ogallala Aquifer: Land Use Suitability and Regional Economic Impacts. *Agric. Water Manag.* **2020**, *233*, 106061. [CrossRef]
32. Reynolds, S.; Guerrero, B.; Golden, B.; Amosson, S.; Marek, T.; Bell, J.M. Economic Feasibility of Conversion to Mobile Drip Irrigation in the Central Ogallala Region. *Irrig. Sci.* **2020**, *38*, 569–575. [CrossRef]

33. Pinto, G.O.; da Silva Junior, L.C.S.; Assad, D.B.N.; Pereira, S.H.; Mello, L.C.B.d.B. Trends in Global Greywater Reuse: A Bibliometric Analysis. *Water Sci. Technol.* **2021**, *84*, 3257–3276. [CrossRef]
34. Minhas, P.S.; Saha, J.K.; Dotaniya, M.L.; Sarkar, A.; Saha, M. Wastewater Irrigation in India: Current Status, Impacts and Response Options. *Sci. Total Environ.* **2022**, *808*, 152001. [CrossRef]
35. Fedler, C.B. Design of Land Application Systems for Water Reuse. *Water* **2021**, *13*, 2120. [CrossRef]
36. Silva Junior, L.C.S.d.; Fedler, C.B. Pond-in-Pond's Anaerobic Pit Performance over Conventional Anaerobic Ponds—A Computational Fluid Dynamics Comparison. *J. Water Process Eng.* **2023**, *51*, 103444. [CrossRef]
37. Stone, R.F. Waste Stabilization Basins for a Desert Sewage Treatment Plant. *Civ. Eng.* **1960**, *30*, 158–160.
38. Oswald, W.J. Fundamental Factors in Stabilization Pond Design. *Air Water Pollut.* **1963**, *7*, 357–393. [PubMed]
39. Adhikari, K.; Fedler, C.B. Pond-In-Pond: An Alternative System for Wastewater Treatment for Reuse. *J. Environ. Chem. Eng.* **2020**, *8*, 103523. [CrossRef]
40. Deviller, G.; Lundy, L.; Fatta-Kassinos, D. Recommendations to Derive Quality Standards for Chemical Pollutants in Reclaimed Water Intended for Reuse in Agricultural Irrigation. *Chemosphere* **2020**, *240*, 124911. [CrossRef]
41. Rahman, M.M.; Hagare, D.; Maheshwari, B.; Dillon, P. Impacts of Prolonged Drought on Salt Accumulation in the Root Zone Due to Recycled Water Irrigation. *Water Air Soil Pollut.* **2015**, *226*, 90. [CrossRef]
42. Dos Santos Ferreira, M.; Gomes de Siqueira, J.; De Paulo Santos de Oliveira, V.; De Andrade Costa, D. Analysis of Municipal Public Policies for Payment for Water Environmental Services through the Public Policy Assessment Index: The State of Rio de Janeiro (Brazil) as a Study Model. *Agua Y Territ. Water Landsc.* **2023**, *23*, e6976. [CrossRef]
43. State of Texas, Office of the Governor, T.E.D. Texas Economic Snapshot. Available online: https://gov.texas.gov/business/page/texas-economic-snapshot (accessed on 10 January 2024).
44. Texas Water Development Board. *Water Use of Texas Water Utilities*; Texas Water Development Board: Austin, TX, USA, 2015.
45. Texas Commission on Environmental Quality. *Domestic Wastewater Effluent Limitation and Plant Siting*; Texas Commission on Environmental Quality: Austin, TX, USA, 2020.
46. Smith, J.E. *Process Design Manual: Land Treatment of Municipal Wastewater Effluents*; US Environmental Protection Agency: Cincinnati, OH, USA, 2006.
47. Reed, S.C.; Crites, R.W.; Middlebrooks, E.J. *Natural Systems for Waste Management and Treatment*, 2nd ed.; McGraw Hill: New York, NY, USA, 1995.
48. Doorenbos, J.; Pruitt, W.O. *Guidelines for Predicting Crop Water Requirements*; Food and Agriculture Organization of the United Nations: Rome, Italy, 1977; ISBN 0254-5284.
49. Allen, R.G.; Pereira, L.S.; Raes, D.; Smith, M. *Crop Evapotranspiration—Guidelines for Computing Crop Water Requirements*; Food and Agriculture Organization of the United Nations: Rome, Italy, 1998.
50. Sentelhas, P.C.; Gillespie, T.J.; Santos, E.A. Evaluation of FAO Penman–Monteith and Alternative Methods for Estimating Reference Evapotranspiration with Missing Data in Southern Ontario, Canada. *Agric. Water Manag.* **2010**, *97*, 635–644. [CrossRef]
51. Lang, D.; Zheng, J.; Shi, J.; Liao, F.; Ma, X.; Wang, W.; Chen, X.; Zhang, M. A Comparative Study of Potential Evapotranspiration Estimation by Eight Methods with FAO Penman–Monteith Method in Southwestern China. *Water* **2017**, *9*, 734. [CrossRef]
52. Cai, J.; Liu, Y.; Lei, T.; Pereira, L.S. Estimating Reference Evapotranspiration with the FAO Penman–Monteith Equation Using Daily Weather Forecast Messages. *Agric. For. Meteorol.* **2007**, *145*, 22–35. [CrossRef]
53. Krebs, L. *Texas Small Rural Land. Technical Report 2391*; Texas A&M University: College Station, TX, USA, 2023.
54. United States Department of Agriculture (USDA). Database. Field Office Technical Guide. Available online: https://efotg.sc.egov.usda.gov/#/state/TX/search (accessed on 10 January 2024).
55. National Agricultural Statistics Service (NASS). *Acreage Report*; NASS: Washington, DC, USA, 2023.
56. Federal Reserve Bank of St. Louis. Consumer Price Index (CPI) and Producer Price Index (PPI). Available online: https://fred.stlouisfed.org/ (accessed on 10 January 2024).
57. Steiner, J.L.; Devlin, D.L.; Perkins, S.; Aguilar, J.P.; Golden, B.; Santos, E.A.; Unruh, M. Policy, Technology, and Management Options for Water Conservation in the Ogallala Aquifer in Kansas, USA. *Water* **2021**, *13*, 3406. [CrossRef]
58. Aguado, R.; Parra, O.; García, L.; Manso, M.; Urkijo, L.; Mijangos, F. Modelling and Simulation of Subsurface Horizontal Flow Constructed Wetlands. *J. Water Process Eng.* **2022**, *47*, 102676. [CrossRef]
59. David, G.; Rana, M.S.; Saxena, S.; Sharma, S.; Pant, D.; Prajapati, S.K. A Review on Design, Operation, and Maintenance of Constructed Wetlands for Removal of Nutrients and Emerging Contaminants. *Int. J. Environ. Sci. Technol.* **2023**, *20*, 9249–9270. [CrossRef]
60. Garcia, X.; Pargament, D. Reusing Wastewater to Cope with Water Scarcity: Economic, Social and Environmental Considerations for Decision-Making. *Resour. Conserv. Recycl.* **2015**, *101*, 154–166. [CrossRef]
61. Wu, J.; Bian, J.; Wan, H.; Ma, Y.; Sun, X. Health Risk Assessment of Groundwater Nitrogen Pollution in Songnen Plain. *Ecotoxicol. Environ. Saf.* **2021**, *207*, 111245. [CrossRef] [PubMed]
62. Tian, F.; Hou, M.; Qiu, Y.; Zhang, T.; Yuan, Y. Salinity Stress Effects on Transpiration and Plant Growth under Different Salinity Soil Levels Based on Thermal Infrared Remote (TIR) Technique. *Geoderma* **2020**, *357*, 113961. [CrossRef]
63. Chauhan, J.S.; Kumar, S. Wastewater Ferti-Irrigation: An Eco-Technology for Sustainable Agriculture. *Sustain. Water Resour. Manag.* **2020**, *6*, 31. [CrossRef]

64. McLennon, E.; Dari, B.; Jha, G.; Sihi, D.; Kankarla, V. Regenerative Agriculture and Integrative Permaculture for Sustainable and Technology Driven Global Food Production and Security. *Agron. J.* **2021**, *113*, 4541–4559. [CrossRef]
65. Tal, A. Israeli Agriculture—Innovation and Advancement. In *From Food Scarcity to Surplus*; Springer: Singapore, 2021; pp. 299–358.
66. Luthy, R.G.; Wolfand, J.M.; Bradshaw, J.L. Urban Water Revolution: Sustainable Water Futures for California Cities. *J. Environ. Eng.* **2020**, *146*, 04020065. [CrossRef]
67. Mainardis, M.; Cecconet, D.; Moretti, A.; Callegari, A.; Goi, D.; Freguia, S.; Capodaglio, A.G. Wastewater Fertigation in Agriculture: Issues and Opportunities for Improved Water Management and Circular Economy. *Environ. Pollut.* **2022**, *296*, 118755. [CrossRef]
68. Saliu, T.D.; Oladoja, N.A. Nutrient Recovery from Wastewater and Reuse in Agriculture: A Review. *Environ. Chem. Lett.* **2021**, *19*, 2299–2316. [CrossRef]
69. Sulich, A.; Rutkowska, M.; Popławski, Ł. Green Jobs, Definitional Issues, and the Employment of Young People: An Analysis of Three European Union Countries. *J. Environ. Manag.* **2020**, *262*, 110314. [CrossRef]
70. Shoushtarian, F.; Negahban-Azar, M. Worldwide Regulations and Guidelines for Agricultural Water Reuse: A Critical Review. *Water* **2020**, *12*, 971. [CrossRef]
71. Morris, J.C.; Georgiou, I.; Guenther, E.; Caucci, S. Barriers in Implementation of Wastewater Reuse: Identifying the Way Forward in Closing the Loop. *Circ. Econ. Sustain.* **2021**, *1*, 413–433. [CrossRef]
72. Mishra, S.; Kumar, R.; Kumar, M. Use of Treated Sewage or Wastewater as an Irrigation Water for Agricultural Purposes—Environmental, Health, and Economic Impacts. *Total Environ. Res. Themes* **2023**, *6*, 100051. [CrossRef]
73. Al-Hazmi, H.E.; Mohammadi, A.; Hejna, A.; Majtacz, J.; Esmaeili, A.; Habibzadeh, S.; Saeb, M.R.; Badawi, M.; Lima, E.C.; Mąkinia, J. Wastewater Reuse in Agriculture: Prospects and Challenges. *Environ. Res.* **2023**, *236*, 116711. [CrossRef]
74. Kokkinos, P.; Mandilara, G.; Nikolaidou, A.; Velegraki, A.; Theodoratos, P.; Kampa, D.; Blougoura, A.; Christopoulou, A.; Smeti, E.; Kamizoulis, G.; et al. Performance of Three Small-Scale Wastewater Treatment Plants. *A Challenge for Possible Re Use. Environ. Sci. Pollut. Res.* **2015**, *22*, 17744–17752. [CrossRef] [PubMed]
75. Leonel, L.P.; Tonetti, A.L. Wastewater Reuse for Crop Irrigation: Crop Yield, Soil and Human Health Implications Based on Giardiasis Epidemiology. *Sci. Total Environ.* **2021**, *775*, 145833. [CrossRef]
76. Otter, P.; Hertel, S.; Ansari, J.; Lara, E.; Cano, R.; Arias, C.; Gregersen, P.; Grischek, T.; Benz, F.; Goldmaier, A.; et al. Disinfection for Decentralized Wastewater Reuse in Rural Areas through Wetlands and Solar Driven Onsite Chlorination. *Sci. Total Environ.* **2020**, *721*, 137595. [CrossRef]
77. *EPA-660/2-73-006b*; Wastewater Treatment and Reuse by Land Application. United States Environmental Protection Agency (EPA): Washington, DC, USA, 1973.
78. Ofori, S.; Puškáčová, A.; Růžičková, I.; Wanner, J. Treated Wastewater Reuse for Irrigation: Pros and Cons. *Sci. Total Environ.* **2021**, *760*, 144026. [CrossRef]
79. Natasha; Shahid, M.; Khalid, S.; Niazi, N.K.; Murtaza, B.; Ahmad, N.; Farooq, A.; Zakir, A.; Imran, M.; Abbas, G. Health Risks of Arsenic Buildup in Soil and Food Crops after Wastewater Irrigation. *Sci. Total Environ.* **2021**, *772*, 145266. [CrossRef]

sustainability

MDPI

Article

An In-Depth Analysis of Physical, Chemical, and Microplastic Parameters of Landfill Fine Fraction for Biocover Construction

Anastasiia Sholokhova [1,*], Inna Pitak [1], Gintaras Denafas [1], Regina Kalpokaitė-Dičkuvienė [1], Marius Praspaliauskas [2] and Juris Burlakovs [3]

1 Laboratory of Materials Research and Testing, Lithuanian Energy Institute, Breslaujos St. 3, LT-44403 Kaunas, Lithuania; inna.pitak@lei.lt (I.P.); gintaras.denafas@lei.lt (G.D.); regina.kalpokaite-dickuviene@lei.lt (R.K.-D.)
2 Laboratory of Heat-Equipment Research and Testing, Lithuanian Energy Institute, Breslaujos St. 3, LT-44403 Kaunas, Lithuania
3 Mineral and Energy Economy Research Institute of Polish Academy of Sciences, Józefa Wybickiego 7 A, 31-261 Kraków, Poland; juris@meeri.pl
* Correspondence: anastasiia.sholokhova@lei.lt

Abstract: Landfills pose global challenges, notably in terms of greenhouse gas (GHG) emissions, pollution release, and extensive land occupation. The transformative practice of landfill mining has redefined these sites as valuable resource reservoirs. The fine fraction (FF), often constituting the majority of excavated waste, is currently underutilized but holds the potential for biocover construction to mitigate methane emissions. This study comprehensively analyzes the FF from the Kuršenai landfill, collecting samples from various depths, reaching up to 10.5 m. The most suitable layers for biocover construction were determined based on basic physical and chemical parameters, along with the concentration of heavy metals and microplastics. The findings unveil significant parameter variations across different depths. Moderate–high correlations (ranging from 0.5 to 0.84) between several parameters were observed. The layer at a depth of 4.5–6 m emerged as the most suitable for biocover construction. However, this layer is characterized by elevated microplastic concentrations (30,208 ± 273 particles/kg), posing a challenge for its use in biocovers as microplastics can be released into the environment during FF extraction and biocover construction. Additionally, microplastics become finer with depth, increasing the associated risks. Therefore, a balanced approach considering material properties and pollution concentrations is vital for sustainable waste management practices.

Keywords: landfills; biocover; fine fraction; microplastics; methane; landfill mining

Citation: Sholokhova, A.; Pitak, I.; Denafas, G.; Kalpokaitė-Dičkuvienė, R.; Praspaliauskas, M.; Burlakovs, J. An In-Depth Analysis of Physical, Chemical, and Microplastic Parameters of Landfill Fine Fraction for Biocover Construction. *Sustainability* **2023**, *15*, 16914. https://doi.org/10.3390/su152416914

Academic Editor: Kaan Yetilmezsoy

Received: 5 November 2023
Revised: 13 December 2023
Accepted: 15 December 2023
Published: 17 December 2023

1. Introduction

Landfills have long occupied significant areas in Europe, serving as repositories of waste materials that represent both a challenge and an opportunity. Europe hosts more than 500,000 landfills [1]. Landfills, once considered the final resting places for waste, are now being recognized as troves of valuable resources [2]. This paradigm shift has given rise to the resurgence of landfill mining—an increasingly popular practice. Landfill mining involves the extraction of valuable materials and resources from existing landfill sites [3], as well as the reduction in the environmental footprint of these sites [4,5]. Four categories of fractions can be obtained during landfill mining, namely, soil-like material or fine fraction; combustible fraction (including plastic, paper, wood, and textile); inert fraction (stone, glass, ceramic, and metal); and others consisting of the remaining fraction [3]. The fine fraction, the smallest fraction that is usually <10–25.4 mm depending on the study, stands out as a crucial yet often overlooked component, typically accounting for 40% to 70% of the total excavated materials [6–8]. Several studies have investigated the fine fraction's potential, highlighting its applicability in waste-to-energy conversion and the recovery of rare metals [9–13]. Extracting resources from this fraction is usually economically challenging, but this fraction has the potential for biocover construction on landfill sites [14].

Landfills are significant sources of greenhouse gas emissions, mainly methane. While methane collection systems are common in sanitary landfills, capturing high-quality gas with a substantial methane content is technically and economically infeasible in old landfills. Biocover emerges as a viable solution to mitigate low-level methane emissions. Methanotrophs play a crucial role in biocover functionality by utilizing methane as an energy source and microbially oxidizing it into carbon dioxide, thereby substantially reducing methane emissions [15]. This reduction aligns with the European Union's commitment to achieving climate neutrality by 2050 and supports EU decarbonization efforts toward the 2030 climate target plan and zero-pollution ambition [16].

One of the most critical parameters for biocover construction is the organic matter content, which is typically recommended to be more than 15%, according to Huber-Humer et al. [17]. While garden compost is commonly used for biocover construction [18–20], in this case, we consider alternative resources, as compost can be used more sustainably as a fertilizer. Consequently, searching for less valuable materials for biocover construction becomes crucial for sustainable waste management. As a potential biocover material, the fine fraction has received limited attention in previous studies, with practical investigations into its suitability for biocover construction conducted only in Sweden [11] and Estonia [16].

Traditionally, the research has focused on excavating and analyzing fine fraction landfill materials from a single depth or a limited range of depths within landfill sites [11,13,16]. While these studies have contributed valuable insights into these materials' physical and chemical properties, they often overlook the dynamic variations that occur at different depths within the landfill. Beyond examining fundamental parameters, this research incorporates an investigation of microplastics in evaluating the suitability of landfill fine fraction as a biocover material. Numerous studies have substantiated a significant abundance of microplastics within landfill bodies up to 83 particles/g [21–25] and their release with leachate [21,26,27]. Growing environmental concern about microplastic pollution relates to their pervasive presence, continual fragmentation, leaching of toxic additives, and integration into the trophic chain. Additionally, microplastics can absorb and transport organic pollutants, trace elements like heavy metals, and other detrimental agents, such as pharmaceuticals and pathogenic organisms, out of landfills [28].

The authors of this article hypothesized that the physical and chemical properties of fine fraction landfill materials vary significantly with depth within landfill sites, leading to the identification of depth-specific layers that may be more suitable for biocover construction. Additionally, considering previous research findings that indicate high microplastic abundance in landfills and a potential increase with depth [23], it is assumed that some layers, even if they exhibit the best suitability for biocovers, may present challenges due to the emission of microplastics. Therefore, this research has two-fold novelty: first, it seeks to analyze and compare fine fraction materials from varying depths within landfill sites, thereby providing a more holistic understanding of the dynamic properties of these materials throughout the landfill profile and choosing the best layer. Second, this study incorporates a rigorous investigation of microplastics as an integral part of the assessment. This inclusion is motivated by the urgent need to prevent microplastic leakage from biocovers, a concern that has been largely overlooked in previous research efforts.

So, the current article aims to analyze fine fraction landfill materials collected at different depths, focusing on their physical, chemical, and microplastic properties, in order to determine the optimal layer for extracting fine fractions and using them in biocover construction, contributing to sustainable waste management practices and enhancing control over landfill gas. Identifying the optimal layer holds significance for potential future landfill mining and reclamation operations, emphasizing the utilization of layers with the highest suitability, highest methane reduction potential, and minimal pollution levels.

2. Materials and Methods

2.1. Study Object

In the second stage of the project, titled "Assessment and Testing of Resource Recovery from Landfills in Lithuanian Conditions", drilling operations were conducted on 28 April 2022 at the closed Kuršėnai landfill, located in the Šiauliai waste management region, with the coordinates of 55°55′05.4″ N 22°52′55.8″ E (Figure 1). The objective of the drilling was to analyze the morphological composition of the landfilled waste and assess the feasibility of landfill mining operations. The Kuršėnai landfill was operational from 1970 to 2007. No biogas collection system is in place, and methane emissions remain high despite landfill age. Given these circumstances, a decision was made to investigate the suitability of the fine fractions of the landfill for the construction of a biocover.

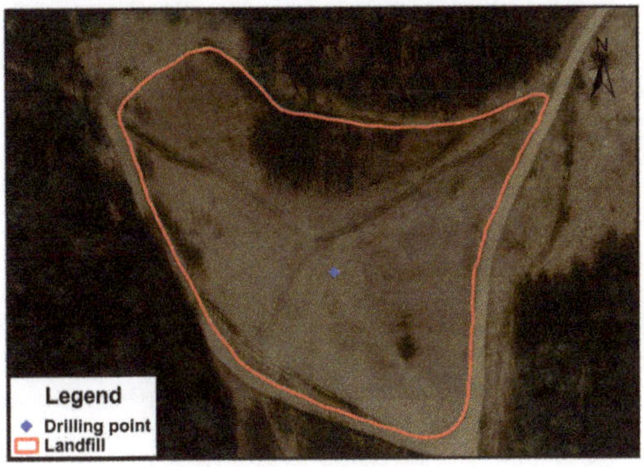

Figure 1. Kuršėnai landfill and drilling point.

Rotary-spoke drilling was carried out using the Wamet MWG-6 (small-size caterpillar drilling rig) at the center of the landfill. Each drill had a length of 1.5 m, and since the drilling was conducted to a depth of 10.5 m, a total of 7 drills were used. Initially, the cover layer was removed, and then every 1.5 m, the entire mass drilled from that interval, typically 10–15 kg as a mixed sample, was collected. Subsequently, this mass was sieved through a 20 mm sieve to separate the fine fractions. The coarse fractions were further sorted into different fractions for morphological analysis. Three replicant samples of fine fractions weighing 1 kg were then collected, transported to the laboratory, and subjected to further analysis. On average, the fine fractions accounted for approximately 74.6% of the total material collected.

2.2. Physical and Chemical Characterization

In this study, the fine fractions from landfill sites were characterized following the methodology outlined by Huber-Humer et al. [17] for assessing their suitability as a methane (CH_4) oxidation substrate. Huber-Humer et al. [17] developed the concepts of biocover and target values for biocover material, which are used as references in many studies [11,16,29]. The characterization process encompassed an array of key parameters, including organic matter (OM) content, moisture content (MC), water-holding capacity (WHC), particle size distribution (PSD), bulk density (BD), pH, and the electrical conductivity (EC) of leachate. In addition to the basic parameters, heavy metals and numerical microplastic concentrations were analyzed.

MC was determined by drying at 105 °C for 24 h and OM by loss on ignition at 550 °C for 2 h. The WHC was calculated using the gravimetric method as the weight difference before and after saturation with water. The bulk density was measured according to ISO 11885. PSD was analyzed in two options: first, following Huber-Humer's recommendation of 6–20 mm, 2–6 mm, and <2 mm; for the second option, categories were chosen as

10–20 mm, 5–10 mm, 2–5 mm, 1–2 mm, 0.5–1 mm, 0.25–0.5 mm, 0.125–0.25 mm, and <0.125 mm.

The electrical conductivity and pH were measured with pH/Cond340i (WTW) and PL-700PVS meters at a dilution ratio of 1:10 (10 g of fine fraction was mixed with 100 mL of distilled water) according to Pansu and Gautheyrou [30] and Kriipsalu [16].

Heavy metals underwent analysis using inductively coupled plasma optical emission spectroscopy (ICP-OES). Fine fraction samples (0.32 g–0.52 g) were mineralized with 6 mL of concentrated nitric acid and 2 mL of hydrofluoric acid, hydrochloric acid, and hydrogen peroxide at 1200 W, 6 MPa, and pRate: 30 kPa$\cdot$s^{-1} (Multiwalve 3000). After the mineralization, the solution was poured into 50 mL flasks and diluted to 40 mL using deionized water. The analysis of the solutions (As, Cd, Cr, Ni, Pb, Zn, Cu, and Fe) was performed using an ICP-OES, Optima 8000 (Perkin Elmer). The scanning of each single sample during element analysis was repeated three times to gather reasonably good results. An analysis was conducted in two replicates of each sample. The mercury concentration was determined according to [31,32].

2.3. *Concentration of Microplastics*

Fine fractions < 20 mm were dried at 80 °C for 24 h to assess the microplastic concentration at each depth. Then, three 10 g samples per depth were sieved through 1 mm; in this article, a concentration of microplastics of 50–1000 µm was analyzed. Microplastic extraction and identification were described in detail in previous articles [23,33]. Microplastics < 50 µm were not studied due to the limitations of the chosen identification method. In short, organic matter was removed using Fenton's reagent with heating. Subsequently, inorganic particles were separated using density separation with potassium formate and vacuum-filtrated on glass fiber filters. Then, microplastic particles were identified by staining the filters with Nile red dye, and they were examined under a fluorescence microscope on a blue excitation scale. The quantity and size of microplastics were analyzed in ImageJ 1.52v software.

2.4. *Data Analysis*

All data are presented as means ± SD (standard deviation). The normality of the data was assessed through a combination of visual observations using histograms and a statistical test, specifically the Shapiro–Wilk test ($p < 0.05$). All datasets deviated from a normal distribution; therefore, the Spearman correlation analysis was subsequently employed to explore relationships between various variables.

3. Results and Discussion

3.1. *Physical Parameters*

Physical parameters of fine fractions from different depths are presented in Figures 2 and 3. Moisture content varied from 20.99 ± 1.14% at a depth of 9–10.5 m to 36.01 ± 0.31% at a depth of 0–1.5 m. According to Huber-Hummer et al. [17], the ideal range of moisture for biocover material should be 30–50%; therefore, except for samples from the deepest depth, all samples met the recommended moisture content. The biogas flowing through the biocover may experience heating due to elevated temperatures, leading to water evaporation. Despite the water produced during the methane oxidation process, there is a possibility of desiccation in the biocover and a decrease in the efficiency of methane reduction [34]. Methanotrophic microorganisms tend to become inactive enzymes at less than 13% water content [17]. All samples demonstrated sufficient WHC in terms of recommended values (50–120%). The water-holding capacity was lowest at a depth of 6–7.5 m, measuring 73.8 ± 3.88%, while it was the highest at a depth of 4.5–6 m. The bulk density ranged from 0.71 ± 0.02 kg/L to 1.01 ± 0.03 kg/L. In samples from a Swedish landfill [11], the BD was lower (0.69 kg/L) than found in this article and lower than the recommended value of 0.8–1.1 kg/L. In our study, the BD in only two layers was lower than the recommended value, highlighting the importance of an in-depth analysis of fine fraction parameters. Bulk density is a critical parameter for efficient biocover functionalization, as it influences porosity and the movement of air and gases, particularly oxygen and methane [35]. Adequate aeration supports the activity of aerobic microorganisms involved in methane oxidation and impacts the efficiency of the reduction in methane emissions.

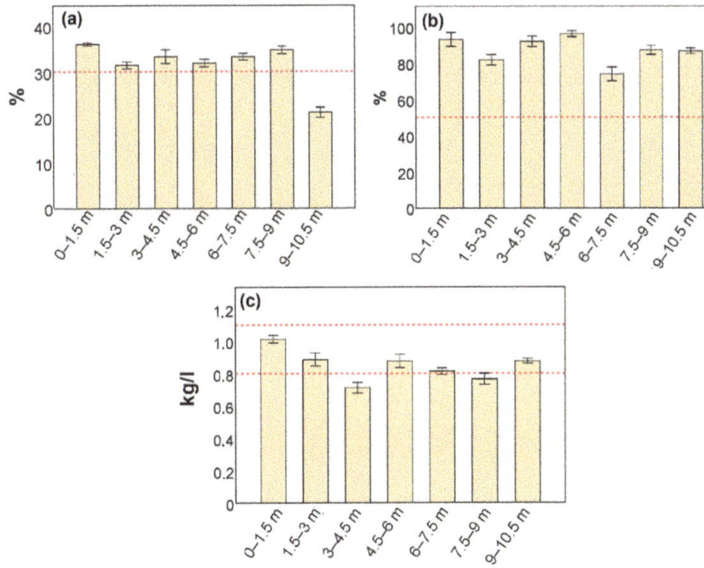

Figure 2. Physical parameters of fine fraction: (**a**) moisture content; (**b**) water-holding capacity; and (**c**) bulk density. Red dotted lines—recommended values. Error bars ± 1 SD.

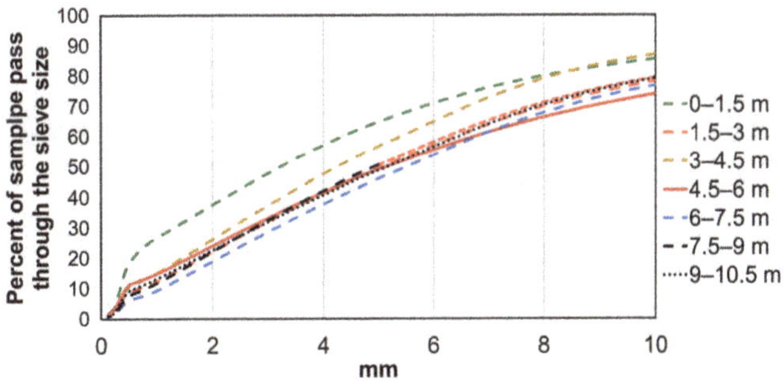

Figure 3. Particle size distribution of fine fractions from different depths.

Particle size distribution is presented in Figure 3 and, according to Huber-Humer et al. [17], in Table S1. For well-balanced materials suitable for biocover applications, it is recommended that approximately 20–30% of the material pass through a 2 mm sieve, roughly 40% pass through a 2–6.3 mm sieve, about 20–40% pass through a 6.3–20 mm sieve, and the remaining 10% consists of particles larger than 20 mm.

In our case, we tested the fine fraction (particles < 20 mm), and the results indicate that all samples exhibit a normal particle size distribution, aligning well with the recommended criteria for biocover materials. However, it is noteworthy that samples from depths of 0–1.5 m exhibited a higher percentage of material passing through the 2 mm sieve, measuring 38.39%. This percentage is slightly above the recommended range and balance for ideal biocover material, which can impact porosity and gas diffusion in the biocover layer. Conversely, samples from depths of 6–7.5 m exhibited a slightly lower percentage, at 16.18%, which falls below the recommended range for the 2 mm sieve fraction.

3.2. Chemical Parameters

Figure 4 presents the OM of fine fractions at different depths. The OM content varied from 13.75 ± 0.5% at a depth of 9–10.5 m to 18.55 ± 0.59% at a depth of 7.5–9 m. As in our study, an old landfill was investigated, and such a value of OM is usual. Mönkäre et al. [13] analyzed fine fractions from two landfills in Finland with ages ranging from 1 to 10 years and 24 to 40 years and found a higher content of OM in the young landfill compared to the old one—15.5–27.3% and 6.0–24.0%, respectively. Our analysis also demonstrates that organic content in samples from the 0–1.5, 3–4.5, 4.5–6, and 7.5–9 m depths met the 15% recommended value, while other depths were characterized with a slightly lower content. The lower level of organic content leads to a reduction in methane removal efficiency. Organic matter is a nutrient source for the microorganisms involved in methane oxidation and is needed for their growth and activity [36,37].

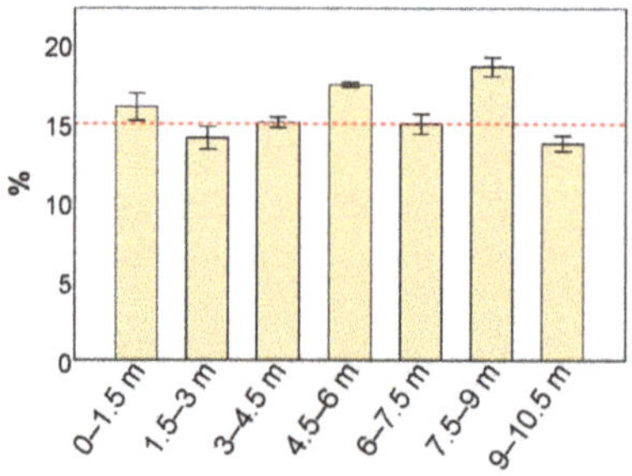

Figure 4. Organic matter content. Red dotted line—recommended value. Error bars ± 1 SD.

pH and conductivity are other important parameters that directly impact the vitality of methanotrophic microorganisms. The pH of leachate varied from 7.99 to 8.44 and was alkaline, which is a common finding and is connected to the age of the landfill (Figure 5) [38]. All samples fall within the recommended pH range for methanotrophs' activity, ranging from 6.5 to 8.5, but for some samples, the pH was close to the upper limit of 8.5.

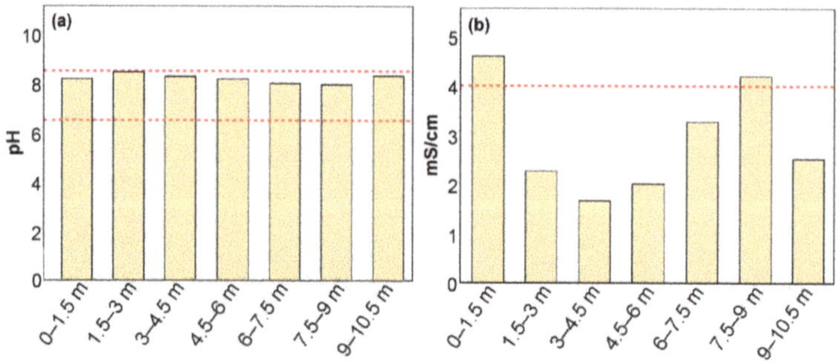

Figure 5. pH (**a**) and conductivity (**b**) of leachate. Red dotted lines—recommended values.

The conductivity values show a significant variation ranging from 1660 µS/cm at a depth of 3–4.5 m to 4592 µS/cm at a depth of 0–1.5 m. According to Huber-Humer et al. [17], the conductivity value for biocover should be no more than 4000 µS/cm. In depths 0–1.5 and 7.5–9 m, the conductivity was slightly higher—4592 µS/cm and 4188 µS/cm, respectively. As methanotrophs are rather tolerant to higher conductivity [17], using fine fractions from these depths as the main material for biocover is limited.

The average concentration of heavy metals at all depths is provided in Table 1. These concentrations were compared with the limit values established by Lithuanian legislation [39]. The table includes the limit values for both industrial soils and forested areas, considering the potential change in land use after biocover construction and remediation.

Table 1. Average concentration of heavy metals in different landfill depths, mg/kg.

	Depths							Soil for Industry Zone	Soil for Forest Areas
	0–1.5 m	1.5–3 m	3–4.5 m	4.5–6 m	6–7.5 m	7.5–9 m	9–10.5 m		
As				<1.35				80	30
Cd				<0.51				120	60
Cr	67.27	37.25	48.98	51.25	44.12	68.39	63.59	600	300
Cu	209.22	89.86	74.16	123.99	91.84	136.30	210.19	200	100
Ni	4.57				<0.42			300	150
Pb	31.63	5.24	<1.20	91.99	<1.20	105.33	88.37	500	150
Zn	281.02	465.09	217.64	367.20	262.19	589.16	244.35	1200	600
Hg	<0.02	0.035	0.11	0.06	0.10	<0.02	<0.02	1.00	0.75
Fe	10,817.1	7517.5	10,665.6	8566.9	7470.1	11,042.7	10,338.2		

From Table 1, it is evident that the concentration of heavy metals was low and was found to be below the regulatory limits set by Lithuanian regulation for both industrial soils and forested areas, except for the concentration of Cu in the top and bottom layers. In addition, in the layer of 7.5–9 m, the concentration of Zn and Pb was the highest among other depths and close to the limit for soil in forested areas. However, heavy metals, with their low concentrations, are well sorbed and do not leak out, as was found in studies by Kaczala et al. [40] and Burlakovs et al. [9]. It should be noted that there does not appear to be a consistent trend in concentration change with depth, likely because the concentration is highly dependent on the composition of waste deposited at a particular time.

3.3. *Concentration of Microplastics and Size Distribution*

Landfill mining involves excavating and disturbing the waste, which could release microplastics into the environment. These microplastics could then contaminate surrounding ecosystems and bodies of water, and they can even enter the food chain, affecting both wildlife and human health.

Numerical concentrations of microplastics are presented in Figure 6. Our study revealed a significant and concerningly high concentration of microplastics in the fine fraction of the landfill, making its use for remediation challenging. The average concentration of microplastics ranges from 4958 $\pm$ 439 particles/kg at a 0–1.5 m depth to 34,917 $\pm$ 1922 particles/kg at a 6–7.5 m depth. The abundance of microplastics in soil [41,42], green and food compost [43], and stabilized organic output from MBT [44] is usually lower than the abundance found in landfill bodies.

It was noted that the concentration of microplastics increases up to a depth of 6–7.5 m, and this rise can be attributed to the migration of microplastics with leachate from the upper layers that are deeper into the landfill profile. As a result, this migration and redistribution of microplastics cause their accumulation at intermediate depths. However, beyond the depth of 6–7.5 m, a decrease in the concentration of microplastics is observed. Several factors can contribute to this decline. Firstly, the morphological analysis shows that the concentration of plastic content (Table S2) in the landfill waste also increases and then decreases as we move deeper into the landfill. Correlation analysis shows a medium–high positive relationship between the concentration of microplastics and plastic content (r = 0.757). Secondly, the landfill body is not homogenous, and localized barriers may form within the waste profile. These barriers can obstruct the further migration of microplastics, causing them to accumulate at certain depths.

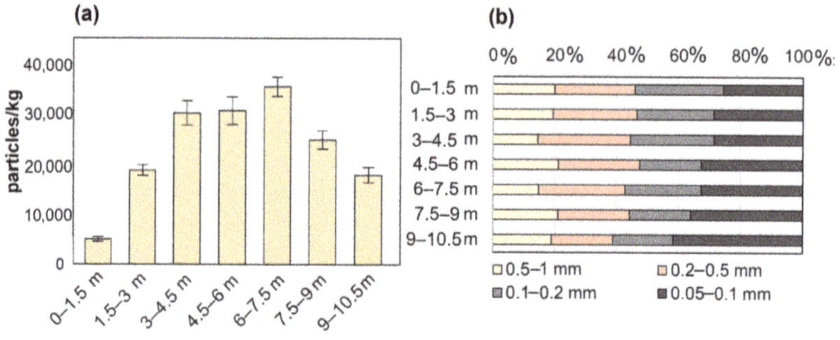

Figure 6. (**a**) Numerical concentration of microplastics. Error bars ± 1 SD; (**b**) size distribution of identified microplastics.

Consequently, the accumulation of microplastics in specific locations due to these barriers could result in decreased concentrations in other sections of the landfill. Moreover, microplastic degradation might occur more rapidly under landfill conditions. This degradation process could contribute to the overall reduction in the concentration of microplastics in landfills. However, the faster degradation alone does not fully explain the sharp decrease observed after the initial increase.

The size distribution of microplastics presented in Figure 6b reveals an intriguing trend: the percentage of the finest microplastics, measuring between 0.05 and 0.1 mm, appears to increase with depth. This phenomenon could be attributed to the degradation of plastic materials over time. However, it is essential to underscore that finer microplastics pose a more significant threat to the environment due to their ease of entry into our trophic chain. Consequently, when selecting the depth for fine fraction extraction and its subsequent use in biocover construction, special attention should be devoted to these finer microplastics to ensure the overall effectiveness of environmental management strategies. Therefore, the first two layers, in the context of the concentration of microplastics and their size distribution, have a lesser environmental impact during biocover construction.

A Spearman correlation analysis was also conducted to examine the relationship between microplastics and heavy metal concentrations. The results revealed a strong positive correlation with mercury (Hg) concentration ($r = 0.82$, $p < 0.05$). Previous studies revealed that microplastics can absorb heavy metals and become more toxic [45,46]. Therefore, opting for a layer with a high concentration of microplastics for biocover construction entails a dual risk, as microplastics can adsorb and transport heavy metals and other pollutants.

3.4. *Comprehensive Assessment of Fine Fraction Suitability for Biocover*

The utilization of fine fraction landfill materials for biocover construction serves as an exemplary embodiment of circular economy and sustainable waste management. Circular economy emphasizes the reduction in waste, the efficient use of resources, and the repurposing of materials to create a closed-loop system [47]. In this context, repurposing fine fraction materials from landfills aligns perfectly with these principles. By transforming a previously discarded waste fraction into a valuable resource for mitigating environmental impacts, such as landfill gas emissions, we embrace a circular approach.

Samples of the fine fraction collected from seven different depths were analyzed for several key parameters to assess their suitability as biocover construction materials. The comprehensive results of this analysis are presented in Table 2. In terms of factors such as moisture content and water-holding capacity, samples from all depths met the requirements. However, it is noteworthy that the organic matter content, one of the most important parameters, only met the criteria in four layers.

Table 2. Summary table of values for various parameters of the fine fractions (bold—under/above the recommended values).

	Parameters							
	pH	**Conductivity, mS/cm**	**Organic Content, %**	**Moisture Content, %**	**Bulk Density, kg/l**	**WHC, %**	**Microplastics, Particle/kg**	**Heavy Metals, mg/kg**
target value	6.5–8.5	4000	15	30–50	0.8–1.1	50–130	as low as possible	separate per each metal
0–1.5 m	8.16	**4592**	16.02	36.01	1.01	92.9	4958	Cu exceeds the limit
1.5–3 m	8.44	2260	**14.06**	31.41	0.89	81.6	18,500	below the limit
3–4.5 m	8.26	1660	15.05	33.27	**0.71**	91.8	29,708	
4.5–6 m	8.16	2000	17.42	31.82	0.88	96	30,208	
6–7.5 m	7.99	3265	**14.95**	33.22	0.81	73.8	34,917	
7.5–9 m	7.93	**4188**	18.55	34.70	**0.76**	87	24,500	
9–10.5 m	8.29	2520	**13.75**	**20.99**	0.87	86.6	17,667	Cu exceeds the limit

Considering all these parameters, it becomes evident that the layer at a depth of 4.5–6 m appears to be the most favorable choice for biocover construction. This depth range exhibits a promising combination of suitable characteristics. However, it is worth noting that the concentration of microplastics in this layer is relatively high.

On the other hand, the top layer also demonstrates good suitability for biocover construction, maintaining a lower concentration of microplastics. However, the conductivity is the highest, and the concentration of copper exceeds the limits. As microplastics are intolerant of high conductivity, the use of this layer for biocover construction is limited. The layer of 1.5–3 m is also suitable according to all parameters except for organic matter content. It is essential to highlight that parameter values can be managed and controlled via a strategic mixing of the fine fractions with appropriate amendments. Therefore, the suitability of the layer of 1.5–3 m can be increased by mixing the fine fraction with sewage sludge or green compost to decrease the concentration of microplastics. In a relevant study, Kriipsalu et al. [16] examined fine fraction materials from an Estonian landfill, and, to increase organic content, the fine fraction was mixed with sewage sludge compost in a proportion of 3:1.

The layer of 3–4.5 m has low bulk density, which can be increased by mixing it with soil, sand, or bottom ash. However, this layer was also characterized by a relatively high concentration of microplastics. The layer of 6–7.5 m had the highest concentration of microplastics and, therefore, is not recommended for use as a material for biocover construction. The layer of 7.5–9 m had low bulk density and high conductivity, so its use for biocover is limited due to the intolerance of methanotrophs. The last layer had a low organic and moisture content, but it can be increased by mixing it with compost or sludge. A correlation analysis revealed several medium–high (0.5–0.8) relationships between tested parameters, such as the relationship between the pH and OM content ($r = -0.74$), pH and MC ($r = -0.65$), OM and MC ($r = 0.71$), pH and conductivity ($r = 0.52$), WHC and OM ($r = 0.64$), and bulk density and microplastics ($r = -0.61$); however, only the first two were statistically significant ($p < 0.05$) (Table 3).

Table 3. Correlation coefficients (r) between analyzed parameters.

	pH	Conductivity	Organic Content	Moisture Content	Bulk Density	WHC	Microplastics
pH		**0.52, $p > 0.05$**	**−0.74, $p < 0.05$**	**−0.65, $p > 0.05$**	0.32	−0.13	−0.39
Conductivity	**0.52, $p > 0.05$**		0.21	0.5	0.32	−0.14	−0.45
Organic content	**−0.74, $p < 0.05$**	0.21		**0.71, $p < 0.05$**	−0.14	**0.64, $p > 0.05$**	0.18
Moisture content	**−0.65, $p > 0.05$**	0.5	**0.71, $p < 0.05$**		−0.11	0.39	−0.04
Bulk density	0.32	0.32	−0.14	−0.11		0.18	**−0.61, $p > 0.05$**
WHC	−0.13	−0.14	**0.64, $p > 0.05$**	0.39	0.18		−0.18
Microplastics	−0.39	−0.45	0.18	−0.04	**−0.61, $p > 0.05$**	−0.18	

The positive correlation between moisture and organic matter content is often observed among other studies [10]. Organic matter, such as food waste, paper, and plant materials, can be hygroscopic, meaning it has an inherent ability to absorb and retain moisture from the surrounding environment. This fact can also explain the positive correlation between OM and WHC. Also, some organic waste categories possess a higher moisture content and, during decomposition, can produce leachate [48].

A negative correlation between pH and organic content, as well as pH and moisture, can be connected to leachate formation. As organic matter decomposes, it releases moisture into the landfill, creating leachate. Leachate can have a lower pH due to organic acids forming during decomposition [49]. A negative correlation between OM and pH was also confirmed by Lindamulla et al. [50]. The correlation between the concentration of microplastics and bulk density is not statistically significant, and there is no explanation for this in the literature. However, according to the authors' presumption, lower bulk density, associated with higher porosity, provides more pore space availability for microplastics to occupy, leading to a negative correlation. Additionally, lower bulk density may facilitate better water infiltration, potentially carrying microplastics downward and resulting in their accumulation in lower layers.

4. Conclusions

The fine fraction from the diverse depths of the Kursenia landfill underwent an extensive assessment across various parameters to ascertain its aptness as a construction material for a biocover. The findings delineated substantial variations in the characteristics of fine fractions at different depths. After an extensive analysis of all key parameters, it was found that some depths are more suitable for fine fraction extraction. In our study, the fine fraction from the 4.5–6 m depth at the Kuršėnai landfill emerged as the most suitable material for biocover construction based on its physical and chemical attributes. However, this layer revealed the second-highest concentration of microplastics, totaling 30,208 ± 273 particles/kg. This significant microplastic content poses a substantial challenge, restricting immediate application and necessitating solutions to enhance other fractions with lower concentrations of microplastics. Elevated concentrations of microplastics raise environmental concerns, particularly during landfill mining operations, which may release microplastics into the ecosystem.

Considering both material properties and pollution concentrations, the layer of 1.5–3 emerges as a preferable option for biocover construction, meeting essential requirements, except for organic content, and demonstrating significantly lower microplastic levels. The shortfall in organic content can be addressed by blending it with organic-rich materials like green compost or sludge.

Within this research, several moderate correlations between parameters were observed, including the crucial relationship between microplastics and mercury concentration (r = 0.82). These findings underscore the dual risk associated with microplastic pollution, emphasizing its capacity to absorb and transport heavy metals, increasing overall toxicity.

Our study underscores the vital importance of extensive investigations into material parameters and pollution concentrations before undertaking landfill mining and biocover construction. This meticulous analysis is essential for selecting the most effective layer for methane degradation while mitigating additional environmental impact.

The utilization of fine fraction as a biocover material is a notable example of a circular economy, repurposing waste fractions and minimizing landfill methane emissions with minimal reliance on additional materials. Moreover, this aligns with the European Union's commitment to reduce greenhouse gas emissions. However, there remain unexplored aspects, and future research is warranted to delve into areas such as the release of microplastics from biocover, monitoring the efficiency of biocovers made of fine fractions on actual landfills, and enhancing the efficiency of fine fractions in methane oxidation.

Supplementary Materials: The following supporting information can be downloaded at https://www.mdpi.com/article/10.3390/su152416914/s1, Table S1: Particle size distribution; Table S2: Plastic content in different depths according to morphological analysis.

Author Contributions: Conceptualization, A.S., I.P., G.D. and R.K.-D.; methodology, A.S. and J.B.; validation, I.P., R.K.-D. and J.B.; formal analysis, A.S.; investigation, A.S., I.P. and M.P.; resources, I.P., G.D. and M.P.; data curation, A.S.; writing—original draft preparation, A.S.; writing—review and editing, I.P., G.D., R.K.-D. and J.B.; visualization, A.S.; supervision, I.P. and G.D.; project administration, G.D.; funding acquisition, G.D. and J.B. All authors have read and agreed to the published version of the manuscript.

Funding: The research was funded through project 01.2.1-LVPA-V-835-03-0005 "Model of Resource Recovery from Landfills Assessment and Its Testing in Lithuanian Conditions".

Informed Consent Statement: Not applicable.

Acknowledgments: The work was also supported by the PASIFIC program "GeoReco" project funding from the European Union's Horizon 2020 research and innovation program under the Marie Sklodowska-Curie grant agreement No. 847639 and from the Ministry of Education and Science (Poland).

Conflicts of Interest: The authors declare no conflict of interest.

References

1. EUROLCO Data Launched on the Landfill Situation in the EU-28. Available online: https://eurelco.org/2018/09/30/data-launched-on-the-landfill-situation-in-the-eu-28/ (accessed on 30 September 2023).
2. Krook, J.; Svensson, N.; Eklund, M. Landfill Mining: A Critical Review of Two Decades of Research. *Waste Manag.* **2012**, *32*, 513–520. [CrossRef]
3. Jain, M.; Kumar, A.; Kumar, A. Landfill Mining: A Review on Material Recovery and Its Utilization Challenges. *Process Saf. Environ. Prot.* **2023**, *169*, 948–958. [CrossRef]
4. Joseph, K.; Nagendran, R.; Palanivelu, K.; Thanasekaran, K.; Visvanathan, C. *Dumpsite Rehabilitation and Landfill Mining. ARRPER Report*; Anna University: Chennai, India, 2004; pp. 1–53.
5. Burlakovs, J.; Kriipsalu, M.; Klavins, M.; Bhatnagar, A.; Vincevica-Gaile, Z.; Stenis, J.; Jani, Y.; Mykhaylenko, V.; Denafas, G.; Turkadze, T.; et al. Paradigms on Landfill Mining: From Dump Site Scavenging to Ecosystem Services Revitalization. *Resour. Conserv. Recycl.* **2017**, *123*, 73–84. [CrossRef]
6. Hernández Parrodi, J.C.; Höllen, D.; Pomberger, R. Characterization of Fine Fractions from Landfill Mining: A Review of Previous Investigations. *Detritus* **2018**, *2*, 46. [CrossRef]
7. Hull, R.M.; Krogmann, U.; Strom, P.F. Composition and Characteristics of Excavated Materials from a New Jersey Landfill. *J. Environ. Eng.* **2005**, *131*, 478–490. [CrossRef]
8. Quaghebeur, M.; Laenen, B.; Geysen, D.; Nielsen, P.; Pontikes, Y.; Van Gerven, T.; Spooren, J. Characterization of Landfilled Materials: Screening of the Enhanced Landfill Mining Potential. *J. Clean. Prod.* **2013**, *55*, 72–83. [CrossRef]
9. Burlakovs, J.; Jani, Y.; Kriipsalu, M.; Vincevica-Gaile, Z.; Kaczala, F.; Celma, G.; Ozola, R.; Rozina, L.; Rudovica, V.; Hogland, M.; et al. On the Way to 'Zero Waste' Management: Recovery Potential of Elements, Including Rare Earth Elements, from Fine Fraction of Waste. *J. Clean. Prod.* **2018**, *186*, 81–90. [CrossRef]
10. Chandana, N.; Goli, V.S.N.S.; Mohammad, A.; Singh, D.N. Characterization and Utilization of Landfill-Mined-Soil-Like-Fractions (LFMSF) for Sustainable Development: A Critical Appraisal. *Waste Biomass Valorization* **2021**, *12*, 641–662. [CrossRef]

11. Jani, Y.; Kaczala, F.; Marchand, C.; Hogland, M.; Kriipsalu, M.; Hogland, W.; Kihl, A. Characterisation of Excavated Fine Fraction and Waste Composition from a Swedish Landfill. *Waste Manag. Res. J. Sustain. Circ. Econ.* **2016**, *34*, 1292–1299. [CrossRef]
12. Parrodi, J.C.H.; Höllen, D.; Pomberger, R. Potential and Main Technological Challenges for Material and Energy Recovery from Fine Fractions of Landfill Mining: A Critical Review. *Detritus* **2018**, *3*, 19–29. [CrossRef]
13. Mönkäre, T.J.; Palmroth, M.R.T.; Rintala, J.A. Characterization of Fine Fraction Mined from Two Finnish Landfills. *Waste Manag.* **2016**, *47*, 34–39. [CrossRef]
14. Pehme, K.M.; Orupõld, K.; Kuusemets, V.; Tamm, O.; Jani, Y.; Tamm, T.; Kriipsalu, M. Field Study on the Efficiency of a Methane Degradation Layer Composed of Fine Fraction Soil from Landfill Mining. *Sustainability* **2020**, *12*, 6209. [CrossRef]
15. Scheutz, C.; Duan, Z.; Møller, J.; Kjeldsen, P. Environmental Assessment of Landfill Gas Mitigation Using Biocover and Gas Collection with Energy Utilisation at Aging Landfills. *Waste Manag.* **2023**, *165*, 40–50. [CrossRef]
16. Kriipsalu, M.; Somani, M.; Pehme, K.; Tamm, O.; Truu, J.; Truu, M.; Orupold, K. Performance of Biocover in Controlling Methane Emissions from Landfill: A Decade of Full-Scale Investigation. *Process Saf. Environ. Prot.* **2023**, *172*, 486–495. [CrossRef]
17. Huber-Humer, M.; Röder, S.; Lechner, P. Approaches to Assess Biocover Performance on Landfills. *Waste Manag.* **2009**, *29*, 2092–2104. [CrossRef] [PubMed]
18. Niemczyk, M.; Berenjkar, P.; Wilkinson, N.; Lozecznik, S.; Sparling, R.; Yuan, Q. Enhancement of CH4 Oxidation Potential in Bio-Based Landfill Cover Materials. *Process Saf. Environ. Prot.* **2021**, *146*, 943–951. [CrossRef]
19. Pedersen, G.B.; Scheutz, C.; Kjeldsen, P. Availability and Properties of Materials for the Fakse Landfill Biocover. *Waste Manag.* **2011**, *31*, 884–894. [CrossRef]
20. Scheutz, C.; Pedersen, R.B.; Petersen, P.H.; Jørgensen, J.H.B.; Ucendo, I.M.B.; Mønster, J.G.; Samuelsson, J.; Kjeldsen, P. Mitigation of Methane Emission from an Old Unlined Landfill in Klintholm, Denmark Using a Passive Biocover System. *Waste Manag.* **2014**, *34*, 1179–1190. [CrossRef]
21. Su, Y.; Zhang, Z.; Wu, D.; Zhan, L.; Shi, H.; Xie, B. Occurrence of Microplastics in Landfill Systems and Their Fate with Landfill Age. *Water Res.* **2019**, *164*, 114968. [CrossRef]
22. Puthcharoen, A.; Leungprasert, S. Determination of Microplastics in Soil and Leachate from the Landfills. *Thai Environ. Eng. J.* **2019**, *33*, 39–46.
23. Sholokhova, A.; Denafas, G.; Ceponkus, J.; Omelianenko, T. Microplastics in Landfill Bodies: Abundance, Spatial Distribution and Effect of Landfill Age. *Sustainability* **2023**, *15*, 5017. [CrossRef]
24. Goli, V.S.N.S.; Singh, D.N. Extraction and Characterization of Microplastics in Landfill-Mined-Soil-like-Fractions: A Novel Methodology. *Chem. Eng. J.* **2023**, *452*, 139217. [CrossRef]
25. Sima, J.; Wang, J.; Song, J.; Du, X.; Lou, F.; Pan, Y.; Huang, Q.; Lin, C.; Wang, Q.; Zhao, G. Dielectric Barrier Discharge Plasma for the Remediation of Microplastic-Contaminated Soil from Landfill. *Chemosphere* **2023**, *317*, 137815. [CrossRef] [PubMed]
26. He, P.; Chen, L.; Shao, L.; Zhang, H.; Lü, F. Municipal Solid Waste (MSW) Landfill: A Source of Microplastics? -Evidence of Microplastics in Landfill Leachate. *Water Res.* **2019**, *159*, 38–45. [CrossRef] [PubMed]
27. van Praagh, M.; Hartman, C.; Brandmyr, E. *Microplastics in Landfill Leachates in the Nordic Countries*; Nordic Council of Ministers: Copenhagen, Denmark, 2019. [CrossRef]
28. Wojnowska-Baryła, I.; Bernat, K.; Zaborowska, M. Plastic Waste Degradation in Landfill Conditions: The Problem with Microplastics, and Their Direct and Indirect Environmental Effects. *Int. J. Environ. Res. Public Health* **2022**, *19*, 13223. [CrossRef] [PubMed]
29. Bajwa, T.M.; Fall, M.; Alshawmar, F. Experimental Characterization of the Engineering Properties of Landfill Compost-Biocover. *Appl. Sci.* **2022**, *12*, 4276. [CrossRef]
30. Pansu, M.; Gautheyrou, J. *Handbook of Soil Analysis: Mineralogical, Organic and Inorganic Methods*; Springer: Berlin/Heidelberg, Germany, 2006; pp. 1–993. [CrossRef]
31. *ISO 16968:2015*; Solid Biofuels—Determination of Minor Elements. International Organization for Standardization: Geneva, Switzerland, 2015.
32. *ISO 12846:2012*; Water Quality—Determination of Mercury. International Organization for Standardization: Geneva, Switzerland, 2012.
33. Sholokhova, A.; Ceponkus, J.; Sablinskas, V.; Denafas, G. Abundance and Characteristics of Microplastics in Treated Organic Wastes of Kaunas and Alytus Regional Waste Management Centres, Lithuania. *Environ. Sci. Pollut. Res. Int.* **2022**, *29*, 20665–20674. [CrossRef]
34. Cossu, R.; Stegmann, R. *Solid Waste Landfilling: Concepts, Processes, Technologies*; Elsevier: Amsterdam, The Netherlands, 2018; ISBN 0128183365.
35. Gebert, J.; Huber-Humer, M.; Cabral, A.R. Design of Microbial Methane Oxidation Systems for Landfills. *Front. Environ. Sci.* **2022**, *10*, 907562. [CrossRef]
36. Hrad, M. *Quantification of Landfill Gas Emissions in Biocovers—An Experimental Simulation in Lysimeters*; Universitätsbibliothek Bodenkultur: Wien, Austria, 2010.
37. Pariatamby, A.; Cheah, W.Y.; Shrizal, R.; Thamlarson, N.; Lim, B.T.; Barasarathi, J. Enhancement of Landfill Methane Oxidation Using Different Types of Organic Wastes. *Environ. Earth Sci.* **2015**, *73*, 2489–2496. [CrossRef]
38. Singh, A.; Chandel, M.K. Physicochemical and FTIR Spectroscopic Analysis of Fine Fraction from a Municipal Solid Waste Dumpsite for Potential Reclamation of Materials. *Waste Manag. Res. J. Sustain. Circ. Econ.* **2020**, *39*, 374–385. [CrossRef]

39. D1-230 Dėl Cheminėmis Medžiagomis Užterštų Teritorijų Tvarkymo Aplinkos/Apsaugos Reikalavimų Patvirtinimo. Available online: https://e-seimas.lrs.lt/portal/legalAct/lt/TAD/TAIS.319604/aWhXQOotGB (accessed on 11 October 2023).
40. Kaczala, F.; Orupõld, K.; Augustsson, A.; Burlakovs, J.; Hogland, M.; Bhatnagar, A.; Hogland, W. Fractionation of Pb and Cu in the Fine Fraction (<10 Mm) of Waste Excavated from a Municipal Landfill. *Waste Manag. Res.* **2017**, *35*, 1175–1182. [CrossRef]
41. Büks, F.; Kaupenjohann, M. Global Concentrations of Microplastics in Soils—A Review. *SOIL* **2020**, *6*, 649–662. [CrossRef]
42. Medynska, A.; Medynska-Juraszek, M.; Szczepá, A. Microplastic Pollution in EU Farmland Soils: Preliminary Findings from Agricultural Soils (Southwestern Poland). *Agriculture* **2023**, *13*, 1733. [CrossRef]
43. Porterfield, K.K.; Hobson, S.A.; Neher, D.A.; Niles, M.T.; Roy, E.D. Microplastics in Composts, Digestates, and Food Wastes: A Review. *J. Environ. Qual.* **2023**, *52*, 225–240. [CrossRef]
44. Sholokhova, A.; Denafas, G.; Mykhaylenko, V. Microplastics Generation and Concentration during Mechanical-Biological Treatment of Mixed Municipal Solid Waste. *Environ. Res.* **2022**, *214*, 113815. [CrossRef]
45. Khalid, N.; Aqeel, M.; Noman, A.; Khan, S.M.; Akhter, N. Interactions and Effects of Microplastics with Heavy Metals in Aquatic and Terrestrial Environments. *Environ. Pollut.* **2021**, *290*, 118104. [CrossRef]
46. Liu, Q.; Wu, H.; Chen, J.; Guo, B.; Zhao, X.; Lin, H.; Li, W.; Zhao, X.; Lv, S.; Huang, C. Adsorption Mechanism of Trace Heavy Metals on Microplastics and Simulating Their Effect on Microalgae in River. *Environ. Res.* **2022**, *214*, 113777. [CrossRef]
47. Korhonen, J.; Honkasalo, A.; Seppälä, J. Circular Economy: The Concept and Its Limitations. *Ecol. Econ.* **2018**, *143*, 37–46. [CrossRef]
48. Jain, M.S.; Daga, M.; Kalamdhad, A.S. Variation in the Key Indicators during Composting of Municipal Solid Organic Wastes. *Sustain. Environ. Res.* **2019**, *29*, 9. [CrossRef]
49. Abdel-Shafy, H.I.; Ibrahim, A.M.; Al-Sulaiman, A.M.; Okasha, R.A. Landfill Leachate: Sources, Nature, Organic Composition, and Treatment: An Environmental Overview. *Ain Shams Eng. J.* **2023**, *15*, 102293. [CrossRef]
50. Lindamulla, L.; Nanayakkara, N.; Othman, M.; Jinadasa, S.; Herath, G.; Jegatheesan, V. Municipal Solid Waste Landfill Leachate Characteristics and Their Treatment Options in Tropical Countries. *Curr. Pollut. Rep.* **2022**, *8*, 273–287. [CrossRef]

sustainability MDPI

Article

Sustainable Water Management with Design and Economic Evaluation of Recycling Greywater at Abu Dhabi University—A Case Study on Decentralization

Chandra Mouli R. Madhuranthakam *, Malak AbuZaid, Omar Chaalal and Tala Ghannam

Chemical Engineering Department, Abu Dhabi University, Abu Dhabi P.O. Box 59911, United Arab Emirates; omar.chaalal@adu.ac.ae (O.C.)

* Correspondence: chandra.mouli@adu.ac.ae; Tel.: +971-2-5015304

Abstract: Wastewater can be segregated as greywater and blackwater separately. The greywater generated in malls, restaurants, and university buildings is generally dilute, while it will later become concentrated when it is merged into the main sewage collection line. It would be more economical and environmentally friendly if the greywater is treated locally using a modular wastewater treatment facility that produces treated water amenable for other uses such as irrigation or horticulture. The objective of this article is to study the economic feasibility and design a decentralized plant that produces fresh water from greywater generated at the Abu Dhabi university campus located in the United Arab Emirates. The proposed unit will consist of a compact design of filtration, chemical treatment and disinfection processes that would generate treated wastewater that can be used for horticulture in and around the local campus or can be stored and supplied for irrigation purposes. Several parameters such as total suspended solids, biological oxygen demand, and chemical oxygen demand are measured and monitored throughout the entire process and are regulated by appropriate operations performed for each unit. This study shows that decentralization of greywater treatment is not only economical but also essential for the management of fresh water, which in turn assures environmental sustainability. By using coagulation, flocculation and chlorination with a 30 mg/L alum dosage, 0.6 mg/L of polyacrylamide and 0.12 mg/L of sodium hypochlorite, respectively, greywater is treated to meet the water specification for reusing it for horticulture. Further, a modular plant with an investment of USD 8 M is proved to process 90,000 tons of greywater with a 34% discounted rate of return.

Keywords: decentralization; economic feasibility; environmental sustainability; greywater treatment; suspended solids; water conservation

Citation: Madhuranthakam, C.M.R.; AbuZaid, M.; Chaalal, O.; Ghannam, T. Sustainable Water Management with Design and Economic Evaluation of Recycling Greywater at Abu Dhabi University—A Case Study on Decentralization. *Sustainability* **2023**, *15*, 16208. https://doi.org/10.3390/su152316208

Academic Editors: Steve W. Lyon and Marc A. Rosen

Received: 6 October 2023
Revised: 3 November 2023
Accepted: 17 November 2023
Published: 22 November 2023

1. Introduction

With the current rate of consumption, it is estimated that the world's water supply will fall short of demand by 40% by 2030. Numerous global regions are already witnessing demand outstripping supply, and in other areas, water scarcity is impeding economic development [1]. As economic growth and unpredictable weather patterns intensify competition for water resources, the threat of a worldwide food crisis looms, potentially affecting businesses, governments, citizens, and farmers. The World Resources Institute (WRI) predicts that 33 countries will face severe water stress by 2040 when assessing future water stress situations. Many of these countries, 14 to be exact, are in the Middle East, with 9 scoring the highest possible score of 5, including the United Arab Emirates (UAE), highlighting a particularly severe water stress situation [2]. While water makes up 70% of our Earth's surface, only 3% of it is not saltwater. Saline water, which accounts for 97% of the total, is unsuitable for human consumption, agriculture, or industrial cleaning without substantial energy expenditure and desalination processes. However, this saline water has limited use, such as in certain types of industrial cooling. Over the past century, our water sources have remained static, but the global population has skyrocketed [3]. A staggering 80% of all freshwater drawn globally is used for irrigating food crops, which establishes agriculture as the primary consumer of water. Additionally, pollution is swiftly contaminating both surface and groundwater, restricting their use for irrigation. For example, over 80% of wastewater from human activities is discharged into oceans and rivers without any prior treatment. This practice results in considerable water pollution, rendering these bodies of water unfit for irrigation [4–7]. Wastewater treatment is an increasingly recognized crucial approach to water resource management. The goal is to remove pollutants from the wastewater, rendering it safe enough to be discharged back into the environment. The different methods for treatment vary depending on the source of wastewater such as irrigation wastewater, industrial wastewater, etc. For instance, utilizing treated wastewater can be seen as a resource recovery strategy in small to medium-sized agricultural areas. This is because it not only fulfills irrigation needs, but it could also potentially serve as a source of nutrients for crops [3,4,8]. Wastewater can be divided into five categories based on what it contains: blackwater, which consists of fecal sludge, urine, and feces; greywater, comprised of water from washing and bathing; blue water, which is stormwater and urban runoff; and green water and red water, which refer to agricultural and industrial waste, respectively. Among the different types of wastewater, greywater treatment is the main focus of this article. Greywater treatment contributes to water demand management by encouraging the preservation of premium freshwater and lowering both environmental pollution and total supply costs [9]. Greywater is used–untreated water from appliances such as bathtubs, showers, washbasins in bathrooms, clothes washers, and laundry tubs [10]. Greywater generated in malls, restaurants, and university buildings is generally dilute and it will later become concentrated when it is merged into the main sewage collection line. It would be more economical and environmentally friendly if the greywater is treated locally using small modular wastewater treating units that produce treated water amenable for irrigation or horticulture [11]. Recycling domestic wastewater like greywater using centralized and decentralized treatment methods is a highly effective strategy to reduce worldwide water demand. The choice between these systems is guided by location accessibility, economic circumstances, and the availability of treatment facilities [12]. Figure 1a,b show a schematic of centralized and decentralized treatment facilities used for wastewater treatment, respectively.

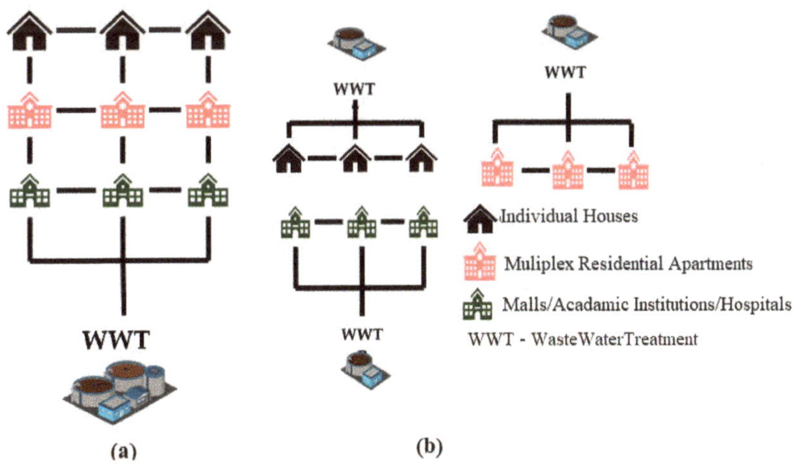

Figure 1. Schematic of different types of greywater treatment facilities: (**a**) centralized treatment; (**b**) decentralized treatment.

In many developed and developing countries, conventional centralized wastewater treatment plants are regarded as primary wastewater management solutions in which various types of wastewater, such as domestic, commercial, industrial, and hospital wastewater, together with storm and runoff water, are processed at a central treatment facility that is planned, developed, and run by a government or private organizations. On the other hand, a decentralized wastewater treatment facility treats wastewater very near the generation point on a small scale utilizing cheaper and simpler technologies [13–15]. In developed nations, chemical processes often have their own wastewater treatment facilities to manage the industrial waste generated by their activities. However, in developing countries such as Yemen, all wastewater from different industries is typically collected by a single sewage treatment plant. This can lead to an overload of the plant's capacity, potentially compromising its effectiveness and resulting in lower-quality treated effluents [16–19]. Due to centralized systems' high capital and operational costs, decentralized systems appear vital, cost-effective, reliable, and environmentally sustainable, especially for developed countries that cannot afford these massive expenses [20,21]. Owing to its benefits, the number of decentralized wastewater systems has grown significantly; more than 1000 systems have been constructed in China [22]. Brazil, for instance, has acknowledged decentralized wastewater treatment plants by law and is a component of the plan of the National Sanitation Strategy to treat 86.5% of all produced wastewater by 2023, increasing the current treatment index by 49% [18]. For small and medium-sized agricultural regions, using reclaimed wastewater can be a resource recovery alternative since it not only meets irrigation demands but also has the potential to be a source of nutrients for plants [23,24]. The substantial distance between treatment facilities and agricultural zones can pose a significant challenge in reusing water treated at centralized wastewater plants. This is because it necessitates the establishment of a distribution network for the treated water, which could result in duplicating the existing infrastructure for treatment, reclamation, and discharge. Decentralized wastewater treatment plants (DWWTPs) not only ease the load on centralized wastewater treatment plants (CWWTPs) and cut down on pumping costs, but they can also tackle the hurdles associated with reusing treated wastewater. This is possible by utilizing the treated water in nearby green spaces and agricultural lands [25–28]. Treated greywater has a variety of uses, including toilet flushing, car washing, fire prevention, and irrigating green spaces like parks and schoolyards. Its use for irrigation is becoming increasingly common, especially in desert regions, where it could reduce drinking water consumption by as much as 50% [29,30].

In this research article, the potential for treating greywater generated at Abu Dhabi University (ADU) located in the United Arab Emirates, is evaluated by conducting experiments at a lab scale and scaling it up to the design requirements. ADU consumes almost 100,000 m^3 of fresh water annually, costing around USD 225,000. Although this water is used for multiple purposes, a significant amount becomes greywater. This greywater is merged into a sewer, connected to the city's main pipeline leading to the CWWTP located in the Al Wathba region of the Abu Dhabi Emirate. The greywater produced at ADU is relatively low in concentration and can be treated in the proposed decentralized wastewater treatment plant (DWWTP) on campus. The treated water from this process can then be employed for campus horticultural needs or marketed to external vendors. Using the results obtained from the lab tests, a scaled-up version of the DWWTP is designed, and an economic analysis is run to confirm the profitability and the environmental benefits associated with the proposed treatment facility.

2. Materials and Methods

Greywater samples obtained from Abu Dhabi University are placed in three 150 mL flasks. Alum, polyacrylamide, and sodium hypochlorite are purchased from Green Water Treatment Solutions LLC, Abu Dhabi, UAE. pH is measured with a benchtop pH meter supplied by Thermo Scientific Orion Star A211, TDS is measured with a conductivity meter and using the APHA method 2540 C, TSS is measured using HACH LXV322.99.00001 and APHA method 2540D, BOD (5 day) and COD are measured based on titration using APHA 5210B and APHA 5220 C methods, respectively. After measuring the pH, total suspended solids (TSS), total dissolved solids (TDS), biological oxygen demand (BOD_5, referred to as BOD in general), and chemical oxygen demand (COD) of the initial sample, three different amounts of alum (20, 25 and 30 mg/L) are added to the three flasks containing greywater to achieve coagulation. Uniform and rapid mixing is conducted by using a magnetic stirrer bar. After one minute, three samples of 25 mL are collected for analysis, and the experiment is continued by adding polyacrylamide (PAM) at three levels (0.6, 0.8, and 1.0 mg/L), facilitating flocculation. After taking the samples from this step, sodium hypochlorite is added at three levels (0.04, 0.08, and 0.12 mL/L) for the chlorination step. After chlorination, the corresponding samples are collected. With alum, polyacrylamide, and chlorine (as sodium hypochlorite) at three levels each, a factorial design of 3^3 = 27 experiments is conducted. The measured quantities (output variables) are pH, TDS, TSS, BOD, and COD. Three-factor levels 1, 2, 3 are used for alum, polyacrylamide and chlorine. Figure 2 shows the complete design of a 3-level, 3-factor full factorial design.

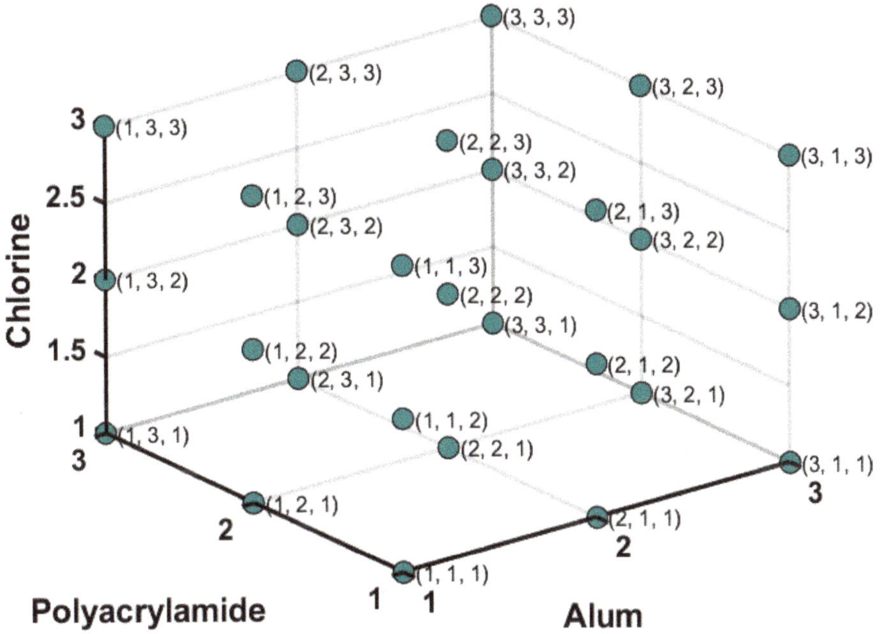

Figure 2. Three-dimensional representation of 3-level 3-factor full factorial design of experiments.

The three independent variables represented by X_i (i = 1, 2 and 3 for alum, polyacrylamide, and chlorine, respectively) and the five dependent variables represented by Y_i, (i = 1, 2, 3, 4 and 5 for pH, TDS, TSS, BOD and COD, respectively) used in this study are summarized as shown in Table 1.

Table 1. Experimental design factors and their levels.

Factors	Variable	Unit	Levels		
			Low	Medium	High
X_1	Alum	mg/L	20	25	30
X_2	Polyacrylamide	mg/L	0.6	0.8	1.0
X_3	Chlorine	ml/L	0.04	0.08	0.12
Y_1	pH	-	Dependent variable		

Table 1. *Cont.*

Factors	Variable	Unit	Levels		
			Low	Medium	High
Y_2	TDS reduction	%		Dependent variable	
Y_3	TSS reduction	%		Dependent variable	
Y_4	BOD reduction	%		Dependent variable	
Y_5	COD	mg/L		Dependent variable	

The obtained data are analyzed using response surface plots in MATLAB (R2023a). Using the lab-scale data results, a scale-up decentralized greywater treatment plant is designed, and its economic feasibility is analyzed with a discounted rate of return method.

3. Results and Discussion

The factorial design for the three factors at three levels yields 27 experiments with all possible combinations of X_1, X_2, and X_3. The response or output variables Y_1, Y_2, Y_3, Y_4, and Y_5 are measured in each combination. The experimental values obtained for this 3^3-factorial design are shown in Table 2.

Table 2. Complete set of factorial levels and the experimental values of X_1, X_2, X_3, Y_1, Y_2, Y_3, Y_4 and Y_5.

Experiment Number	X_1	X_2	X_3	Y_1	Y_2	Y_3	Y_4	Y_5
1	1	1	1	8.4	−47	43	−1	1132
2	1	1	2	8.4	−70	57	12	984
3	1	1	3	8.4	−14	43	8	1032
4	1	2	1	8.5	−27	43	10	1012
5	1	2	2	8.3	−58	71	0	1120
6	1	2	3	8.7	−79	0	3	1092
7	1	3	1	8.4	−25	57	9	1018
8	1	3	2	8.2	−48	71	12	986
9	1	3	3	8.4	−45	43	−18	1321
10	2	1	1	8.3	−63	43	21	890
11	2	1	2	8.3	−16	0	−30	1462
12	2	1	3	8.5	−68	71	13	983
13	2	2	1	8.6	−49	43	−1	1132
14	2	2	2	8.3	−73	71	−2	1145
15	2	2	3	8.5	−21	71	3	1092
16	2	3	1	8.4	−53	43	11	997
17	2	3	2	8.5	−47	71	9	1023
18	2	3	3	8.5	−41	57	2	1098
19	3	1	1	8	5	43	12	985
20	3	1	2	8.2	1	43	18	923
21	3	1	3	8.2	4	−14	62	429
22	3	2	1	8.3	3	57	18	920
23	3	2	2	8.3	3	−14	50	564
24	3	2	3	8.4	4	57	40	673
25	3	3	1	8.2	2	14	−10	1234
26	3	3	2	8.3	4	14	9	1021
27	3	3	3	8.2	4	14	13	980

3.1. Effect on pH

The initial pH of the untreated greywater was 8, and the pH varied between 8 and 8.8 when alum, PAM, and Cl were added. The desired pH for using the treated greywater for horticulture is 8 to 8.5. Figure 3a–f shows the response surface plots and corresponding contour plots for pH at different levels of alum.

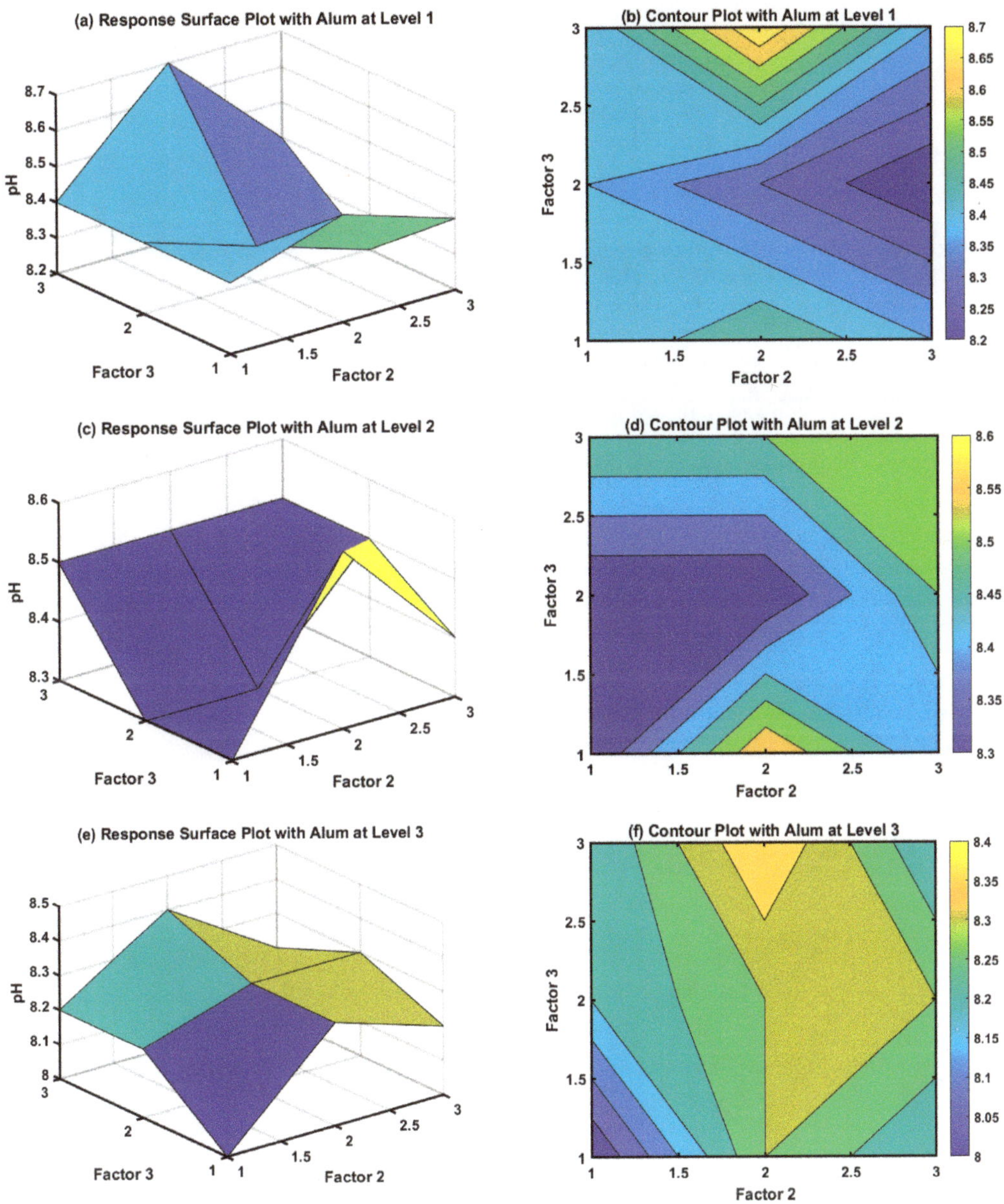

Figure 3. Response surface plot for pH vs. Factor 2 and Factor 3 levels with alum at (**a**) level 1, (**c**) level 2; (**e**) level 3. (**b**,**d**,**f**) are contour plots corresponding to (**a**,**c**,**e**), respectively.

From the contour plots shown in Figure 3b,d,f, it can be observed that as the alum dosage increased, the pH level reduced for the individual effects. Thammasane and Kaosol [31] and Wei et al. [32] reported similar changes when alum and polyaluminium chloride were used as coagulants in treating wastewater. When alum is added to water, it undergoes hydrolysis to form aluminum hydroxide and lowers the pH of water due to the generation of acidic species. Adding PAM does not significantly affect the increase or decrease in pH levels. However, chlorination led to an increase in pH, which could be due to the formation of sodium hydroxide, which is one of the products when sodium hypochlorite is used as a source of chlorine during the disinfection stage of water treatment. For all combinations of alum, PAM, and chlorine levels, the measured pH was in the range of the target 8.0 to 8.5. This facilitates choosing optimal factor levels based on the effect on other measured variables.

3.2. Effect on TDS

The initial TDS of the greywater sample used in this study is 894 mg/L. The allowable TDS for the treated greywater in Abu Dhabi is 2000 mg/L. Although the initial TDS of the greywater sample is less than that of the allowable TDS level, adding alum, PAM, and chlorine led to an increase in the TDS level. For individual effects, with alum at levels 1 and 2, TDS increased, while at level 3, TDS decreased. However, this increase/decrease did not exceed the maximum allowable TDS of 2000 mg/L for the treated greywater reused for horticulture. Figure 4a–f show the response surface plots and the corresponding contour plots for the reduction in the TDS level with the three different alum levels.

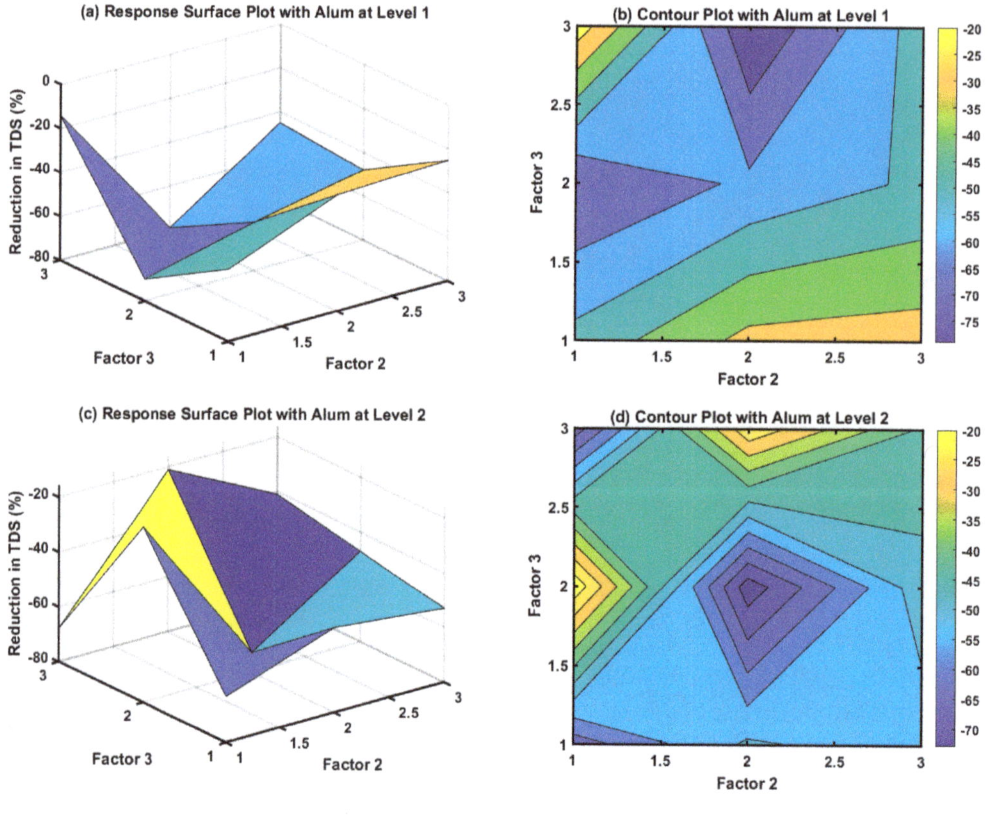

Figure 4. *Cont.*

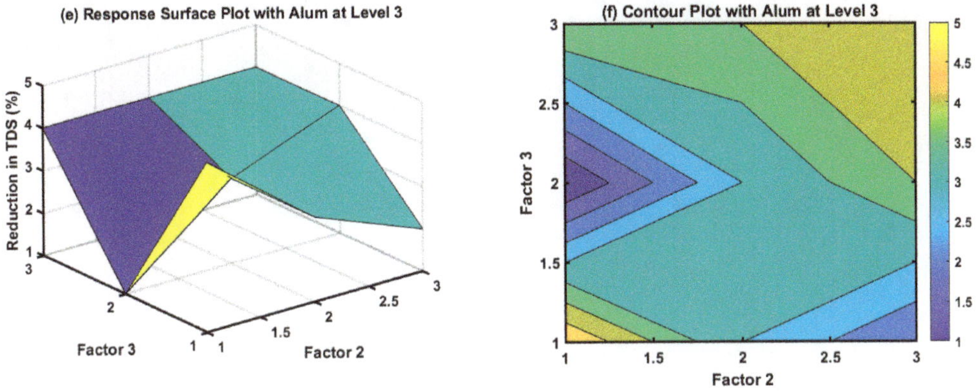

Figure 4. Response surface plot for the reduction level in TDS (%) vs. Factor 2 and Factor 3 levels with alum at (**a**) level 1, (**c**) level 2; (**e**) level 3. (**b**,**d**,**f**) are contour plots corresponding to (**a**,**c**,**e**), respectively.

The contour plots of Figure 3b,d,f, show that TDS increased with the first and second levels of alum (the negative values on the color bar mean an increase in the TDS level) for all levels of PAM and chlorine. However, a further increase in the alum level led to a positive TDS reduction (see Figure 3f). If the objective is to reduce TDS to a significant level, then by adding more alum, one can achieve this goal.

3.3. *Effect on TSS*

Out of the many water treatment operations (such as sedimentation, filtration, floatation) that can reduce TSS, the main operation in this study that reduces the TSS is the coagulation step achieved with alum and flocculation achieved with the PAM. Figure 5a–f shows the response surface plots and the contour plots for the reduction in TSS for all factorial experiments conducted in this study. The target TSS in the treated greywater is 10 mg/L as per the Abu Dhabi municipality requirements. It is observed that the maximum TSS achieved in the set of all 27 experimental results was 8 mg/L. The minimum TSS level achieved in this study is 2 mg/L.

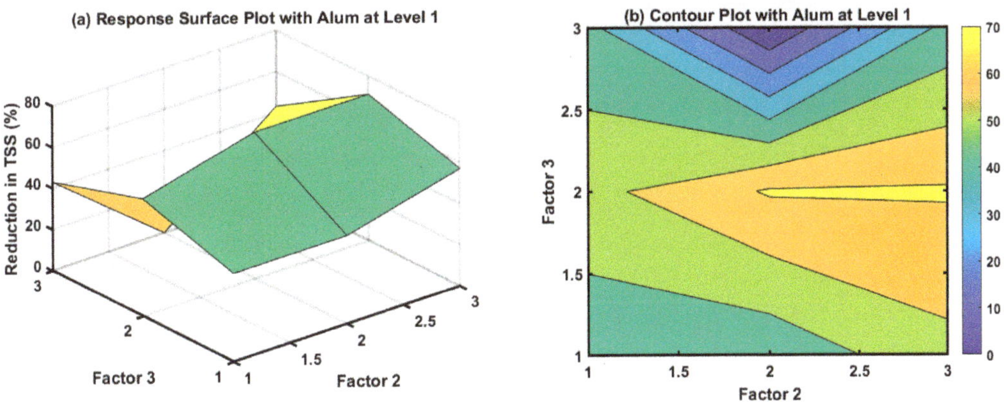

Figure 5. *Cont.*

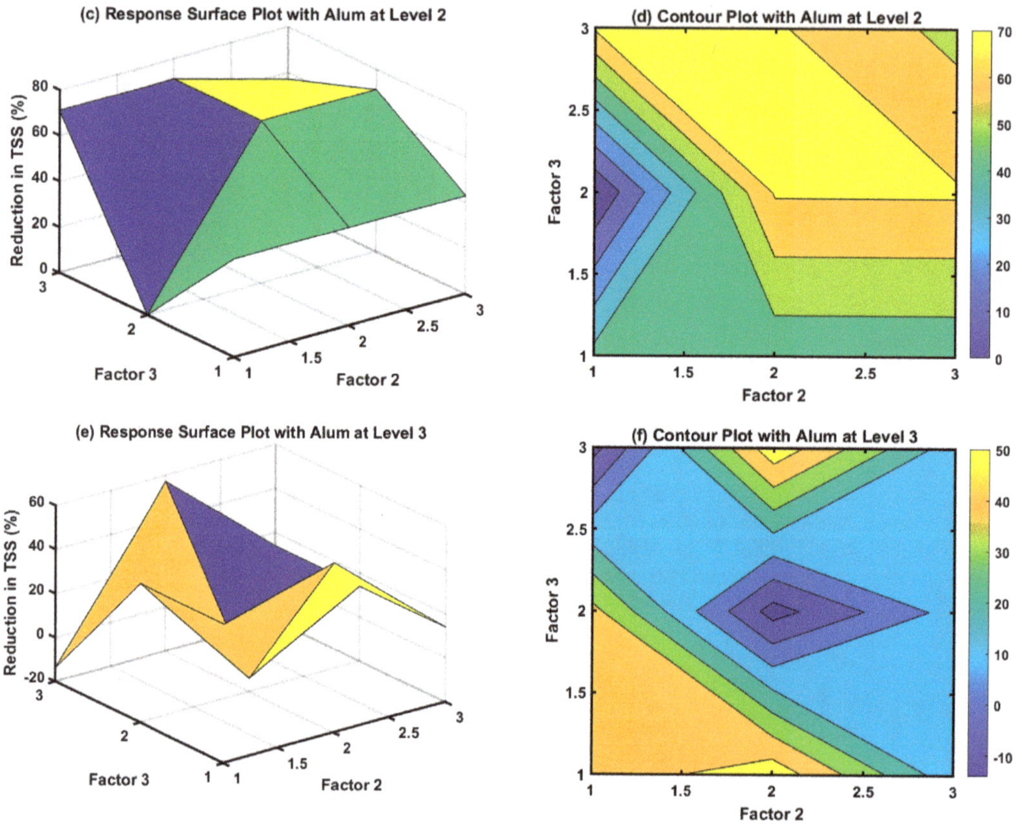

Figure 5. Response surface plot for the reduction level in TSS (%) vs. Factor 2 and Factor 3 levels with alum at (**a**) level 1, (**c**) level 2; (**e**) level 3. (**b**,**d**,**f**) are contour plots corresponding to (**a**,**c**,**e**), respectively.

3.4. *Effect on BOD and COD*

The initial BOD and COD of the untreated greywater sample were 337 mg/L and 1124 mg/L, respectively. Figure 6a–f shows the response surface and contour plots for the reduction in BOD for different levels of alum, PAM, and chlorine used to treat greywater in this study. From Figure 6b,d, it shows that the BOD increased with alum, PAM, and chlorine at levels 1 and 2 but from Figure 6f, it showed that the BOD decreased with further addition of alum and chlorine. Assessing the COD to BOD ratio is crucial in evaluating the biodegradability of greywater. This ratio for all the treated samples was observed to be 3.3. According to Khajvand et al. [33], the COD/BOD ratio typically ranges between 1.4 and 3.3, signifying that a substantial portion, almost half, of the organic matter is biodegradable. The highest ratio is due to elevated COD from xenobiotic organic compounds (XOCs). These compounds are commonly found in household chemicals and pharmaceuticals or can be produced during greywater treatment. It is well known that XOCs are recalcitrant against biological degradation and toxic to aquatic organisms, even at exceedingly low concentrations [34]. Figure 7 shows the relationship between COD and BOD of the treated greywater samples. Due to the elevated usage of detergents, surfactants are one of the main chemical components of greywater. In developed countries, the commonly used surfactants are biodegradable components. Therefore, the ratio of COD to BOD, encompassing organic substances such as fats, oil, food, and surfactants, tends to reach 2. On the other hand, greywater in developing countries is characterized by a relatively higher COD/BOD ratio reaching 3 or 4, as illustrated in Khanam and Patidar's [35] study and in this research. This is attributed mainly to the prevalent use of non-biodegradable detergents in such regions.

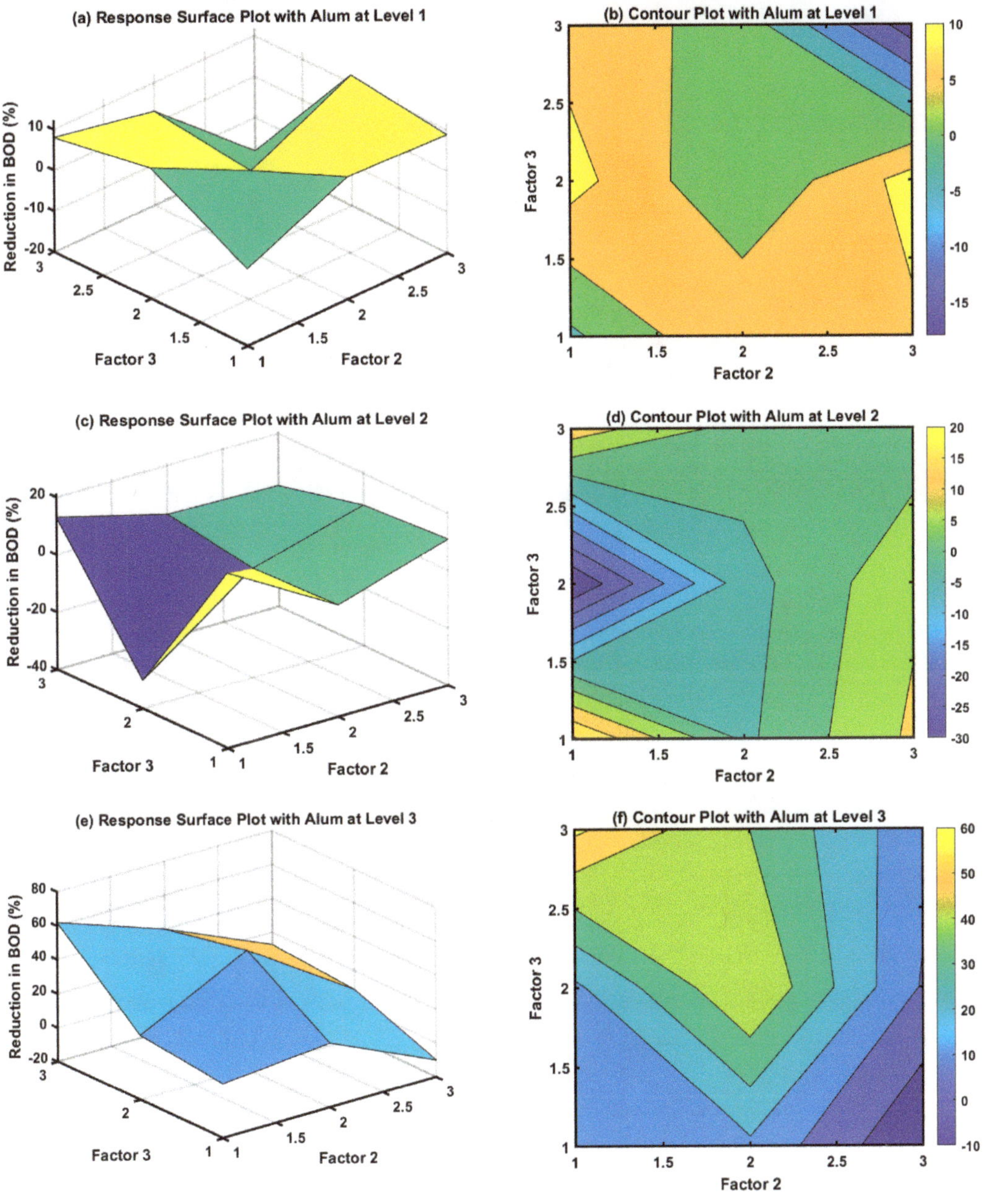

Figure 6. Response surface plot for the reduction level in BOD (%) vs. Factor 2 and Factor 3 levels with alum at (**a**) level 1, (**c**) level 2; (**e**) level 3. (**b**,**d**,**f**) are contour plots corresponding to (**a**,**c**,**e**) respectively.

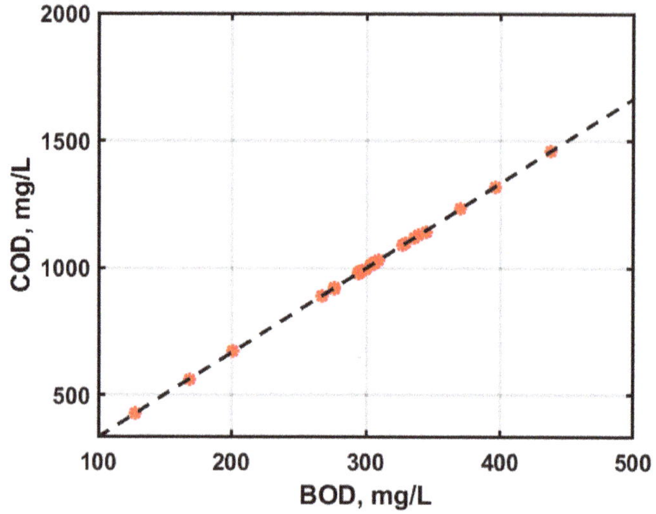

Figure 7. COD vs. BOD of the treated greywater samples (see Table 2 for different factors).

A nonlinear statistical model is obtained by fitting the experimental values of X_4 using least square nonlinear regression. MATLAB (Ver 2023a) is used for modeling and analysis of the results. The model obtained for X_4 is given by Equation (1).

$$\begin{aligned} Y_4 = {} & \beta_0 + \beta_1 X_1 + \beta_2 X_2 + \beta_3 X_3 + \beta_{12} X_1 X_2 + \beta_{23} X_2 X_3 + \beta_{31} X_3 X_1 + \beta_{11} X_1^2 + \beta_{22} X_2^2 + \beta_{33} X_3^2 + \beta_{210} X_1^2 X_2 \\ & + \beta_{201} X_1^2 X_3 + \beta_{120} X_2^2 X_1 + \beta_{103} X_3^2 X_1 + \beta_{013} X_3^2 X_2 \end{aligned} \quad (1)$$

where Y_4 is BOD and β's are the corresponding coefficients of each term in Equation (1). The values of β's for a 95% confidence limit is given in Table 3, along with their confidence interval. Table 4 shows the ANOVA obtained for the corresponding model shown in Equation (1), while Table 5 shows the ANOVA obtained for the parameters (β's) of the model. Table 4 shows that removal of BOD is strongly affected by the alum dosage and the interaction of alum with chlorine factors. This is also apparent and confirmed in Figure 5f where the interaction effect between the alum dosage (X_1) and chlorine dosage (X_3) reveals a maximum of 60% BOD removal when alum dosage is 30 mg/L and the chlorine dosage is at 0.12 mg/L. The R^2 for this fit is 0.78. Several modifications to this model in terms of adding cubic terms or removing interaction terms led to a fit with an R^2 value less than 0.77. Hence, this model (shown by Equation (1)) is retained to model removal of BOD for all the experimental data used in this study.

Table 3. Coefficients and their confidence interval corresponding to Equation (1) for Y_4.

Coefficient	Value	Lower Limit	Upper Limit
β_0	180.11	−82.88	443.10
β_1	−91.83	−283.70	100.03
β_2	−118.33	−310.20	73.53
β_3	−88.16	−280.03	103.70
β_{12}	74.66	−10.58	159.91
β_{23}	50.66	−34.58	135.91
β_{31}	−28	−113.25	57.25
β_{11}	16.83	−27.30	60.97

Table 3. *Cont.*

Coefficient	Value	Lower Limit	Upper Limit
β_{22}	13	−31.13	57.13
β_{33}	27.16	−16.97	71.30
β_{210}	−11	−25.99	3.99
β_{201}	8	−6.99	22.99
β_{120}	−9	−23.99	5.99
β_{103}	1.5	−13.49	16.49
β_{013}	−14	−28.99	0.99

Table 4. Analysis of variance (ANOVA) corresponding to the model (shown in Equation (1)).

	Sum Square Error	Degrees of Freedom	Mean Square Error	F Value	*p* Value
Total	8922.7	26	343.18		
Model	6881	15	458.73	2.47	0.0676
Linear	2259	3	753	4.06	0.0362
Nonlinear	4622	12	385.17	2.07	0.1184
Residual	2041.7	11	185.61		

Table 5. Analysis of variance for BOD removal (the underline shows the *p*-values < 0.05).

Parameter	Sum Square Error	Degrees of Freedom	Mean Square Error	F Value	*p* Value
x1	1740.5	1	1740.5	9.3774	0.010808
x2	338	1	338	1.8211	0.20431
x3	180.5	1	180.5	0.97249	0.34525
x1:x2	341.33	1	341.33	1.839	0.20225
x2:x3	341.33	1	341.33	1.839	0.20225
x1:x3	1200	1	1200	6.4653	0.027348
x1^2	704.17	1	704.17	3.7939	0.077409
x2^2	150	1	150	0.80816	0.38792
x3^2	28.167	1	28.167	0.15176	0.7043
(x1^2):x2	484	1	484	2.6077	0.13464
(x1^2):x3	256	1	256	1.3793	0.26502
x1:(x2^2)	324	1	324	1.7456	0.21325
x1:(x3^2)	9	1	9	0.04849	0.82975
x2:(x3^2)	784	1	784	4.224	0.064392
Error	2041.7	11	185.61		

Figure 8a–d shows the validation indicator plots for the model and Figure 9 shows the comparison of the fit to the experimental values of Y_4.

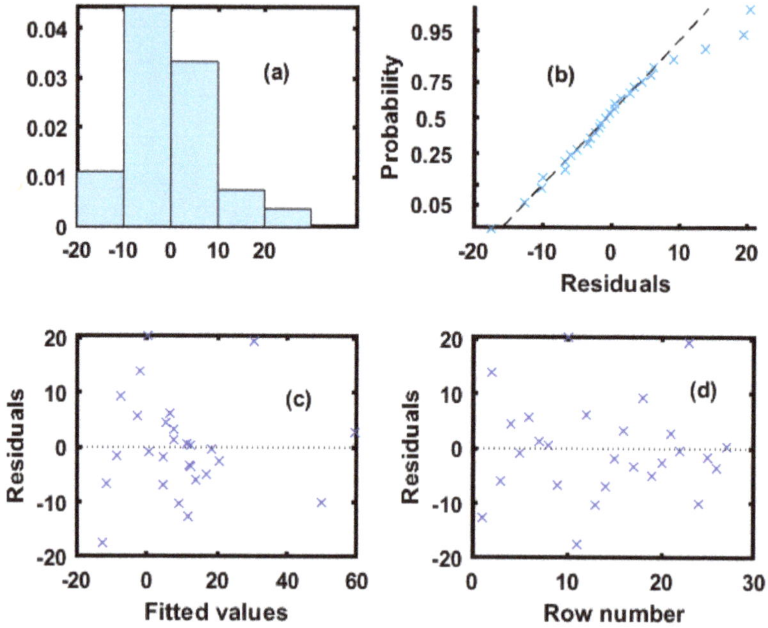

Figure 8. Validation indicator plots for the model: (**a**) histogram of the residuals; (**b**) normal probability of residuals; (**c**) residuals versus fitted values and (**d**) case order plot of residuals.

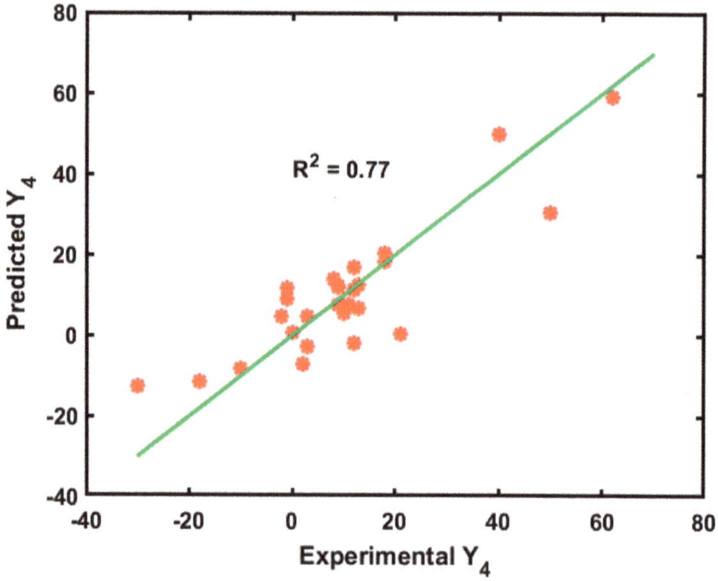

Figure 9. Comparison of the predicted reduction of BOD, Y_4 to the experimental reduction in BOD.

Figure 8a shows the histogram of residuals (difference between the experimental and fitted value) with a bell shape skewed a little bit to the right side, while Figure 8b shows the normal probability of the residuals. Except for few points (which are also responsible for the slight skewness seen in Figure 8a), most of the points fall along the straight line, suggesting that the predicted BOD values and the experimental BOD values are in reasonable agreement and further confirms the normal distribution of the data. This is also confirmed from Figure 8c where the residuals are symmetrically aligned around zero, except for a few outliers. Figure 8d shows that the residuals are scattered randomly around zero. Figure 9 shows that 77% of the variability can be captured from the statistical model specified by Equation (1). The R^2 value obtained for this data set can be further improved by fitting nonlinear models or neural networks (by using more experiment data). A summary of the analysis of the treated and untreated greywater with respect to the optimal levels of alum (30 mg/L), PAM (0.6 mg/L) and chlorine (0.12 mg/L) is shown in Table 6.

Table 6. Greywater characteristics before and after treatment.

Property	Untreated Greywater	Treated Greywater	Deviation
pH	8	8.2	−0.2
TDS (mg/L)	894	861	33
BOD (mg/L)	337	128	209
COD (mg/L)	1124	429	695

3.5. Economic Feasibility of a Pilot-Scale Plant

From the lab-scale experimental analysis, alum at level 3, PAM at level 2 or 3, and chlorine at level 3 ensures treated greywater with desired properties (targeted for use in horticulture). A small modular plant (that can be installed in two car parking lots, 16 ft × 18 ft) in Abu Dhabi University is conceptually designed to produce 98,000 tons of treated greywater per year, targeted for horticulture reuse. A process flow diagram with the equipment used and the sequence of operations is shown in Figure 10.

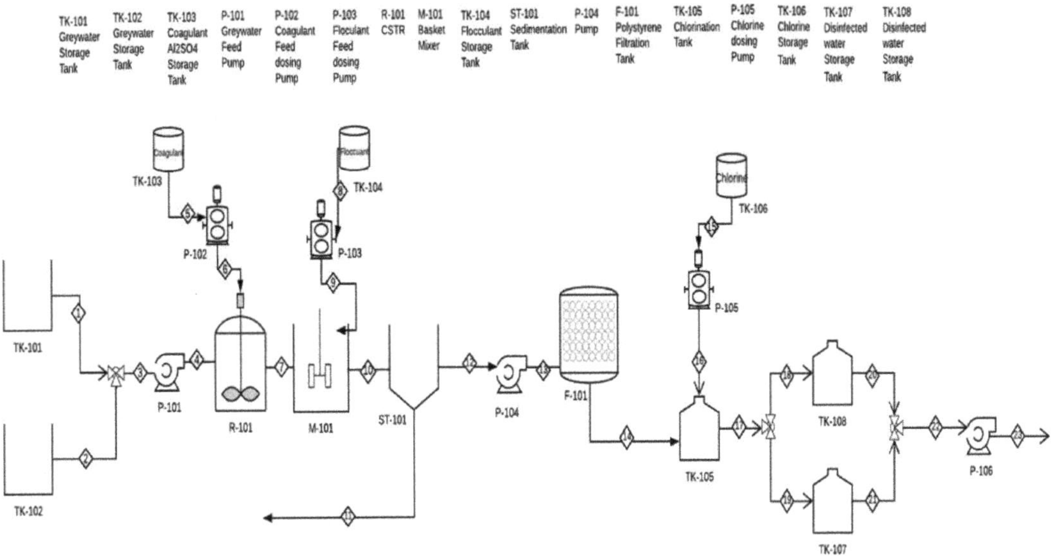

Figure 10. Process flow diagram (PFD) of pilot-scale plant used for treating the greywater generated in Abu Dhabi University.

This process consists of mainly three stages: chemical treatment, filtration, and disinfection. Initially, greywater is collected and stored in dual storage tanks, as a critical measure to avoid potential overflow. The untreated water is then pumped to a continuously stirred tank reactor (R-101) where the key coagulant, alum, is added to mainly remove turbidity by neutralizing the charge of suspended particles and allowing for aggregation. To further enhance the effectiveness of coagulation and optimize the process, a coagulant-aid or flocculant, polyacrylamide is added to the greywater as it enters the basket mixer (M-101). This step involves slow mixing, enabling the agglomeration and formation of bigger flocs. Following chemical treatment, the treated effluent is pumped to a sedimentation tank (ST-101), where the resulting flocs settle down due to gravity. The sludge formed at the bottom of the tank containing impurities can be disposed of appropriately. To ensure the elimination of any remaining contaminants, the treated greywater is directed to a polystyrene filtration unit (F-101). Polystyrene (PS), which can be recycled, effectively remediates organic matter and dyes [36–38]. Finally, the filtered water goes to the disinfection tank (TK-105) where sodium hypochlorite (which releases chlorine) is added. This step is very important to remove any harmful pathogens in the water. Finally, twin storage tanks receive the disinfected water from where it can be pumped to designated sources. The detailed design and parameters of each piece of equipment are provided in the Supplementary Information (see Tables S1–S7).

To estimate the economic feasibility and viability of the project, a profitability analysis was conducted, employing the discounted cash flow method DCF at a discount rate of 10%. This approach considers the time value of money [39] and involves the discounted fixed capital investment DFCI, discounted cash flow rate of return DCFROR, and discounted payback period DPBP. The rate of return or discount rate is when the project's net present value becomes zero; however, a higher rate of return indicates a more substantial level of profitability associated with the project [40]. According to cost analysis (see Supplementary Information), this project was found to be economically feasible and has a DCFROR of 34%, showing its potential on a financial basis and in addressing environmental issues.

4. Conclusions

A factorial three-level design of lab-scale experiments for treating greywater that originated from Abu Dhabi University showed that by using coagulation, flocculation, and chlorination steps, treated water amenable for horticulture can be obtained. It was found that alum at 30 mg/L, PAM at 0.6 mg/L, and chlorine (in the form of sodium hypochlorite) at 1 mg/L can be used to treat greywater to reduce TDS, TSS, BOD, and COD levels while maintaining the desired pH range. Using the experimental results, the design of a conceptual modular greywater treatment unit was performed with a target of processing 90,000 tons of water. Detailed design and economic calculations showed that this plant can be constructed in a size equivalent to two car parking lots with a speculative USD 8 M for the fixed capital investment and a discounted rate of return of 34%. The main limitation of this study would be the variation in the greywater characteristics on a continuous basis. Nevertheless, an empirical model can be used to relate the optimal dosage levels of the coagulant, flocculant, and sodium hypochlorite with respect to different initial greywater properties. The proposed concept of decentralization of greywater treatment will not only reduce the load on the regular centralized wastewater treatment but also minimize the treatment costs and constitute a sustainable method for closing the loop of water consumption and wastewater generation.

Supplementary Materials: The following supporting information can be downloaded at: https://www.mdpi.com/article/10.3390/su152316208/s1, The detailed equipment sizes and cost analysis is shown in the supplementary tables provided along with this article. Table S1: Design of Continuous Stirred Tank Reactor R-101; Table S2: Design of flocculation basin M-101; Table S3: Design of pumps; Table S4: Tanks Design; Table S5: Capital cost for process equipment; Table S6: Process cost; Table S7: Cost of manufacturing.

Author Contributions: Conceptualization, C.M.R.M.; methodology, C.M.R.M.; software, C.M.R.M.; validation, C.M.R.M., M.A. and T.G.; formal analysis, C.M.R.M. and O.C.; investigation, C.M.R.M., M.A.; resources, C.M.R.M.; data curation, C.M.R.M., M.A. and T.G.; writing—original draft preparation, C.M.R.M., M.A.; writing—review and editing, C.M.R.M.; visualization, C.M.R.M., M.A.; supervision, C.M.R.M.; project administration, C.M.R.M.; funding acquisition, C.M.R.M. All authors have read and agreed to the published version of the manuscript.

Funding: This research received no external funding.

Institutional Review Board Statement: Not Applicable.

Informed Consent Statement: Not Applicable.

Data Availability Statement: The data presented in this study are available in this article and Supplementary Materials.

Acknowledgments: We thank Abu Dhabi University for providing the infrastructure to conduct this research project.

Conflicts of Interest: The authors declare no conflict of interest.

References

1. Ensuring Sustainable Water Management for All by 2030 | World Economic Forum. Available online: https://www.weforum.org/impact/sustainable-water-management/ (accessed on 9 January 2023).
2. Maddocks, A.; Young, R.S.; Reig, P. Ranking the World's Most Water-Stressed Countries in 2040. 2015. Available online: https://www.wri.org/insights/ranking-worlds-most-water-stressed-countries-2040 (accessed on 24 November 2022).
3. Landels, A.; Cagney, N.; Bauer, L.; Beacham, T.A.; Balabani, S.; Allen, M.J. Development of Vortex Bioreactor Technology for Decentralised Water Treatment. In *Vortex Structures in Fluid Dynamic Problems*; InTech: Vienna, Austria, 2017. [CrossRef]
4. Estévez, S.; González-García, S.; Feijoo, G.; Moreira, M.T. How decentralized treatment can contribute to the symbiosis between environmental protection and resource recovery. *Sci. Total Environ.* **2022**, *812*, 151485. [CrossRef] [PubMed]
5. Velasco-Muñoz, J.F.; Aznar-Sánchez, J.A.; Belmonte-Ureña, L.J.; Román-Sánchez, I.M. Sustainable Water Use in Agriculture: A Review of Worldwide Research. *Sustainability* **2018**, *10*, 1084. [CrossRef]
6. Singh, S.; Pradhan, N.; Ojha, N.; Roy, B.; Bose, S. Grey Water Treatment and Its Application in Cultivation of Plants. 2016. Available online: https://www.researchgate.net/publication/314116189 (accessed on 29 August 2023).
7. United Nations. The Sustainable Development Goals Report. Available online: https://unstats.un.org/sdgs/report/2022/ (accessed on 3 January 2023).
8. Poustie, A.; Yang, Y.; Verburg, P.; Pagilla, K.; Hanigan, D. Reclaimed wastewater as a viable water source for agricultural irrigation: A review of food crop growth inhibition and promotion in the context of environmental change. *Sci. Total Environ.* **2020**, *739*, 139756. [CrossRef] [PubMed]
9. Al-Hamaiedeh, H.; Bino, M. Effect of treated grey water reuse in irrigation on soil and plants. *Desalination* **2010**, *256*, 115–119. [CrossRef]
10. U.S. Green Building Council. Glossary. Available online: https://www.usgbc.org/glossary (accessed on 11 January 2023).
11. Difference between Blackwater and Greywater | Difference between. Available online: http://www.differencebetween.net/miscellaneous/difference-between-blackwater-and-greywater/ (accessed on 11 January 2023).
12. Athirah, A.; Al-Gheethi, A.A.S.; Noman, E.A.; Mohamed, R.M.S.R.; Kassim, A.H.M. Centralised and Decentralised Transport Systems for Greywater and the Application of Nanotechnology for Treatment Processes. In *Management of Greywater in Developing Countries: Alternative Practices, Treatment and Potential for Reuse and Recycling*; Springer: Berlin/Heidelberg, Germany, 2019; pp. 227–244. [CrossRef]
13. Zhang, D.Q.; Tan, S.K. Decentralized wastewater management and its application in an urban area of Beijing, China. In Proceedings of the 2010 4th International Conference on Bioinformatics and Biomedical Engineering, iCBBE 2010, Chengdu, China, 18–20 June 2010. [CrossRef]
14. Chirisa, I.; Bandauko, E.; Matamanda, A.; Mandisvika, G. Decentralized domestic wastewater systems in developing countries: The case study of Harare (Zimbabwe). *Appl. Water Sci.* **2017**, *7*, 1069–1078. [CrossRef]
15. Massoud, M.A.; Tarhini, A.; Nasr, J.A. Decentralized approaches to wastewater treatment and management: Applicability in developing countries. *J. Environ. Manag.* **2009**, *90*, 652–659. [CrossRef]
16. Mladenov, N.; Dodder, N.G.; Steinberg, L.; Richardot, W.; Johnson, J.; Martincigh, B.S.; Buckley, C.; Lawrence, T.; Hoh, E. Persistence and removal of trace organic compounds in centralized and decentralized wastewater treatment systems. *Chemosphere* **2022**, *286*, 131621. [CrossRef]
17. Al-Gheethi, A.A.; Mohamed, R.M.; Jais, N.M.; Efaq, A.N.; Halid, A.A.; Wurochekke, A.A.; Amir-Hashim, M.K. Influence of pathogenic bacterial activity on growth of *Scenedesmus* sp. and removal of nutrients from public market wastewater. *J. Water Health* **2017**, *15*, 741–756. [CrossRef]

18. Ferreira, M.M.; Fiore, F.A.; Saron, A.; da Silva, G.H.R. Systematic review of the last 20 years of research on decentralized domestic wastewater treatment in Brazil: State of the art and potentials. *Water Sci. Technol.* **2021**, *84*, 3469–3488. [CrossRef]
19. Ahmed, M.; Arora, M. Suitability of Grey Water Recycling as decentralized alternative water supply option for Integrated Urban Water Management. *IOSR J. Eng.* **2012**, *2*, 31–35. [CrossRef]
20. Capodaglio, A.G.; Callegari, A.; Cecconet, D.; Molognoni, D. Sustainability of decentralized wastewater treatment technologies. *Water Pract. Technol.* **2017**, *12*, 463–477. [CrossRef]
21. Muzioreva, H.; Gumbo, T.; Kavishe, N.; Moyo, T.; Musonda, I. Decentralized wastewater system practices in developing countries: A systematic review. *Util Policy* **2022**, *79*, 101442. [CrossRef]
22. Liang, X.; Van Dijk, M.P. Financial and economic feasibility of decentralized wastewater reuse systems in Beijing. *Water Sci. Technol.* **2010**, *61*, 1965–1973. [CrossRef] [PubMed]
23. Jaramillo, M.F.; Restrepo, I. Wastewater Reuse in Agriculture: A Review about Its Limitations and Benefits. *Sustainability* **2017**, *9*, 1734. [CrossRef]
24. Helmecke, M.; Fries, E.; Schulte, C. Regulating water reuse for agricultural irrigation: Risks related to organic micro-contaminants. *Environ. Sci. Eur.* **2020**, *32*, 4. [CrossRef]
25. Qureshi, A.S. Challenges and Prospects of Using Treated Wastewater to Manage Water Scarcity Crises in the Gulf Cooperation Council (GCC) Countries. *Water* **2020**, *12*, 1971. [CrossRef]
26. Cirelli, G.L.; Consoli, S.; Di Grande, V. Long-term storage of reclaimed water: The case studies in Sicily (Italy). *Desalination* **2008**, *218*, 62–73. [CrossRef]
27. Capodaglio, A.G. Integrated, Decentralized Wastewater Management for Resource Recovery in Rural and Peri-Urban Areas. *Resources* **2017**, *6*, 22. [CrossRef]
28. Malila, R.; Lehtoranta, S.; Viskari, E.L. The role of source separation in nutrient recovery—Comparison of alternative wastewater treatment systems. *J. Clean. Prod.* **2019**, *219*, 350–358. [CrossRef]
29. Lu, W.; Leung, A.Y.T. A preliminary study on potential of developing shower/laundry wastewater reclamation and reuse system. *Chemosphere* **2003**, *52*, 1451–1459. [CrossRef]
30. DHWA (Department of Health Western Australia). Draft Guidelines for the Reuse of Greywater in Western Australia. Available online: https://www.yumpu.com/en/document/view/28027174/draft-guidelines-for-the-reuse-of-greywater-in-western-australia (accessed on 31 January 2023).
31. Thammasane, S.; Kaosol, T. Impact of Chemical Coagulants for *Oscillatoria* sp. Removal from Raw Water on Chemical Coagulation Process. *Am. J. Environ. Sci.* **2019**, *14*, 257–265. [CrossRef]
32. Wei, N.; Zhang, Z.; Liu, D.; Wu, Y.; Wang, J.; Wang, Q. Coagulation behavior of polyaluminum chloride: Effects of pH and coagulant dosage. *Chin. J. Chem. Eng.* **2015**, *23*, 1041–1046. [CrossRef]
33. Khajvand, M.; Mostafazadeh, A.K.; Drogui, P.; Tyagi, R.D.; Brien, E. Greywater characteristics, impacts, treatment, and reclamation using adsorption processes towards the circular economy. *Environ. Sci. Pollut. Res.* **2022**, *29*, 10966–11003. [CrossRef] [PubMed]
34. Fatta-Kassinos, D.; Kalavrouziotis, I.K.; Koukoulakis, P.H.; Vasquez, M.I. The risks associated with wastewater reuse and xenobiotics in the agroecological environment. *Sci. Total Environ.* **2011**, *409*, 3555–3563. [CrossRef] [PubMed]
35. Khanam, K.; Patidar, S.K. Greywater characteristics in developed and developing countries. *Mater. Today Proc.* **2022**, *57*, 1494–1499. [CrossRef]
36. de Assis, G.C.; Skovroinski, E.; Leite, V.D.; Rodrigues, M.O.; Galembeck, A.; Alves, M.C.; Eastoe, J.; de Oliveira, R.J. Conversion of "Waste Plastic" into Photocatalytic Nanofoams for Environmental Remediation. *ACS Appl. Mater. Interfaces* **2018**, *10*, 8077–8085. [CrossRef]
37. Arrosyid, B.H.; Zulfi, A.; Nur'aini, S.; Hartati, S.; Rafryanto, A.F.; Noviyanto, A.; Hapidin, D.A.; Feriyanto, D.; Khairurrijal, K. High-Efficiency Water Filtration by Electrospun Expanded Polystyrene Waste Nanofibers. *ACS Omega* **2023**, *8*, 23664–23672. [CrossRef]
38. Kik, K.; Bukowska, B.; Sicińska, P. Polystyrene nanoparticles: Sources, occurrence in the environment, distribution in tissues, accumulation and toxicity to various organisms. *Environ. Pollut.* **2020**, *262*, 114297. [CrossRef]
39. Garrett, D.E. Profitability Analysis; Discounted Cash Flow (DCF). In *Chemical Engineering Economics*; Springer: Berlin/Heidelberg, Germany, 1989; pp. 81–106. [CrossRef]
40. González-Arias, J.; Sánchez, M.E.; Cara-Jiménez, J. Profitability analysis of thermochemical processes for biomass-waste valorization: A comparison of dry vs wet treatments. *Sci. Total Environ.* **2022**, *811*, 152240. [CrossRef]

sustainability

MDPI

Article

Prediction of Water Level in Lakes by RNN-Based Deep Learning Algorithms to Preserve Sustainability in Changing Climate and Relationship to Microcystin

Serkan Ozdemir * and Sevgi Ozkan Yildirim

Department of Information Systems, Middle East Technical University, 06800 Ankara, Turkey; sevgiozk@metu.edu.tr

* Correspondence: serkano@metu.edu.tr

Abstract: In recent years, intensive water use combined with global climate change has increased fluctuations in freshwater lake levels, hydrological characteristics, water quality, and water ecosystem balance. To provide a sustainable management plan in the long term, deep learning models (DL) can provide fast and reliable predictions of lake water levels (LWLs) in challenging future scenarios. In this study, artificial neural networks (ANNs) and four recurrent neural network (RNN) algorithms were investigated to predict LWLs that were applied in time series such as one day, five days, ten days, twenty days, one month, two months, and four months ahead. The results show that the performance of the Long Short-Term Memory (LSTM) model with a prediction of 60 days is in the very good range and outperforms the benchmark, the Naïve Method, by 78% and the ANN at the significance level ($p < 0.05$) with an RMSE = 0.1762 compared to other DL algorithms. The RNN-based DL algorithms show better prediction performance, specifically, for long time horizons, 57.98% for 45 days, 78.55% for 60 days, and 58% for 120 days, and it is better to use a prediction period of at least 20 days with an 18.45% performance increase to take advantage of the gated RNN algorithms for predicting future water levels. Additionally, microcystin concentration was tightly correlated with temperature and was most elevated between 15 and 20 m water depths during the summer months. Evidence on LWL forecasting and microcystin concentrations in the context of climate change could help develop a sustainable water management plan and long-term policy for drinking water lakes.

Keywords: sustainable water level management; deep learning models; microcystin; climate change

Citation: Ozdemir, S.; Ozkan Yildirim, S. Prediction of Water Level in Lakes by RNN-Based Deep Learning Algorithms to Preserve Sustainability in Changing Climate and Relationship to Microcystin. *Sustainability* **2023**, *15*, 16008. https://doi.org/10.3390/su152216008

Academic Editor: Andreas Angelakis

Received: 12 October 2023
Revised: 10 November 2023
Accepted: 14 November 2023
Published: 16 November 2023

1. Introduction

Depending on climatic, geographic, geological, social, and economic factors, each location has its own water quality and quantity challenges. Additionally, ongoing global warming and meteorological patterns are likely to disrupt the temporal and spatial balance of water, leading to freshwater scarcity and impeding the achievement of the United Nations Sustainable Development Goals around the world. Modeling studies suggest that there will be a paradigm shift in the distribution of freshwater on the planet by 2050 [1,2]. Therefore, a sound water management plan, developed using reliable forecasting models, is essential for implementing sustainable water use and conserving water resources in a given basin or region.

Turkey experiences frequent droughts that significantly reduce surface and groundwater resources, including wetlands and lakes [3,4]. Drought conditions affect standing water bodies when there is a reduction in surface runoff and in stream inputs. Droughts typically coincide with hot weather, which causes evaporation to increase significantly during dry periods. The effects of drought include a decrease in water levels in what is usually a very fertile littoral zone. This can leave aquatic fauna (e.g., mussels, snails, and flora) stranded in the area. The increased water temperature associated with drought can lead to stratification, increased salinity, and reduced oxygen levels. In some cases, the combination of high temperatures with low oxygen may lead to the extinction of fish species [5].

Uncontrolled drinking water supplies and inadequately managed water reservoirs pose a significant threat to developing and densely populated cities. Lake Sapanca, for example, is an important source of fresh water supply for the cities of Sakarya and Kocaeli and is also used by several bottled water companies for commercial purposes. The prospects of the reservoir appear to be affected by climate change and recent droughts, which could negatively impact several parts of the region and its ecosystems [6], as well as water quality associated with cyanobacteria and microcystin [7]. Because of the multitude of factors that affect the surface area of a lake, one of the most critical hydrologic problems is estimating the water level of a lake before it reaches its threshold. Hydrological models have certain limitations in terms of providing accurate predictive results [8] due to the complex nature of hydrological and meteorological variables as well as the temporal and spatial properties of an individual catchment. Therefore, it is vital to develop more reliable predictive models that can accurately and reliably estimate the future water level of a lake.

There are two different approaches for LWL prediction in the literature. The most prominent approach follows the physical process, and the emerging approach focuses on data-driven methodologies, which focus on historical datasets to predict future values. Data-driven methods simulate the LWL in addition to the factors affecting it using scientific computer models. Different types of models have been developed to promote specific cases. For instance, Chang and Chang evaluated the model with Support Vector Regression (SVR) and an Adaptive Neuro-Fuzzy Inference System (ANFIS) [9]. Liu et al. presented a multivariate conditional model based on copulas to predict water level and improve spatial precipitation estimation [10]. Wang et al. applied SVR to simulate the causality between LWL and the quantity of water discharged from the reservoir [11]. Statistical methods and Artificial Intelligence (AI) techniques are two common data-driven approaches to solving LWL prediction problems [12]. These methods include multiple regression, pattern recognition, neural network techniques, time series methods, and probability features [13].

During the recent decade, a variety of contemporaneous techniques have been applied to compare the predictive performance of the algorithms. For example, Ghorbani et al. investigated the ability of the Genetic Programming (GP) and ANN models to predict LWLs in Australia, and reported accurate predictions with good agreement [14]. To predict LWLs at Lake Urmia in Iran, Talebizadeh and Moridnejad employed the ANN and ANFIS [15]. The ANFIS algorithm has better accuracy compared to the ANN model, as shown by the uncertainty analysis. In another study, neural network, neural fuzzy, and GP models were applied to estimate the LWL on a daily basis [16]. The results showed that each of the three models accurately predicted the LWL. Buyukyildiz et al. developed a series of AI models, Multilayer Perceptron (MLP), hybridized SVR with Particle Swarm Optimization (PSO), a Radial Basis Neural Network, and ANFIS, to predict LWLs [17]. Their results show that the hybrid model SVR-PSO is a reliable predictive model. Similarly, for three upstream rivers on the east coast of Malaysia, water levels for the next five hours were successfully estimated using an ANN [18]. To predict the LWL, Yadav and Eliza used a Support Vector Machine (SVM) and Wavelet [19]. The results of the study showed that the model implemented to predict future values of the reservoir was more accurate compared to regression models. Despite the successful attempts to use machine learning (ML) methods in these studies, there are certain inherent limitations in the algorithms used in the literature [20]. For instance, in ANNs, the rules that could explain underlying methods are not given. In terms of fuzzy logic, setting precise, fuzzy enrollment limitations and parameters can be difficult and the fuzzy justification is not always correct. Regression models show that as the number of variables increases, their accuracy decreases. The regression models work better when there are fewer variables. Lastly, training a deep learning model requires a lot of computing power, which leads to the need for powerful GPUs and a large amount of RAM. Another potential drawback is an overfitting issue that arises when a model performs poorly on newly untrained data after being overtrained on training data.

The most used time series prediction model with statistical analysis that is conducted by scholars for lake level prediction is the Autoregressive Integrated Moving Average (ARIMA) model [21,22]. It can be expressed in several ways, including as Moving Average (MA), Autoregressive (AR), or hybrid AR or MA, known as Autoregressive Moving Average (ARMA) or Seasonal Autoregressive Integrated Moving Average (SARIMA) [23]. The SARIMA model, on the other hand, has the advantage of requiring fewer model features to explain the structure of time series that exhibit nonstationarity in seasons and between seasons [24]. Unlike ML methods, which often require multiple features as input, this is an important simplification [22]. The artificial neural network (ANN) algorithm is a widely used ML method for water flow modeling, water quality assessment, and water level prediction in the field of hydrology and water resources [18,25–27]. In addition, some research papers have presented a hybrid ANN-ARIMA model [28,29].

Review of the aforementioned studies shows that various models for LWL prediction have different findings and highlight their estimation uncertainty. Some scholars have used time series techniques for predicting various areas such as energy price, stock price, and corporate sales forecasts, which are critical to the global economy [30,31], including weather, environment, hydrology, and geological phenomena [32,33] in recent years. Nearly all of them concluded that the time series forecasting methods provide more accurate results compared to the benchmark models.

The recurrent neural network (RNN)-based deep learning (DL) approach is proposed in this paper as a state-of-art technique for examining the LWL that would improve the prediction performance. DL networks, which differ from conventional approaches in that they allow computer models consisting of numerous layers to learn representations of data consisting of multiple levels of abstraction, replicate the functioning of the human brain [34]. The approach of DL has been used for object recognition, speech recognition, and visual object recognition including genomics and drug discovery [35]. The extraordinary success of supervised RNN-based DL algorithms for conducting recognition studies directed the use of RNN-based algorithms in multivariate time series studies. The LWL studies also have time series data due to their nature and attract hydrologists to exploit the power of these DL algorithms in their future time series prediction studies. However, the application of DL models for LWLs is limited and is the focus of this study in order to overcome several drawbacks of the available approaches to predict LWLs, such as the large number of input variables and their uncertainty. The motivation behind this study is to provide an effective prediction technique for water managers to handle drinking water supply availability in lakes before reaching an alarming level. The limited water supply in lakes not only causes frequent drought experiences and water shortages, but also causes a decrease in water quality.

In this work, novel gated RNN-based algorithms are used to build a model that can predict the future LWL to support drought mitigation and reservoir management. In addition, this study aims to help fill the gap in the literature regarding the selection of DL models and the evaluation of the performance of LWL prediction algorithms by using Naïve Benchmarks and the Diebold–Mariano test. As far as the authors are aware, there is no other study that focuses on the comparison between algorithms for multivariate prediction studies with different time lags.

2. Materials and Methods

2.1. Case Study Area

Lake Sapanca extends between the latitudes of 40°41′–40°44′ E and the longitudes of 30°09′–30°20′ N in the northwestern part of Turkey (Figure 1). It is located between two cities: the western part of the lake is in Kocaeli and the eastern end is within the provincial border of Sakarya. It is a 16 km (east–west) and 5 km (north–south) long tectonic fresh water source that provides the drinking water needs of both cities. It has a surface area of 46 km^2 and a reasonable depth of 30 m. It has a volume of about 1.3 billion m^3. The greatest depth of the lake basin is 54 m, and its catchment area is 250 km^2 [36]. The Lake is surrounded by southern mountains and northern hills.

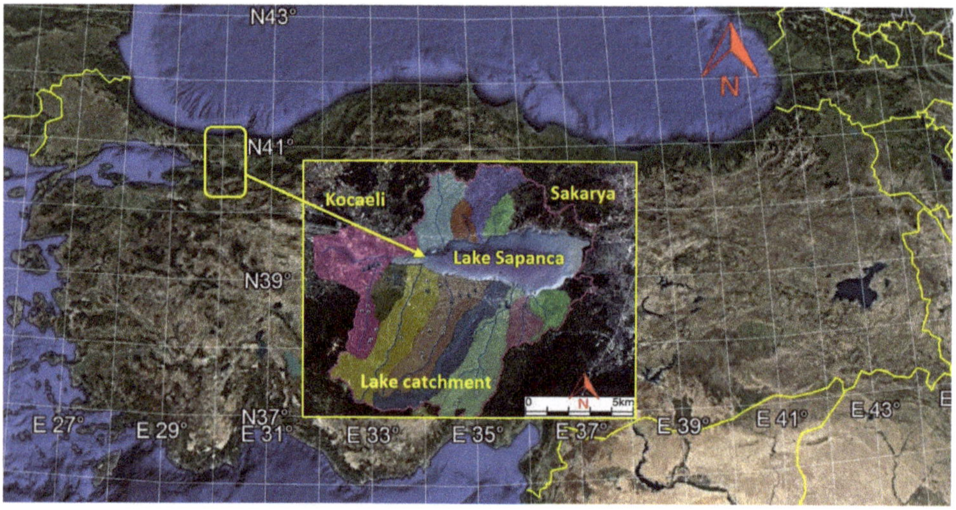

Figure 1. Lake Sapanca area and its catchment with river basins.

The transitional climate found in the Sapanca basin is influenced by the Black Sea and Mediterranean climates. While the basin exhibits characteristics of both the Black Sea and Mediterranean climates, it may also display elements of a continental climate due to its interaction with an intermediary air system. Despite the warm and rainy winters experienced in the basin, summers are comparatively less hot and dry than what is typically observed in the Mediterranean region.

Figure 1 depicts the catchment area of Lake Sapanca and its sub-basins, which consist of 12 streams that inflow to Lake Sapanca. The lake has a controlled outflow with Cark Creek, which regulates the maximum LWL to prolong water retention in the lake. The seasonal precipitation, water withdrawal, and surface outflow results in inter-annual LWL variations of 2.28 m, between 29.90 and 32.18 m above sea level. The lake is noteworthy because it supplies potable water to the provinces of Sakarya and Kocaeli. It is also believed that the lake basin will eventually meet the bottled water needs of Istanbul. Although the basin area does not include any industrial regions, 23% of the basin area is used as cultivated land mainly covered by ornamentals and fruit orchards, and 9.5% is used as settlement land. The remaining basin area is covered by 65% of forest land and 2.5% as natural land. The water needs of urban, agricultural, and industrial sectors have caused Lake Sapanca basin's water quantity and quality to worsen. Whereas the mean growth rate of the population in Turkey is 0.8%, the population growth rate of the basin has increased from 1.5% to 3.5% in the last 20 years [37]. The rapid growth in the population of the basin is adversely affecting the quantity and quality of the water. Despite the fact that the lake is in a transitional stage from oligotrophic to mesotrophic, its ecological status is deteriorating as the water level drops below the lake's surface discharge during droughts, and point and nonpoint runoff flows in from numerous sources [20]. The lake's ecological state deteriorates primarily as a result of unchecked agricultural operations and household wastewater leakages in the vicinity. In addition, the droughts that periodically occur cause the lake's water quality and quantity to deteriorate [6].

2.2. Dataset Description

Several characteristics are used in the literature to evaluate future LWLs. The most commonly used features in the literature are precipitation (17%), LWL, and evaporation [20]. Other major features used by researchers include discharge [38], temperature [39], inflow [40], streamflow and humidity [41], wind speed and solar radiation [42], and volume and area [43].

The State Hydraulic Works and Turkish State Meteorological Service, through their river monitoring program for Lake Sapanca, provided the data examined for this study. LWL, maximum temperature, minimum temperature, average temperature, precipitation, and withdrawal were the features that were supplied (Table 1). Among these, withdrawal feature includes water withdrawal for industrial, agricultural, and domestic use. Measurements were taken daily between 2012 and 2023, with occasional missing data. The interpolation technique was used to complete the missing data.

Table 1. Dataset Features.

Inputs:	Output:
Maximum Temperature Minimum Temperature Average Temperature Precipitation Withdrawal	LWL

Time series data are a collection of values generated over a period of time in continuous or discrete time units. Numerous research has demonstrated the effectiveness of time series prediction as a control and early warning system. Time series prediction seeks to forecast upcoming changes across time at observation locations. The dataset employed in this research, as shown in Figure 2, is a typical multivariate time series that typically contains real-valued LWL and meteorological information in addition to water removal from the reservoir. One can spot abnormalities in the LWL, meteorological, and hydrological data by carefully examining the graph. However, only annual and seasonal patterns of change are seen in the temperature data. Distribution data from LWL are compatible to meteorological data, especially in annual precipitation. Additionally, water withdrawals show an increasing trend over time (Figure 2).

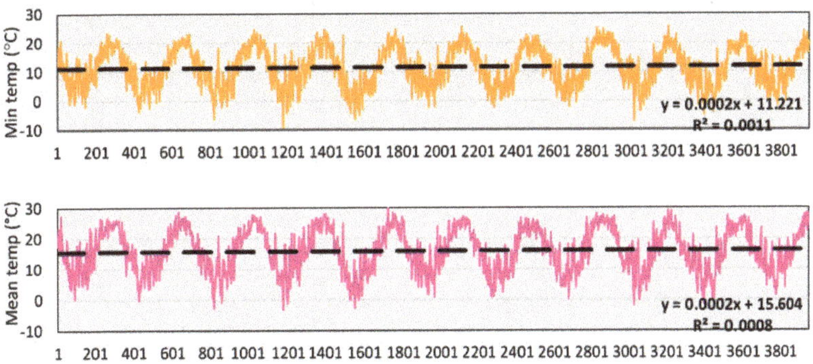

Figure 2. *Cont.*

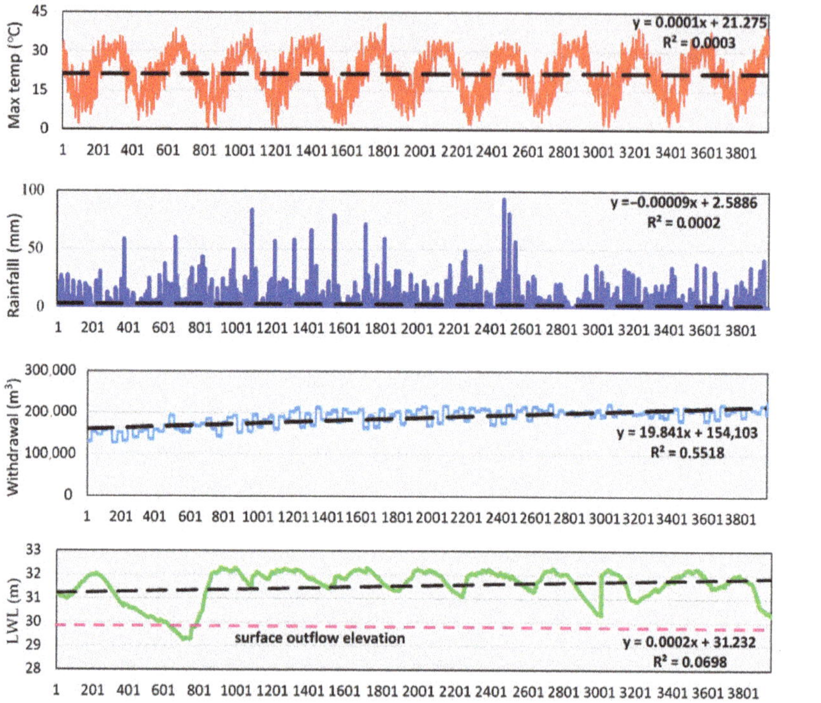

Figure 2. Time series plots of daily meteorological data, water withdrawals, and lake water level for Lake Sapanca from 11 October 2012 through 4 August 2023. (x-axis: data rows in sequence.)

As shown in Figure 2, during prolonged drought, the LWL drops below the discharge elevation at the surface, which is 29.90 m above sea level in Lake Sapanca. Data from LWL indicate that in years of low precipitation, LWL decreases. Higher precipitation in the last decade (2015–2018) coincides with LWLs above the lake's discharge elevation. In addition, higher maximum temperatures and low precipitation in recent years reduce LWLs to the surface runoff elevation during dry periods (Figure 2). Low precipitation also increases water demand, while low temperatures decrease water use. In addition, increasing population and industrialization are related to water withdrawal from the lake. Therefore, multivariate time series data that include freshwater demand and meteorological characteristics are critical for predicting lake water levels.

2.3. Data Preprocessing

The dataset was created on a daily basis with monthly stacks and converted to a time series format to be used as a predictive model. The dataset contains several missing points that prevent the model from running. Although the dataset has small gaps, some columns contain large blanks. The large gaps that are located either at the beginning or at the end of the dataset were removed from the dataset. Other missing data were interpolated using the linear method. Among the features, maximum temperature, minimum temperature, average temperature, and withdrawal do not have any missing values. The only missing values are included in the features of precipitation (0.7%) and water level (3.03%). Therefore, the dataset can be used for an RNN-based neural network study with minimal bias with the interpolation method due to its negligible missingness rate. No outlier was detected using the interquantile range method and expert opinions. The data were used after the necessary preprocessing steps had been performed.

Due to the sequential nature of time series, the data should be used to assign the training, validation, and test sets. The daily hydrological and meteorological data collected at the lake basin from 11 October 2012 to 4 August 2023 were used to train, validate, and test the algorithms. The dataset was divided into training, validation, and test sets with 60%, 20%, and 20% proportions, respectively, to cover the high/low values in the training and test subsets to determine the optimal pattern for the data and to improve the model's validity. To avoid overfitting and to include all seasons in the dataset, dry and wet seasons were included in all sets created. The lag values used in the study are 1, 5, 10, 20, 30, 45, 60, and 120. The lag values were determined considering the seasonal cycles in terms of wet and dry seasons. After 120 days, the LWL values arrive for the next seasonal cycle, which causes the Naïve Method to become overly optimistic.

2.4. Model Descriptions

The RNN is the ancestor of gated recurrent-based networks. Due to time delays during the backpropagation error in the learning stage of the RNN-based network, it was established as a remedy for the gradient explosion issue. At each time step, gated RNNs predict the label of an activity. To predict an activity label, any number of previous time steps can be merged. The gated RNN model networks have been shown to be a significant model in the past and are capable of learning from sequential inputs. They can effectively learn from sequences of different lengths and capture long-term dependencies.

There are 4 different gated RNN networks used in this study: LSTM, GRU, Stacked LSTM, and Bidirectional LSTM.

The gates and feedback loops used by LSTM are self-trained using the input data. By incorporating a gate mechanism, the LSTM network, a particular architecture created to simulate dynamic temporal and spatial sequences, is able to more precisely resolve long-range dependencies [12]. The LSTM network is made up of a number of memory blocks connected by layers made up of a collection of memory cells with recurrent connections (Figure 3). LSTM has three multiplicative units: input, output, and forget gates. Through the hyperbolic tangent function, sigmoid function, and regulatory filter, the input gate transforms the information. The forget gate erases the less important information. The output gate selects the pertinent data from the active cell. The LSTM layer uses the following mathematical operation to determine the output variable [20]:

$$\sigma(t) = \frac{1}{1+e^{-t}}\tanh(t) = \left(\frac{e^t - e^{-t}}{e^t + e^{-t}}\right) \tag{1}$$

$$f_t = \sigma\left(Wf(h_{t-1}, X_t) + B_f\right) \tag{2}$$

$$i_t = \sigma(Wi(h_{t-1}, X_t) + B_i) \tag{3}$$

$$o_t = \sigma(Wo(h_{t-1}, X_t) + B_o) \tag{4}$$

where f_t is the forgotten variable, $\mathrm{i_t}$ the input variable, and o_t the output variable. $\mathrm{X_t}$ indicates the values that the feature receives at t time, and h_{t-1} is the output cell of the previous cell. Inside the LSTM cell, memory is indicated by c_{t-1}. W is the weight matrix, and B is the term bias. The sigmoid function (σ), the hyperbolic tangent function (tanh), processes the X_t variable and the h variable from the previous learning.

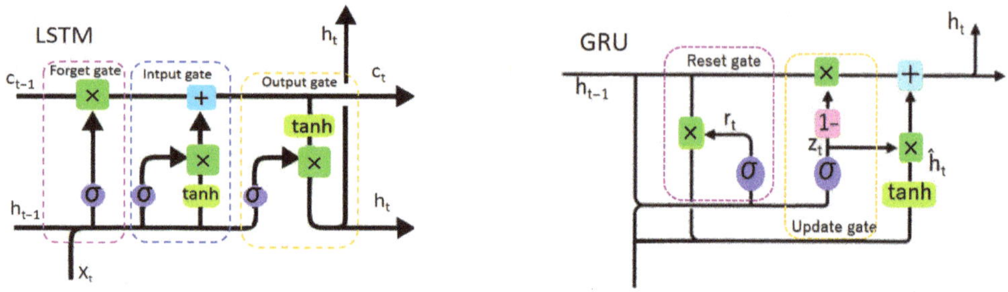

Figure 3. Basic structure of LSTM and GRU algorithms.

While GRU has two gates, it requires less memory and runs faster than LSTM. GRU is computationally more efficient than LSTM because the structure is simpler and more straightforward. The input gate and the forgetting gate are combined into one update gate and simplified (Figure 3). GRU has two activation functions and one tanh function. Therefore, GRU is able to build a long-term memory similar to the LSTM, but has the advantage of having fewer parameters and a faster training speed than the LSTM. GRU uses the following equations to determine the output variables [20]:

$$r = \sigma(W_r(h_{t-1}, X_t) + U_r X_t) \tag{5}$$

$$z = \sigma(W_z(h_{t-1}, X_t) + U_z X_t) \tag{6}$$

$$c = tanh(W_c(h_{t-1} \times r) + U_c X_t\ ) \tag{7}$$

$$h_t = (z * c) + ((1 - z) \times h_{t-1}) \tag{8}$$

where *tanh* and σ are the hyperbolic tangent and logistic sigmoid functions, respectively, and r and z are vectors for the activation values of the update and reset gates, respectively. W_r, U_r, W_z, U_z, W_c, U_c represent the weight matrix.

The RNN is the ancestor of gated recurrent-based networks. Due to time delays during backpropagation error in the learning process of the RNN model network, gated structures were established as a remedy for the gradient explosion problem [44]. At each time step, gated RNN networks predict the label of an activity. To predict an activity label, any number of previous time steps can be merged [45]. The gated RNN model networks have proven to be a significant model in the past and are capable of learning from sequential inputs. They can effectively learn from sequences of different lengths and capture long-term dependencies [46]. Stacked LSTM is a variant of LSTM with multiple LSTM layers containing multiple memory cells that give the model the ability to capture the structure of time series and combine the learned representation of previous layers while providing a higher level of abstraction for the final results [47]. This structure contributes to the model's ability to learn higher-level temporal representations, but can lead to degradation problems due to the low convergence rate of the LSTM layers, although this error is different from the vanishing gradient problem. Another variation of the LSTM is the Bidirectional LSTM, in which the input currents of the LSTM flow in both directions so that information from both the input and output sides can be used [48]. The use of both forward and reverse information improves the accuracy of this model and supports improved learning for long-term dependency data.

$$h_n = \text{LSTMforward}\left(in, \underset{h_{n-1}}{\rightarrow}\right) \otimes \text{LSTMbackward}\left(in, \underset{h_{n+1}}{\leftarrow}\right) \tag{9}$$

where h_n = new state, in = input, h_{n-1} = output of past state, h_{n+1} = output of future state, and the $\otimes$ symbol represents the concatenation operation.

Lastly, the RNN algorithms are also compared against ANN, which is the most used neural network algorithm for LWL studies [20]. A massively parallel-distributed information processing system called an ANN mimics the function of the neuron network in the human brain. Human learning is a result of neurons, and ANNs employ this important feature for ML.

An NN is made up of several nodes, or basic processing units. The mathematical functions and network architecture make up the ANNs. The architecture is made up of the arrangement of nodes in a specific way. Typically, the nodes are organized in layers that facilitate the flow of information from the input layer to the output layer. Between the input and output layers, there may be multiple hidden levels. The network's capacity to represent more complicated events is enhanced by the hidden layers.

2.5. Hyperparameters

This study uses the Tensorflow Keras (2.11.0) libraries to implement the proposed different RNN-based networks, with Tensorflow as the backend [49]. The implementation of ANN, LSTM, GRU, Stacked LSTM, and Bidirectional LSTM layers in the algorithm uses the sequential approach. The loss function is set as "MAE" and the optimizer as "Adam" since these hyperparameters do not have a significant impact on the performance of the algorithm. However, the hyperparameters such as neuron number, epoch, batch size, number of previous time steps, and number of layers are optimized. The hyperparameters in the algorithms that have the best performance are briefly listed in Table 2.

Table 2. Optimized hyperparameter values of algorithms.

	ANN	LSTM	GRU	Stacked LSTM	Bidirectional LSTM
Neuron number	128	128	64	128	32
Epoch	250	100	100	100	50
Batch size	64	128	128	128	128
Number of layers	1	2	2	2	2
Prediction period	45	60	60	60	60

The optimized hyperparameter values in Table 2 are different for the different algorithms. However, all RNN-based algorithms performed the best when the number of layers was 2 and the prediction period was 60 days.

2.6. Evaluation Metrics

A significant number of researchers in the literature prefer Root Mean Squared Error (RMSE), Mean Squared Error, Mean Absolute Percentage Error (MAPE), Mean Absolute Error (MAE), R^2, or R as evaluation metrics to compare their algorithms with the base model or with other algorithms [20]. These metrics account for more than 50% of the evaluation metrics in the literature. On the other hand, there are less favorable evaluation metrics used by some researchers, including Nash–Sutcliffe Efficiency [42], accuracy [50], Mean Relative Error [20], and Percent Bias [51].

The goal of model performance determination is to verify the accuracy and precision of the proposed model and to determine the difference rate so that it can be used with confidence [52]. The evaluation metrics chosen for this study are RMSE and MAPE. The RMSE value shows the root of the squares of the average differences between predicted and observed values. Lower *RMSE* values indicate higher model performance and better correlation between observed and predicted values. Equation (10), discussed in more detail below, was used as a performance measure in the evaluation of the model.

$$RMSE = \sqrt{\frac{\sum_{i-1}^{n}(P_i - Q_i)}{n}} \tag{10}$$

where "Σ" stands for sum, P_i is the expected value for the dataset's ith observation, Q_i is the actual value for that observation, and "n" denotes sample size.

The exactness of a forecasting approach is determined by a statistic called Mean Absolute Percentage Error (MAPE). In order to determine how accurate the predicted quantities are in relation to the actual numbers, the value represents the mean of the absolute percentage errors for every value in a dataset. MAPE necessitates the usage of data values apart from zero and is frequently useful for large-scale data analysis. MAPE is a simple metric; a 10% MAPE indicates that, irrespective of whether the variance was positive or negative, the average difference between the expected and actual amounts was 10%. Lower MAPE values indicate higher model performance and better correlation between observed and predicted values. The MAPE formula used to calculate the error rate is in Equation (11)

$$\mathrm{MAPE} = \frac{1}{m}\sum_{i=1}^{m}\left|\frac{Y_i - X_i}{Y_i}\right| \tag{11}$$

where "Σ" stands for sum, X_i is the expected value for the dataset's ith observation, Y_i is the actual value for that observation, and "m" denotes sample size.

Although the performance of the model is evaluated using the RMSE and MAPE values, the comparison of forecast accuracy is made using the Naïve Method. The Naïve Benchmark is one of the most commonly used method for comparing time series forecasting models because it is easy to compute and understand [53]. In this approach, each forecast is equated to the last observed value for the intended time step. The performance of the algorithm is considered successful if the RMSE or MAPE value is lower than the RMSE or MAPE results of the Naïve Method. The reason for such a comparison is that the RMSE or MAPE values for earlier time steps are always lower than for further time steps due to their proximity to the actual values. For this reason, the performance of further time periods (i.e., 60 days and 120 days) cannot be compared using only the RMSE or MAPE values themselves. To compare all time values, performance is evaluated using the percentage increase in RMSE compared to the RMSE score or difference in MAPE values over the Naïve Method.

$$Y_t = Y_{t-n} \tag{12}$$

where Y_t is the forecast value at time t and Y_{t-n} is the value at the previous nth day

To determine whether the proposed algorithms are successful enough to be used as a prediction method, the prediction results of the algorithms are also compared. The Diebold–Mariano significance test was used to control the algorithm differences and their significance at p-value < 0.05, as described by Van der Heijden et al. [53]. If the p-value of the test is less than 0.05, the prediction accuracies are significantly different from each other. This approach gives the impression that the sophisticated procedure is only recommended if it is significantly better than the benchmark, not only if it has better accuracy statistics. This also implies that there is a significant difference between the prediction results and that the proposed algorithms cannot be used interchangeably.

A graphical representation of the entire modelling process with the flowchart applied to predict LWL in this study is shown in Figure 4.

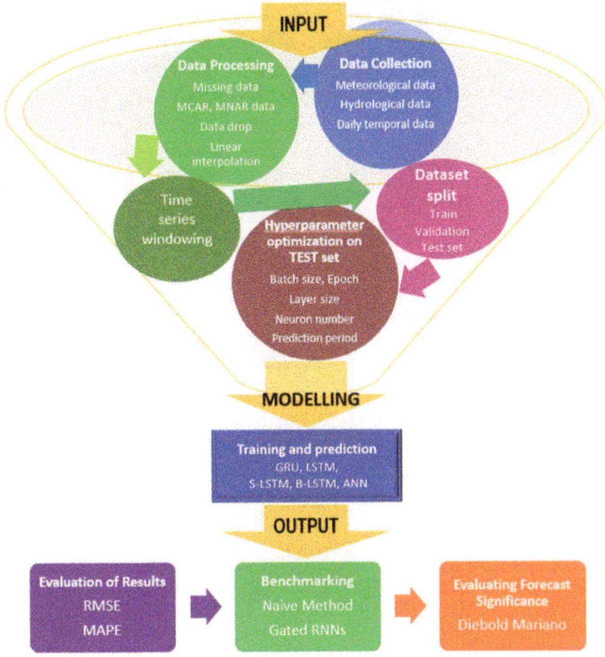

Figure 4. Flowchart of LWL modelling.

2.7. *Water Quality Indicator*

In freshwater habitats, temperature and light intensity are the most important meteorological factors determining algal photosynthesis and algal blooms [54]. Toxin production behavior of freshwater algal species is strongly influenced by environmental conditions. As an indicator of biological water quality, monthly microcystin measurement data at various depths from the surface to 20 m during the period from March 2019 to April 2023 were subjected to statistical analysis. Computer-based models require long-term data to make more reliable and accurate predictions for the future. Therefore, the Mann–Kendall trend analysis test is used with environmental time series. In this test, the null hypothesis assumes that there is no trend and the alternative hypothesis assumes that there is a trend. Furthermore, Spearman rank correlation analysis was applied to determine the relationship between key meteorological parameters and the concentration of the cyanobacterial growth byproduct microcystin since microcystin concentrations did not follow a normal distribution.

3. Results

This study uses the ANN and four different RNN-based deep learning algorithms to compare their forecasting accuracy from day 1 to day 120 ahead, based on RMSE and MAPE values, the Naïve Method, and Diebold–Mariano test results. The ANN, LSTM, GRU, Stacked LSTM, and Bidirectional LSTM algorithms were successfully trained and validated, and compared with test data consisting of 3004 lines to evaluate the model's reliability for the unknown dataset. Table 3 presents the performance of LWL prediction of the investigated ANN and RNN algorithms from day 1 to day 120 ahead forecasting. These results show that all investigated ANN and RNN algorithms showed excellent prediction accuracy in the 1 day to 10 day ahead prediction scenario with RMSE values of <0.1 m. On the other hand, in the 60 day ahead scenarios, the LSTM algorithm had the best performance value for training and testing with an RMSE = 0.1762 m, while in the 120 day ahead scenarios, GRU showed the best performance with an RMSE score of 0.3838 m (Table 3). In contrast, the Stacked LSTM and Bidirectional LSTM models did not show additional performance in terms of prediction accuracy over LSTM. In summary, the LSTM model is significantly efficient considering its high accuracy among other advanced models, specifically for long-term predictions such as a 60 days ahead forecast, due to the architectural benefits of the process of parameter tuning and its migration to different tasks. GRUs are easier to train and faster to run than LSTMs, but they may not be as effective at storing and accessing long-term dependencies. Since it is necessary to know the next timestamp in advance for a Bidirectional LSTM, it is more appropriate for offline applications [55]. On the other hand, the performance difference between Stacked LSTM and LSTM comes from additional dimensions for next value prediction other than the time dimension.

Table 3. The performance of ANN and RNN-based algorithms for predicting lake water level with increasing time intervals, RMSE results. (Metric is based on m.)

Algorithm/Prediction Period	Naïve Method	ANN	LSTM	GRU	Stacked LSTM	Bidirectional LSTM
1 day	0.0134	0.0131	0.0162	0.0134	0.0171	0.0156
5 days	0.0484	0.0445	0.0514	0.0429	0.0494	0.0563
10 days	0.0875	0.0815	0.0799	0.0732	0.0890	0.0875
20 days	0.1551	0.1271	0.1227	0.1070	0.1289	0.1257
30 days	0.2168	0.1540	0.1356	0.1316	0.1221	0.1226
45 days	0.3139	0.1918	0.1775	0.1728	0.1769	0.1947
60 days	0.4041	0.2627	0.1762	0.2203	0.1976	0.1985
120 days	0.6973	0.4810	0.4586	0.3838	0.4275	0.3873

Additionally, the RMSE values describing the prediction error rates of time series algorithms were compared using the Naïve Method, and the algorithms that performed better than the Naïve Method were identified as successful algorithms for predicting future LWL values. The Naïve Benchmark comparison results of algorithms are presented in Table 4 from day 1 to day 120 forecasting. The higher value for each investigated algorithm to each prediction period indicates higher performance and good predictive power. Based on the Naïve Method benchmark, the algorithm performances increased up to the 60 days ahead predictions, then decreased for the 120 days ahead predictions. As an average, the performance of GRU was higher for all investigated periods, whereas Stacked LSTM had a lower average performance value, followed by the Bidirectional LSTM algorithms.

Table 4. Benchmark performance comparison of algorithms; figures indicate improvement in RMSE values over Naïve Method.

Algorithm/Prediction Period	ANN	LSTM	GRU	Stacked LSTM	Bidirectional LSTM
1 day	2.26%	−18.92%	0.00%	−24.26%	−15.17%
5 days	8.40%	−6.01%	12.05%	−2.04%	−15.09%
10 days	7.10%	9.08%	17.80%	−1.70%	0.00%
20 days	19.84%	23.33%	36.70%	18.45%	20.94%
30 days	33.87%	46.08%	48.91%	55.89%	55.51%
45 days	48.29%	55.51%	57.98%	55.83%	46.87%
60 days	42.41%	78.55%	58.87%	68.64%	68.24%
120 days	36.71%	41.30%	58.00%	47.97%	57.16%

The variabilities between Naïve Benchmark comparison scores are much more apparent than for RMSE values (Table 4). The decreasing performance goes down to −24.26%, which indicates that it would be disadvantageous to use the RNN-based algorithm for predicting that specific period. The results also show that the RMSE results of some algorithms are close to those of the Naïve Method, especially for the predictions of 5 and 10 days. Therefore, the algorithms were tested even more to find if it is necessary to use these algorithms for future LWL values. The results show an increase in performance of at least 18.45% (Stacked LSTM) when the prediction horizon is set to 20 days or more. Based on the Naïve Method comparison, LSTM showed the highest performance with a 78.55% improvement over the Naïve Method at 60 day ahead forecasting. It is also worth noting that ANN is the only algorithm that performed better than the Naïve Method in the 1 day prediction period.

The performances of ANN and RNN-based algorithms were also tested using MAPE as an evaluation metric. A similar pattern was observed in the MAPE results when considering the results in the RMSE values because as the time horizon extends to further time periods, the model performance decreases. This pattern indicates there needs to be additional evaluation criteria for model performance results between different time periods. For this reason, the results in Table 5 were calculated further by taking differences between the Naïve Method results and the algorithms' results (Table 6).

Table 5. The performance of ANN and RNN-based algorithms for predicting lake water level with increasing time intervals, MAPE results. (%).

Algorithm/Prediction Period	Naïve Method	ANN	LSTM	GRU	Stacked LSTM	Bidirectional LSTM
1 day	0.03%	0.09%	0.17%	0.37%	0.12%	0.13%
5 days	0.13%	0.27%	0.23%	0.30%	0.24%	0.34%
10 days	0.24%	0.22%	0.46%	0.58%	0.54%	0.44%
20 days	0.42%	0.94%	0.47%	0.68%	0.47%	0.38%
30 days	0.60%	0.43%	0.76%	0.88%	0.53%	0.46%
45 days	0.90%	0.91%	0.59%	0.84%	0.85%	0.78%
60 days	1.20%	0.75%	1.09%	0.90%	0.85%	0.91%
120 days	2.09%	2.19%	1.50%	1.24%	1.55%	1.40%

Table 6. Benchmark performance comparison of algorithms; figures indicate difference of MAPE values compared with Naïve Method.

Algorithm/Prediction Period	ANN	LSTM	GRU	Stacked LSTM	Bidirectional LSTM
1 day	−0.06	−0.14	−0.34	−0.09	−0.10
5 days	−0.14	−0.10	−0.17	−0.11	−0.21
10 days	0.02	−0.22	−0.34	−0.30	−0.20
20 days	−0.52	−0.05	−0.26	−0.05	0.04
30 days	0.17	−0.16	−0.28	0.07	0.14
45 days	−0.01	0.31	0.06	0.05	0.12
60 days	0.45	0.11	0.30	0.35	0.29
120 days	−0.10	0.59	0.85	0.54	0.69

Table 6 reveals the performance differences that indicate the performance improvement in models in terms of the Naïve Method. As can be seen in the table, none of the models perform better when compared with the Naïve Method in the 1 day and 5 day prediction periods. However, as the time period increases, the performance improvement also increases. The best performance is observed for the GRU algorithm in the 120 day prediction period with a 0.85 points performance increase in the MAPE. The MAPE results are occasionally compatible with the RMSE results, but the MAPE results indicate that in order to achieve the advantage of RNN-based algorithms, the models should focus on at least a 30 day ahead prediction. In addition, the ANN algorithm is advantageous when used in the 30 day and 60 day prediction period according to the results.

The performance of ANN, LSTM, GRU, Stacked LSTM, and Bidirectional LSTM for LWL, and their observed and estimated values compared with the Naïve Method for the day 1 to day 120 ahead scenarios are presented in Figures A1–A4. It can be seen from Figure A1 that the observed and simulated lines are generally distributed closely for each investigated model, showing that all ANN and RNN algorithms have high simulation performance at day 1. However, as the forecasting time extends from day 1 to day 120, the observed, estimated, and Naïve Method lines diverge for each of the algorithms.

Figure A1 shows the 1 day and 5 days prediction results of the ANN and gated RNN algorithms and the comparison with the observed and Naïve Method values. The prediction results of all the studied algorithms are quite similar to each other and to the Naïve Method for the 1 day and 5 days prediction (Figure A1), indicating good training, validation, and prediction. Figure A2 shows the prediction results of the ANN and gated RNN algorithms for 10 and 20 days ahead and the comparison with the observed values and the Naïve Method. Compared to the Naïve Method, all tested algorithms had a similar prediction trend for 10 days ahead, but all algorithms outperformed the Naïve Method in their predictions for 20 days ahead. When forecasting 10 and 20 days ahead, the GRU achieved the best results (Figure A2), showing a lower RMSE (Table 3) and a higher performance improvement compared with the Naïve Method (Table 4).

When comparing the performance results of the algorithms for day 30, Stacked LSTM and Bidirectional LSTM produced a similar prediction performance to LSTM and GRU, whereas for day 45 prediction, the GRU, Stacked LSTM, and LSTM algorithms produced a similar performance to Bidirectional LSTM (Figure A3, Table 4).

Figure A4 shows the 60 and 120 day forecast results of the ANN and gated RNN algorithms and the comparison with the observed values and the Naive Method. The 60th day was the culminating point for the prediction performance of the tested algorithms, and LSTM performed better for LWL at the 60 day prediction based on the RMSE and Naïve Method values. Although all tested algorithms performed well in 60 day prediction (Table 4), LSTM provided the closest prediction values to the observed values of LWL 60 days in advance compared with the other methods, as shown in Figure A4. For the 120 day ahead predictions, there was a significant decrease in values for the studied algorithms compared to the Naïve Method, with the exception of GRU. Although the prediction performance was low, the GRU algorithm provided a statistically similar prediction performance for day 60 and day 120. These results show that the GRU algorithm may still be superior to the other algorithms in terms of prediction accuracy with higher Naïve values. However, the degree of agreement between the predicted value and the actual value is not very good and exceeds the actual value.

As a summary, Figures A1–A4 show that the tested algorithms predicted LWL at a statistically acceptable level for up to 120 days. Among the proposed algorithms, the LSTM algorithm was clearly superior in tracking the nonlinear behavior of Lake Sapanca over a 60-day period with the smallest RMSE (0.1762 m) and a higher performance ratio compared to the Naïve Method result (78.55%). Thus, when a model is needed for long-term forecasting LWL, the LSTM-based DL algorithm can help to automate and manage LWL to implement more effective water management strategies. It is optimal for 60-day forecasts of LWL.

The Diebold–Mariano test values to determine the statistical significance of two separate prediction results are summarized in Table 7. It can be noted that the RNN algorithms did not show significant superiority for the 1-day, 5-days, and 10-days LWL forecasting over the Naïve Method. However, the Naïve Method and the GRU algorithm for 5 days gave a p-value of 0.031, indicating the GRU algorithm's superior result is significant compared with the Naïve Method to predict the next 5 days. The same is true for predicting the next 10 days using the LSTM and GRU algorithms. However, when the prediction significances of LSTM and GRU are tested, the p-value is lower than 0.05, indicating that GRU must be used to predict the next 10 days.

Table 7. Forecast difference results of Naïve Method, ANN, and RNN algorithms based on Diebold–Mariano (DM) test for increasing day intervals from day 1 to day 120 (p-value ≤ 0.05 indicates the significance of the DM test results, Green boxes indicate significantly different prediction results with distinct tones, red boxes indicate insignificant results).

	Day-1	Day-5	Day-10	Day-20	Day-30	Day-45	Day-60	Day-120
Naïve Method-ANN	0.578	0.094	0.984	0.055	0.055	0.815	0	0.222
Naïve Method-LSTM	0.122	0.31	0.007	0.612	0.009	0.014	0.003	0
Naïve Method-GRU	0.005	0.031	0	0	0	0.006	0.012	0
Naïve Method-Stacked LSTM	0.485	0.181	0	0.506	0.009	0.253	0.009	0
Naïve Method-Bidirectional LSTM	0.261	0.007	0.011	0.686	0.161	0.923	0.187	0
ANN-LSTM	0.264	0.474	0.008	0	0	0.025	0	0
ANN-GRU	0.011	0.581	0	0	0	0.443	0.046	0
ANN-Stacked LSTM	0.878	0.71	0	0.072	0.072	0.169	0.058	0
ANN-Bidirectional LSTM	0.523	0.233	0.012	0.593	0.593	0.741	0.002	0
LSTM-GRU	0.099	0.21	0.032	0	0.004	0.003	0	0.015
LSTM-Stacked LSTM	0.326	0.728	0.244	0.874	0.006	0	0	0.995
LSTM-Bidirectional LSTM	0.608	0.062	0.878	0.364	0	0.011	0.017	0.752
GRU-Stacked LSTM	0.014	0.358	0.319	0.001	0	0.541	0.917	0.014
GRU-Bidirectional LSTM	0.037	0.516	0.022	0	0	0.662	0.229	0.033
Stacked LSTM-Bidirectional LSTM	0.623	0.122	0.188	0.287	0.202	0.295	0.192	0.747

From Table 4, it can be seen that, for day 20 predictions, the best performance improvement comes from the GRU algorithm. Accordingly, the p-values are significant ($p < 0.05$) based on the Diebold–Mariano test (Table 7), which confirms the superiority of GRU. Regarding the Naïve Method comparison (Table 4) and the Diebold–Mariano (Table 7) test results, only the GRU algorithm should be preferred to predict the LWL for the next 20 days.

Similarly, GRU, LSTM, and Stacked LSTM gave a p-value of less than 0.05 in the Diebold–Mariano tests compared to the Naïve Method for predicting the next 30 days LWL. On the other hand, the predictive performance of Bidirectional LSTM was not significant compared to the Naïve Method as the p-value is greater than 0.05.

According to the Naïve Method comparison, GRU performed better than the other algorithms in the 45-day forecast (Table 4). Table 7 further confirms that the predictions of GRU algorithms have a significant p-value compared to the Naïve Method. Moreover, the p-values are more remarkable than 0.05 when GRU is compared with Stacked LSTM and Bidirectional LSTM, indicating that the GRU algorithm can be used interchangeably with the Stacked LSTM and Bidirectional LSTM algorithms.

The results of the Diebold–Mariano test show that the accuracy of the prediction results and the stability of the performance of the LSTM algorithm are significantly better, with a p-value of less than 0.05 (Table 7). Considering the results of the RMSE, Naïve Method, and Diebold–Mariano test, only the LSTM algorithm should be preferred for predicting the next 60 days to obtain a more reliable and accurate prediction of the future dynamics of LWL.

It is clear that the implemented ANN and RNN algorithms provide a relatively accurate prediction pattern when the prediction values are compared with the observed data for the 120 day prediction (Table 7), even though the magnitude of the Naïve Method benchmark result is reduced compared to the 60 day prediction. In addition to the benchmark, the Naïve Method, the GRU algorithm has the significant best performance for the 120 day forecast considering the Diebold–Mariano test results compared to the other algorithms (Table 7), indicating that the GRU algorithm is more efficient at forecasting the next 120 days of LWL.

From the obtained results for LWL prediction from day 1 to day 120, we can see that: (1) Day 60 predictions provide the most optimized LWL detection based on high Naïve Benchmark performance comparison values. (2) The best performance of the investigated algorithms can change in terms of the selected prediction periods. (3) The LSTM algorithm can better predict LWL for 60 days in advance with higher accuracy, which allows water managers to take action. In addition, it is worth noting that the Bidirectional LSTM and Stacked LSTM algorithms contribute to the forecast with little or no performance increase for the short prediction period of less than 20 days.

Among the features, the most important one to affect the output was determined as withdrawal using the Mutual Information technique. The importance levels can be ordered as withdrawal, average temperature, minimum temperature, maximum temperature, and precipitation (Figure 5).

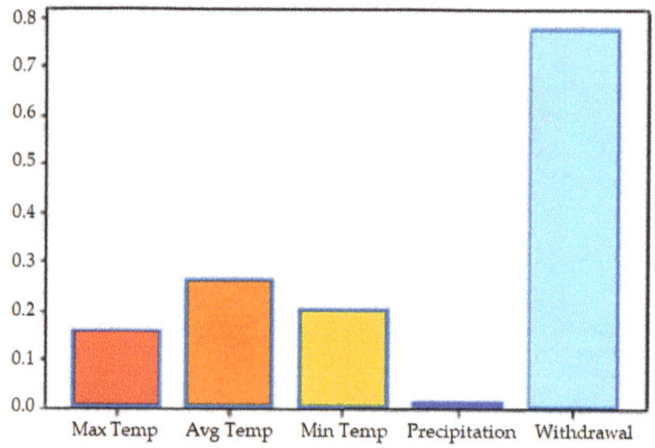

Figure 5. Variable importance.

Accurate LWL prediction is a necessity, not only to prevent possible drought conditions but also possible water quality effects. Therefore, this study conducted extra work to observe the relationship between microcystin concentrations previously observed during low LWL periods. In addition to the LWL effect, this study investigated the relationship for the maximum temperature, mean temperature, minimum temperature, precipitation, light intensity, and evaporation. This experiment was conducted in order to reveal their importance so as to predict LWL in advance and be able to take measurable actions in advance.

To begin with, the microcystin concentrations at the surface, 1 m, 5 m, 10 m, 15 m, and 20 m were measured over the period of 2019–2023 to understand the relationship between the changing meteorological situation and water quality was affected by algal growth. The microcystin concentration in all sampled depths showed approximately the same increasing pattern over time, except for the samples collected from the depth of 15 m (Figure 6). The variations in Figure 6 indicate there is an increasing trend of microcystin for the surface water, 1 m, 5 m, 10 m, and 20 m depths. However, the trend is decreasing for the 15 m depth. The microcystin level was almost similar for each depth of the first 10 m; however, significant differences were recorded in the spring and autumn, specifically, vertically mixing periods. During the summer, the microcystin concentration stayed relatively low (<0.5 µg/L) or at an undetectable level from May to October. The highest concentrations were observed during the winter period from November to April with a significant fluctuation, which coincided with the mixing period. By contrast, the microcystin concentrations were higher at the sampling depths of 15 and 20 m. The microcystin was recorded at all sampling times during the experimental period. In general, the concentrations were below the 2 µg/L for both sampling depths; however, the highest concentrations of around 8 µg/L were recorded during the summer stratification phase (June to August). For the two years 2020 and 2021, the microcystin concentration was the lowest (with <3.31 µg/L), especially for 2021 (<1.61 µg/L).

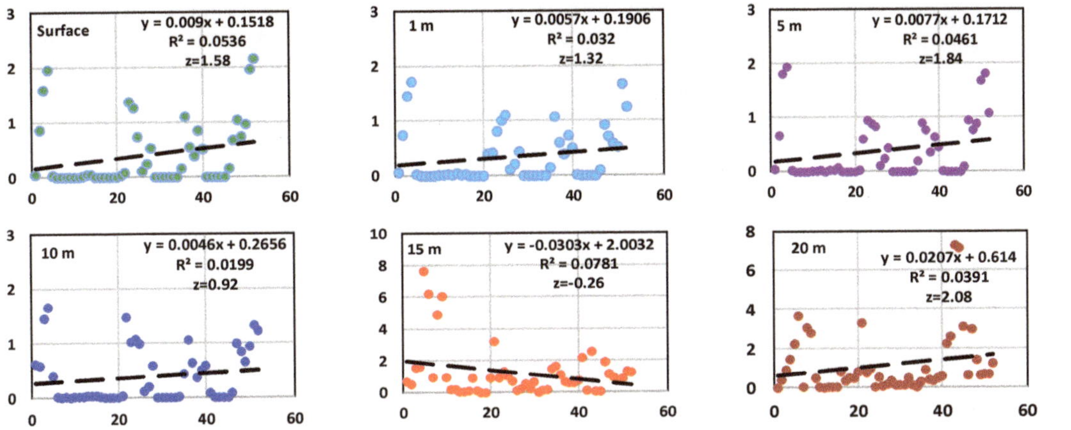

Figure 6. Linear trend of microcystin concentration in the water column at different depths from surface to 20 m from 21 March 2019 to 12 April 2023. (x-axis: data rows in sequence, y-axis: microcystin concentration).

The nonparametric Mann–Kendall test shows that the microcystin concentration decreases monotonically at a depth of 15 m and increases at the other depths. However, only the microcystin concentration at 20 m depth was significant at the 95% confidence limit with a z-value of 2.08 (Figure 6), indicating an increasing positive trend in the microcystin data time series that dominates at this depth.

Due to temporal and spatial variability, it is difficult to obtain sufficient input data needed for data-driven predictive models to analyze and learn the relationships between microcystin and meteorological parameters, i.e., temperature, precipitation associated with algal proliferation, and microcystin concentration. To better understand the changing meteorological parameters on microcystin concentration, Spearman correlations were evaluated using monthly microcystin data collected from raw water before water treatment. From Figure 7, the significant positive contribution of temperature on microcystin concentration is evident. Light intensity also has a positive effect on microcystin concentration. On the other hand, the water level of the lake had no significant effect on the microcystin concentration.

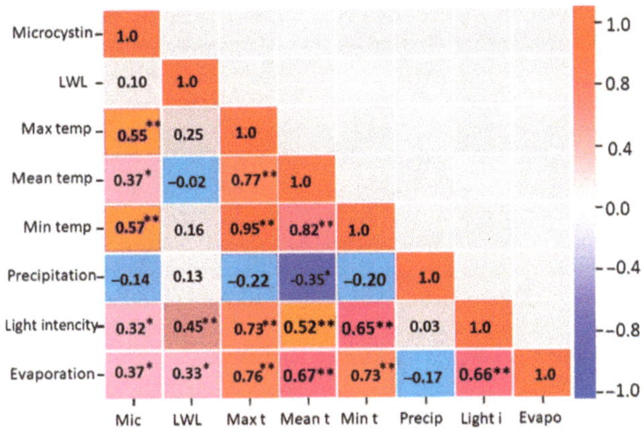

Figure 7. Spearman rank correlation between microcystin and meteorological parameters (** $p < 0.01$, * $p < 0.05$).

The degree of association differs in terms of the features in Figure 7. The minimum temperature, maximum temperature, mean temperature, and evaporation have a moderate correlation with microcystin [56]. In addition, light intensity has a weak correlation. On the other hand, the LWL and precipitation have a very weak correlation with microcystin. The results provide a better understanding that the water quality is rarely affected by the level of water. However, the temperature, which is one of the indicators for predicting LWL, affects the water quality. Thus, it can be concluded that LWL does not directly affect the water quality, but the effect is indirect through the consideration of temperature values.

4. Discussion

Based on the experimental result of this case study that applies ANN and RNN-based deep learning algorithms for lake water level prediction, it is possible to forecast the next 120 days with a smaller RMSE (0.3838 m), reasonable Naïve Benchmark comparison value (58.00%), and significant Diebold–Mariano test results ($p < 0.05$). However, compared with other models, the prediction result based on LSTM proposed in this study is optimal for the next 60 days LWL forecasting with a smaller RMSE (0.1762 m), the highest Naïve Benchmark comparison value (78.55%), and a significant Diebold–Mariano test p-value (<0.003). The goal of this study is to compare the impact of various climates and comprehend how new AI techniques behave and perform on various event forecasting tasks. The prediction performance of the investigated ANN and RNN algorithms aligns with previous research based on the RMSE and the Naïve Method. Using ANN and SVM, Yoon et al. predicted the groundwater levels in the nearshore aquifer in Donghae City, Korea, for two wells with RMSE values of 0.13 m and 0.136 m, respectively [57]. The objective of their research was to create and evaluate data-driven time series forecasting models for the short-term fluctuations in groundwater levels in a coastal aquifer caused by tidal influence and precipitation recharge. However, their study lacks a comparison of the proposed algorithms with the baseline models and other algorithms from DL. Therefore, the performance of the models cannot be evaluated for predicting water levels. The algorithms are also not evaluated against basic benchmark methods such as the Naïve Method, which raises the question of whether it is necessary to create fancy DL algorithms for LWL prediction. Thus, this study could be a milestone for further water level studies that attempt to develop every single DL algorithm available in the field of data science.

Hrnjica, B. and Bonacci found that the LSTM and RNN algorithms performed better than the traditional ANN algorithms on datasets with a given number of features and a time scale of one month [58]. They also found that the feed-forward neural network and LSTM models performed better than the traditional time series forecasting models based on ARIMA and other similar techniques. The objective of their study was motivated by the realization that traditional regression and statistical techniques were insufficiently effective at predicting stochastic events such as water level. In contrast to traditional models, Lee et al. showed that the LSTM model better reproduces the variability and correlation structure of the broader time scale as well as the important statistics of the original time domain [59]. Applying the LSTM into stochastic simulation and determining if the long-term trends of known hydroclimatological indicators can be replicated was the main objective of their work. The improved representation of long-term variability is critical for water managers as they rely on these data to plan and manage future water resources. In the future, the performance improvement over the Naïve Benchmark can be tested with other novel models, such as attention-based algorithms or other derivatives. However, the recent attempt to use an attention-based algorithm showed that it did not perform better than a recurrent network [60].

The main hypothesis of the present study is confirmed by the fact that RNN-based algorithms achieve better predictive performance of LWL when using long-term daily data from a decade and improve predictive accuracy for 60-day forecasts (Table 4). The trends of observations and model predictions in Figure A1 through Figure A4 suggest that the potential performance of RNN algorithms can also be extended beyond 120-day forecasts by incorporating more data into the models. The LSTM model network has demonstrated its ability to learn from sequential data in the past and has been shown to be a useful model. It can effectively learn from sequences of varied durations, capturing long-term dependencies. [35]. To confirm the results of this paper, Zhu et al. studied 69 lakes in Poland for 30 day ahead water level prediction and concluded that the recurrent DL models performed similarly to attention-based recurrent DL models in terms of predictive performance [60]. The results of the LSTM algorithm between its variants, namely the Stacked LSTM and the Bidirectional LSTM, in the present study show that there is no significant difference in predicting less than 30 days ahead. The LSTM algorithm requires long observation datasets and the selection and optimization of hyperparameters, learning rate, and number of epochs to achieve correct prediction results [20]. For example, Morovati et al. reported a better prediction performance of LSTM when using daily recorded data over 20 years [61]. The results obtained for LSTM in this study are consistent with these findings. The findings also show that the LSTM algorithm reflects well when compared to the fluctuation trend of the real LWL value. This is due to the use of a gated structure in the LSTM model, so the LSTM algorithm is good at extracting short-term temporal correlations. However, due to the cyclic periods of water level variations, the performance increase drops when it reaches the next LWL cycle after 60 days. The better performance of RNN algorithms compared to the Naïve Method is also due to the successful optimization of hyperparameters in the RNN networks.

Another important aspect is that although the prevailing opinion suggests using all available DL algorithms to find the algorithm that performs best according to the RMSE or MAPE results, the results of the algorithms do not seem to differ significantly with respect to the Diebold–Mariano test. Therefore, in order to suggest a better performing algorithm, the statistical difference must be shown in addition to the RMSE or MAPE results [53], and in some cases, the ANN and gated RNN derivatives, as indicated in the Results section, do not appear to have statistical significance and can be used interchangeably.

The fluctuations in LWL are associated with meteorological processes and anthropogenic activities, which lead to a nonlinear and complex system. In this context, the study has several limitations due to its nature. One of the limitations is that the results depend on the geographical location. The experiment was conducted at Lake Sapanca in the northeastern Marmara region of Turkey. This location has characteristics of both Black Sea and Mediterranean climates. Therefore, the results may change in regions with different climate characteristics. Another limitation of the study is that the dataset produced by the Turkish Meteorological Service contains several missing data for selected parameters. Although it is possible to interpolate missing data, the results with interpolated data rows may produce a biased LWL value. The results could change with a dataset containing complete records for a longer period without missing data. In addition, there are limitations to the study in that there is an insufficient amount of data, especially for some features. In practice, there may not be a chance to gather all the features from the field. The potential feature(s) may not be represented to the algorithm and the potential feature(s) may even increase the model prediction performance. In addition, in the case of there being very few available features, the prediction performance could be underrepresented. However, in the case of there being too many features, it may cause the model to overfit. Therefore, a balance between overrepresentation and underrepresentation must be provided. Thus, it is further suggested to apply other appropriate preprocessing methods to improve the predictive performance of the RNN DL models with different time horizons. In the future, the LWL prediction could be practiced by using GIS methods with a satellite dataset. The performance difference between time series prediction and prediction with image data could be compared with the

Naïve Method Benchmark. In addition, in terms of the availability of more features, the researchers can conduct sensitivity analysis and uncertainty analysis to eliminate some of the features to prevent possible overfitting issues.

Several well-known nutrient inputs and relatively less known meteorological parameters, together with hydrological disturbances, cause excessive growth of cyanobacteria in freshwater ecosystems, which degrade water quality with their toxins. Extreme heat waves are becoming more common as global and regional warming continues and are expected to become the norm in future scenarios. Microcystin concentration correlated positively with temperature variables (max, min, and mean, $p < 0.01$), including evaporation and light intensity ($p < 0.05$), and not significantly with precipitation (negative correlation), which is directly related to LWL (Figure 7). Significant correlations between meteorological parameters and microcystin concentrations in freshwater bodies have been reported previously [7]. Light intensity in the metalimnion zone leads to greater development of cyanobacteria and the presence of large amounts of microcystins, posing potential problems for the use of water resources [62]. Since freshwater lakes are used as drinking water sources, proper water and algae management is necessary to ensure a clean and safe water supply. The use of tap water is restricted when large amounts of algae are found in water reservoirs because various water treatment problems can occur, such as clogged treatment systems, a bad odor, color in the water, and regulated toxic substances such as microcystin. Predicting the correlation of algal blooms with easily measured meteorological or hydrological parameters in advance and taking rapid response actions to algal growth can minimize damage and ensure uninterrupted production of purified water.

5. Conclusions

Monitoring and forecasting lake water levels is one of the most important tasks to ensure sustainable water resource management, safeguard water quality, and maintain watershed balance in the face of global climate change. The gated RNN-based algorithms are powerful modeling techniques for future forecasting and were tested in this study to obtain more accurate estimates of water level changes in lakes. The gated RNN algorithms correctly adapt to changing input conditions, such as adjustments in water demand policy during reservoir operation. The fact that the gated RNN structure accounts for the nonlinear dynamics of the problem throughout the dataset, means that it can be used to explain why gated RNNs perform better than conventional approaches in predicting reservoir levels. With respect to the RMSE, the results demonstrated here show the ability of the models used to understand the nonlinear behavior of LWLs.

The modeling results support the following findings:

1. The results of the algorithms can be compared, and although there could be different but similar results, the algorithms can be used interchangeably.
2. Overall, the GRU algorithm performs better than other gated RNN algorithms because it has a lower RMSE. However, it does not perform better in all time periods, so the algorithm needs to be replaced by another one to achieve better results for LWL prediction cases further in the future.
3. Gated RNN-based algorithms appear to have higher RMSE results as the prediction horizon increases, indicating poorer performance in lower prediction time periods. A more accurate comparison is possible using the Naïve Method, and the percentage increase could provide a healthier result for comparing algorithm results with different prediction time periods. Although the prediction may differ from the actual values as the time period increases, the performance increase is much higher compared to the Naïve Benchmark, making it more attractive for use in LWL prediction cases.

In addition, this study also examined the relationship between global warming and microcystin levels in freshwater lakes and demonstrated a clear relationship with meteorological data. However, more research is needed in this area to close the gap between LWL predictions with different geographical locations using the same available features, algal growth, and microcystin levels.

Overall, the prediction results suggest that the proposed RNN algorithms can be successfully used to predict the future state of LWL for drinking water resource management leading to the achievement of sustainability under changing climatic conditions.

Author Contributions: Conceptualization, S.O.; methodology, S.O.; software, S.O.; validation, S.O.; formal analysis, S.O.; investigation, S.O.; resources, S.O.; data curation, S.O.; writing—original draft preparation, S.O.; writing—review and editing, S.O. and S.O.Y.; visualization, S.O.; supervision, S.O.Y.; project administration, S.O. and S.O.Y. All authors have read and agreed to the published version of the manuscript.

Funding: This research received no external funding.

Institutional Review Board Statement: Not applicable.

Informed Consent Statement: Not applicable.

Data Availability Statement: Restrictions apply to the availability of these data. Data were obtained from The State Hydraulic Works for water level and withdrawal features, and the Turkish State Meteorological Service for maximum temperature, average temperature, minimum temperature, and precipitation features, which are governmental institutions for water administration and meteorological services in Turkey. The data are available from the authors with the permission of The State Hydraulic Works and Turkish State Meteorological Service.

Conflicts of Interest: The authors declare no conflict of interest.

Appendix A

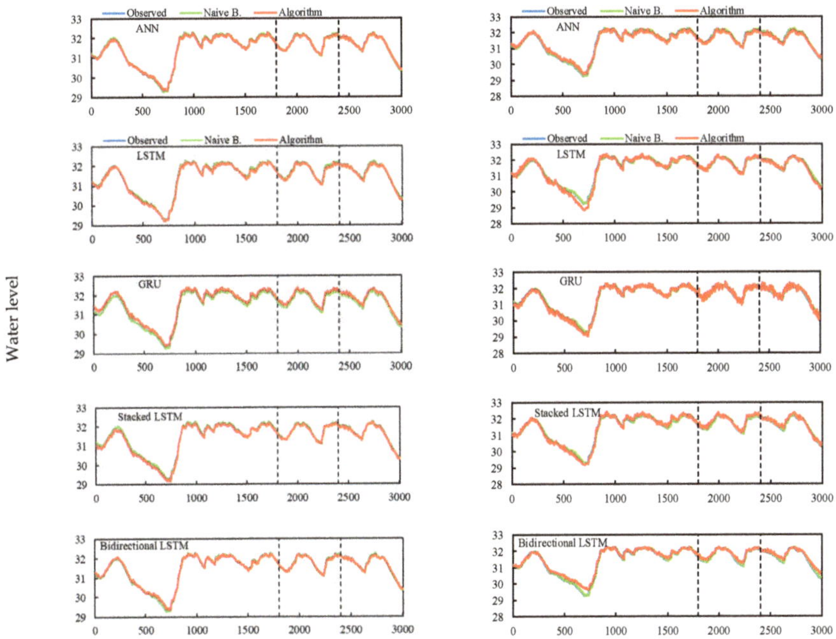

Figure A1. One day (**left side**) and 5 days (**right side**) ahead prediction results. The vertical dashed lines indicate the train set-validation set-test set, respectively.

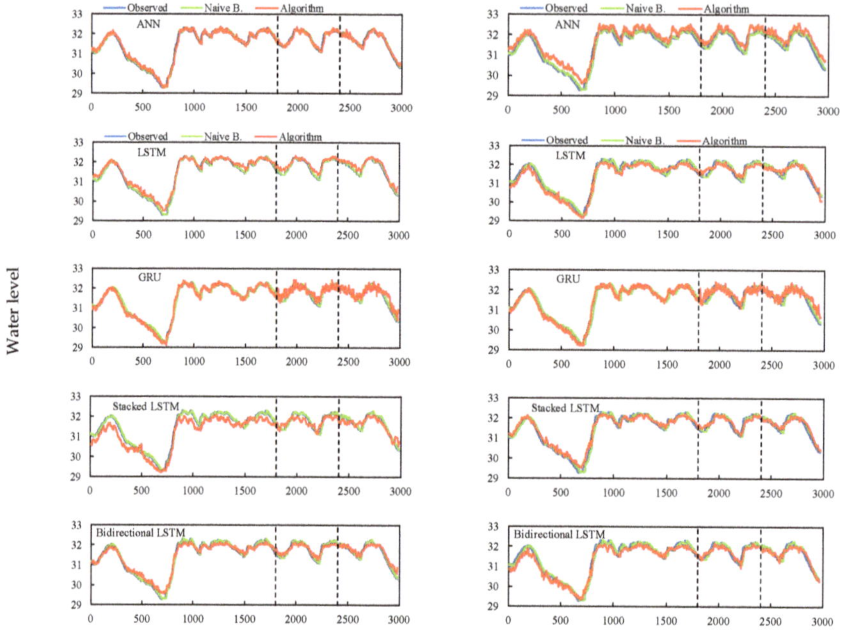

Figure A2. Ten days (**left side**) and 20 days (**right side**) ahead prediction results. The vertical dashed lines indicate the train set-validation set-test set, respectively.

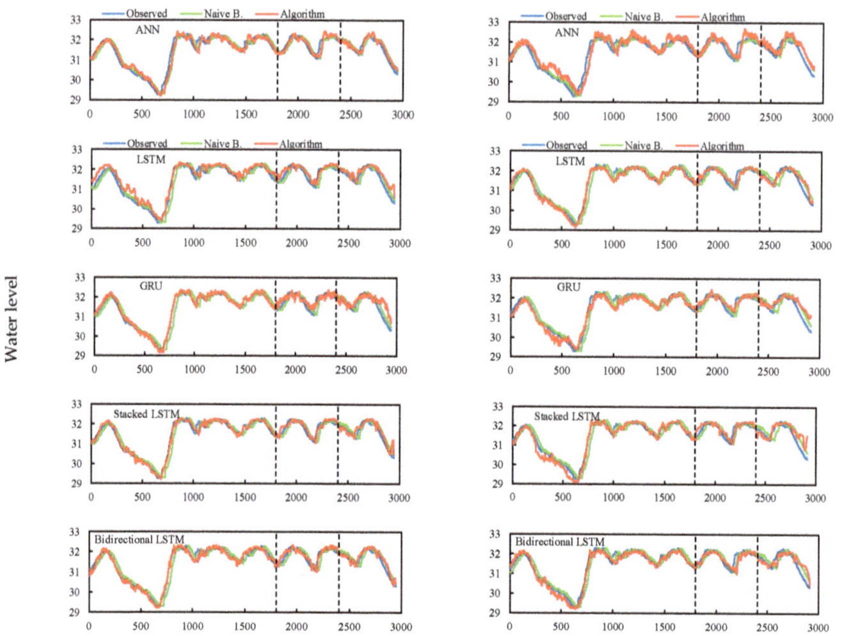

Figure A3. Thirty days (**left side**) and 45 days (**right side**) ahead prediction results. The vertical dashed lines indicate the train set-validation set-test set, respectively.

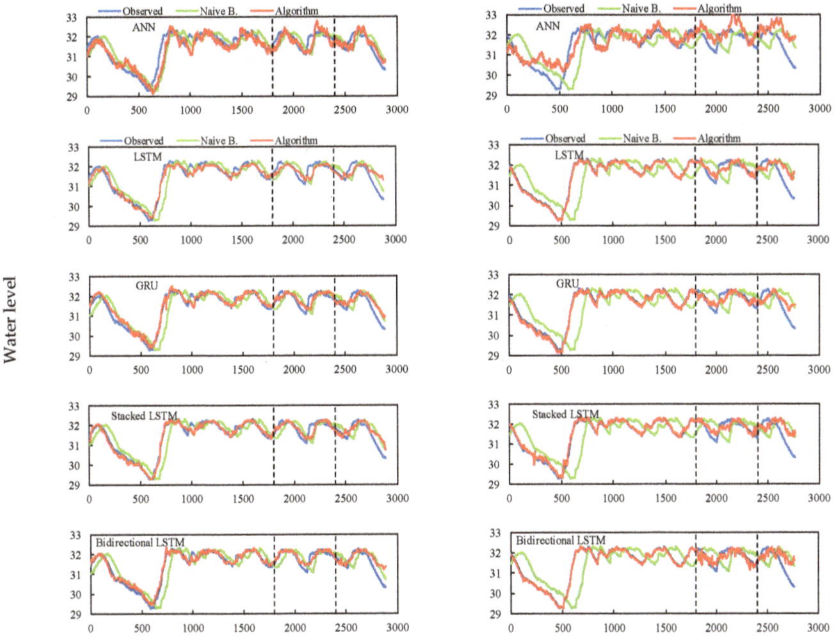

Figure A4. Sixty days (**left side**) and 120 days (**right side**) ahead prediction results. The vertical dashed lines indicate the train set-validation set-test set, respectively.

References

1. Paul, N.; Elango, L. Predicting future water supply-demand gap with a new reservoir, desalination plant and waste water reuse by water evaluation and planning model for Chennai megacity, India. *Groundw. Sustain. Dev.* **2018**, *7*, 8–19. [CrossRef]
2. Castillo-Botón, C.; Casillas-Pérez, D.; Casanova-Mateo, C.; Moreno-Saavedra, L.M.; Morales-Díaz, B.; Sanz-Justo, J.; Gutiérrez, P.A.; Salcedo-Sanz, S. Analysis and prediction of dammed water level in a hydropower reservoir using machine learning and persistence-based techniques. *Water* **2020**, *12*, 1528. [CrossRef]
3. Soylu Pekpostalci, D.; Tur, R.; Danandeh Mehr, A.; Vazifekhah Ghaffari, M.A.; Dąbrowska, D.; Nourani, V. Drought monitoring and forecasting across Turkey: A contemporary review. *Sustainability* **2023**, *15*, 6080. [CrossRef]
4. Yeşilköy, S.; Şaylan, L. Spatial and temporal drought projections of northwestern Turkey. *Theor. Appl. Climatol.* **2022**, *149*, 1–14. [CrossRef]
5. Bond, N.R.; Lake, P.S.; Arthington, A.H. The impacts of drought on freshwater ecosystems: An Australian perspective. *Hydrobiologia* **2008**, *600*, 3–16. [CrossRef]
6. Duru, U. Shoreline change assessment using multi-temporal satellite images: A case study of Lake Sapanca, NW Turkey. *Environ. Monit. Assess.* **2017**, *189*, 385. [CrossRef]
7. Novais, M.H.; Penha, A.M.; Catarino, A.; Martins, I.; Fialho, S.; Lima, A.; Palma, P. The usefulness of ecotoxicological tools to improve the assessment of water bodies in a climate change reality. *Sci. Total Environ.* **2023**, *901*, 166392. [CrossRef]
8. Zounemat-Kermani, M.; Batelaan, O.; Fadaee, M.; Hinkelmann, R. Ensemble machine learning paradigms in hydrology: A review. *J. Hydrol.* **2021**, *598*, 126266. [CrossRef]
9. Chang, F.J.; Chang, Y.T. Adaptive neuro-fuzzy inference system for prediction of water level in reservoir. *Adv. Water Resour.* **2006**, *29*, 1–10. [CrossRef]
10. Liu, Z.; Zhou, P.; Chen, X.; Guan, Y. A multivariate conditional model for streamflow prediction and spatial precipitation refinement. *J. Geophys. Res. Atmos.* **2015**, *120*, 10–116. [CrossRef]
11. Wang, M.; Dai, L.; Dai, H.; Mao, J.; Liang, L. Support vector regression based model for predicting water level of Dongting Lake. *J. Drain. Irrig. Mach. Eng.* **2017**, *35*, 954–961.
12. Zhang, X.; Liu, P.; Zhao, Y.; Deng, C.; Li, Z.; Xiong, M. Error correction-based forecasting of reservoir water levels: Improving accuracy over multiple lead times. *Environ. Model. Softw.* **2018**, *104*, 27–39. [CrossRef]
13. Bourdeau, M.; Zhai, X.q.; Nefzaoui, E.; Guo, X.; Chatellier, P. Modeling and forecasting building energy consumption: A review of data-driven techniques. *Sustain. Cities Soc.* **2019**, *48*, 101533. [CrossRef]
14. Ghorbani, M.A.; Khatibi, R.; Aytek, A.; Makarynskyy, O.; Shiri, J. Sea water level forecasting using genetic programming and comparing the performance with artificial neural networks. *Comput. Geosci.* **2010**, *36*, 620–627. [CrossRef]

15. Talebizadeh, M.; Moridnejad, A. Uncertainty analysis for the forecast of lake level fluctuations using ensembles of ANN and ANFIS models. *Expert Syst. Appl.* **2011**, *38*, 4126–4135. [CrossRef]
16. Kisi, O.; Shiri, J.; Nikoofar, B. Forecasting daily lake levels using artificial intelligence approaches. *Comput. Geosci.* **2012**, *41*, 169–180. [CrossRef]
17. Buyukyildiz, M.; Tezel, G.; Yilmaz, V. Estimation of the change in lake water level by artificial intelligence methods. *Water Resour. Manag.* **2014**, *28*, 4747–4763. [CrossRef]
18. Lukman, Q.A.; Ruslan, F.A.; Adnan, R. 5 Hours ahead of time flood water level prediction modelling using NNARX technique: Case study terengganu. In Proceedings of the 2016 7th IEEE Control and System Graduate Research Colloquium (ICSGRC), Shah Alam, Malaysia, 8 August 2016; IEEE: New York, NY, USA, 2016; pp. 104–108. [CrossRef]
19. Yadav, B.; Eliza, K. A hybrid wavelet-support vector machine model for prediction of lake water level fluctuations using hydro-meteorological data. *Measurement* **2017**, *103*, 294–301. [CrossRef]
20. Ozdemir, S.; Yaqub, M.; Yildirim, S.O. A systematic literature review on Lake water level prediction models. *Environ. Model. Softw.* **2023**, *163*, 105684. [CrossRef]
21. Yu, Z.; Lei, G.; Jiang, Z.; Liu, F. ARIMA modelling and forecasting of water level in the middle reach of the Yangtze River. In Proceedings of the 2017 4th International Conference on Transportation Information and Safety (ICTIS), Banff, AB, Canada, 8–10 August 2017; IEEE: New York, NY, USA, 2017; pp. 172–177.
22. Viccione, G.; Guarnaccia, C.; Mancini, S.; Quartieri, J. On the use of ARIMA models for short-term water tank levels forecasting. *Water Supply* **2020**, *20*, 787–799. [CrossRef]
23. Azad, A.S.; Sokkalingam, R.; Daud, H.; Adhikary, S.K.; Khurshid, H.; Mazlan, S.N.A.; Rabbani, M.B.A. Water level prediction through hybrid SARIMA and ANN models based on time series analysis: Red hills reservoir case study. *Sustainability* **2022**, *14*, 1843. [CrossRef]
24. Fang, T.; Lahdelma, R. Evaluation of a multiple linear regression model and SARIMA model in forecasting heat demand for district heating system. *Appl. Energy* **2016**, *179*, 544–552. [CrossRef]
25. Altunkaynak, A. Forecasting surface water level fluctuations of Lake Van by artificial neural networks. *Water Resour. Manag.* **2007**, *21*, 399–408. [CrossRef]
26. Adhikary, S.K.; Muttil, N.; Yilmaz, A.G. Improving streamflow forecast using optimal rain gauge network-based input to artificial neural network models. *Hydrol. Res.* **2018**, *49*, 1559–1577. [CrossRef]
27. Nouri, H.; Ildoromi, A.; Sepehri, M.; Artimani, M. Comparing three main methods of artificial intelligence in flood estimation in Yalphan catchment. *Geogr. Environ. Plan.* **2019**, *29*, 35–50. [CrossRef]
28. Khandelwal, I.; Adhikari, R.; Verma, G. Time series forecasting using hybrid ARIMA and ANN models based on DWT decomposition. *Procedia Comput. Sci.* **2015**, *48*, 173–179. [CrossRef]
29. Phan, T.T.H.; Nguyen, X.H. Combining statistical machine learning models with ARIMA for water level forecasting: The case of the Red river. *Adv. Water Resour.* **2020**, *142*, 103656. [CrossRef]
30. Sethia, A.; Raut, P. Application of LSTM, GRU and ICA for stock price prediction. In *Information and Communication Technology for Intelligent Systems, Proceedings of ICTIS 2018, Padang, Indonesia, 25–26 July 2018*; Springer Singapore: Singapore, 2019; Volume 2, pp. 479–487. [CrossRef]
31. Anupa, A.; Sugathadasa, R.; Herath, O.; Thibbotuwawa, A. Artificial neural network based demand forecasting integrated with federal funds rate. *Appl. Comput. Sci.* **2021**, *17*, 34–44. [CrossRef]
32. Ebtehaj, I.; Bonakdari, H.; Gharabaghi, B. A reliable linear method for modeling lake level fluctuations. *J. Hydrol.* **2019**, *570*, 236–250. [CrossRef]
33. Xiang, Z.; Demir, I. Distributed long-term hourly streamflow predictions using deep learning–A case study for State of Iowa. *Environ. Model. Softw.* **2020**, *131*, 104761. [CrossRef]
34. Chen, Y.; Fan, R.; Yang, X.; Wang, J.; Latif, A. Extraction of urban water bodies from high-resolution remote-sensing imagery using deep learning. *Water* **2018**, *10*, 585. [CrossRef]
35. LeCun, Y.; Bengio, Y.; Hinton, G. Deep learning. *Nature* **2015**, *521*, 436–444. [CrossRef] [PubMed]
36. Akkoyunlu, A.; Akiner, M.E. Pollution evaluation in streams using water quality indices: A case study from Turkey's Sapanca Lake Basin. *Ecol. Indic.* **2012**, *18*, 501–511. [CrossRef]
37. Population Growth (Annual %)—Turkiye. Available online: https://data.worldbank.org/indicator/SP.POP.GROW?locations=TR (accessed on 8 March 2023).
38. Jiang, F.; Dong, Z.; Wang, Z.A.; Zhu, Y.; Liu, M.; Luo, Y.; Zhang, T. Flood forecasting using an improved NARX network based on wavelet analysis coupled with uncertainty analysis by Monte Carlo simulations: A case study of Taihu Basin, China. *J. Water Clim. Change* **2021**, *12*, 2674–2696. [CrossRef]
39. Nourani, V.; Tootoonchi, R.; Andaryani, S. Investigation of climate, land cover and lake level pattern changes and interactions using remotely sensed data and wavelet analysis. *Ecol. Inform.* **2021**, *64*, 101330. [CrossRef]
40. Tsao, H.H.; Leu, Y.G.; Chou, L.F.; Tsao, C.Y. A method of multi-stage reservoir water level forecasting systems: A case study of Techi hydropower in Taiwan. *Energies* **2021**, *14*, 3461. [CrossRef]
41. Obringer, R.; Nateghi, R. Predicting urban reservoir levels using statistical learning techniques. *Sci. Rep.* **2018**, *8*, 5164. [CrossRef]

42. Guyennon, N.; Salerno, F.; Rossi, D.; Rainaldi, M.; Calizza, E.; Romano, E. Climate change and water abstraction impacts on the long-term variability of water levels in Lake Bracciano (Central Italy): A Random Forest approach. *J. Hydrol. Reg. Stud.* **2021**, *37*, 100880. [CrossRef]
43. Dinka, M.O. Estimation of groundwater contribution to Lake Basaka in different hydrologic years using conceptual netgroundwater flux model. *J. Hydrol. Reg. Stud.* **2020**, *30*, 100696. [CrossRef]
44. Hochreiter, S.; Schmidhuber, J. Long short-term memory. *Neural Comput.* **1997**, *9*, 1735–1780. [CrossRef]
45. Graves, A.; Graves, A. *Supervised Sequaence Labelling*; Springer: Berlin/Heidelberg, Germany, 2012; pp. 5–13.
46. Chemali, E.; Kollmeyer, P.J.; Preindl, M.; Ahmed, R.; Emadi, A. Long short-term memory networks for accurate state-of-charge estimation of Li-ion batteries. *IEEE Trans. Ind. Electron.* **2017**, *65*, 6730–6739. [CrossRef]
47. Ojo, S.O.; Owolawi, P.A.; Mphahlele, M.; Adisa, J.A. Stock market behaviour prediction using stacked LSTM networks. In Proceedings of the 2019 International Multidisciplinary Information Technology and Engineering Conference (IMITEC), Vanderbijlpark, South Africa, 21–22 November 2019; IEEE: New York, NY, USA, 2019; pp. 1–5.
48. Kang, H.; Yang, S.; Huang, J.; Oh, J. Time series prediction of wastewater flow rate by bidirectional LSTM deep learning. *Int. J. Control Autom. Syst.* **2020**, *18*, 3023–3030. [CrossRef]
49. Gulli, A.; Pal, S. *Deep Learning with Keras*; Packt Publishing Ltd.: Birmingham, UK, 2017.
50. Paul, S.; Oppelstrup, J.; Thunvik, R.; Magero, J.M.; Ddumba Walakira, D.; Cvetkovic, V. Bathymetry development and flow analyses using two-dimensional numerical modeling approach for Lake Victoria. *Fluids* **2019**, *4*, 182. [CrossRef]
51. Nhu, V.H.; Shahabi, H.; Nohani, E.; Shirzadi, A.; Al-Ansari, N.; Bahrami, S.; Miraki, S.; Geertsema, M.; Nguyen, H. Daily water level prediction of Zrebar Lake (Iran): A comparison between M5P, random forest, random tree and reduced error pruning trees algorithms. *ISPRS Int. J. Geo-Inf.* **2020**, *9*, 479. [CrossRef]
52. Zheng, F.; Maier, H.R.; Wu, W.; Dandy, G.C.; Gupta, H.V.; Zhang, T. On lack of robustness in hydrological model development due to absence of guidelines for selecting calibration and evaluation data: Demonstration for data-driven models. *Water Resour. Res.* **2018**, *54*, 1013–1030. [CrossRef]
53. Van der Heijden, T.; Lago, J.; Palensky, P.; Abraham, E. Electricity price forecasting in European Day Ahead Markets: A greedy consideration of market integration. *IEEE Access* **2021**, *9*, 119954–119966. [CrossRef]
54. Albay, R.A.; Köker, L.; Gürevin, C.; Albay, M. Planktothrix rubescens: A perennial presence and toxicity in Lake Sapanca. *Turk. J. Bot.* **2014**, *38*, 782–789.
55. Sahar, A.; Han, D. An LSTM-based indoor positioning method using Wi-Fi signals. In Proceedings of the 2nd International Conference on Vision, Image and Signal Processing, Las Vegas, NV, USA, 27–29 August 2018; pp. 1–5. [CrossRef]
56. Fowler, J.; Cohen, L.; Jarvis, P. *Practical Statistics for Field Biology*; John Wiley & Sons: Hoboken, NJ, USA, 2013.
57. Yoon, H.; Jun, S.C.; Hyun, Y.; Bae, G.O.; Lee, K.K. A comparative study of artificial neural networks and support vector machines for predicting groundwater levels in a coastal aquifer. *J. Hydrol.* **2011**, *396*, 128–138. [CrossRef]
58. Hrnjica, B.; Bonacci, O. Lake level prediction using feed forward and recurrent neural networks. *Water Resour. Manag.* **2019**, *33*, 2471–2484. [CrossRef]
59. Lee, T.; Shin, J.Y.; Kim, J.S.; Singh, V.P. Stochastic simulation on reproducing long-term memory of hydroclimatological variables using deep learning model. *J. Hydrol.* **2020**, *582*, 124540. [CrossRef]
60. Zhu, S.; Ji, Q.; Ptak, M.; Sojka, M.; Keramatfar, A.; Chau, K.W.; Band, S.S. Daily water-level forecasting for multiple polish lakes using multiple data-driven models. *Geogr. J.* **2023**, *189*, 357–369. [CrossRef]
61. Morovati, K.; Nakhaei, P.; Tian, F.; Tudaji, M.; Hou, S. A Machine learning framework to predict reverse flow and water level: A case study of Tonle Sap Lake. *J. Hydrol.* **2021**, *603*, 127168. [CrossRef]
62. Boscaini, A.; Brescancin, F.; Cerasino, L.; Fedrigotti, C.; Anna, F.E.; Salmaso, N. Vertical and horizontal distribution of the microcystin producer Planktothrix rubescens (Cyanobacteria) in a small perialpine reservoir. *Adv. Oceanogr. Limnol.* **2017**, *8*, 208–221. [CrossRef]

sustainability

Article

Bacteriophage and Their Potential Use in Bioaugmentation of Biological Wastewater Treatment Processes

Şuheda Reisoglu and Sevcan Aydin *

Division of Biotechnology, Biology Department, Faculty of Science, Istanbul University, Istanbul 34134, Turkey
* Correspondence: sevcan.aydin@istanbul.edu.tr

Abstract: In wastewater treatment plants, the stability of the related microbiota is pivotal for a steady and appropriate operation in biological wastewater treatment (BWT). The interactions between phages, which are high in amounts and diversity, and their particular hosts are significant due to their specificity in this microbial community. As drivers of diversity, phages are capable of changing the microbial structure by affecting host cells. This study aimed to enhance the stability of the functional microbiota, a primary concern of BWT, by evaluating the influence of bioaugmentation with bacteriophage cocktails. The sequencing data revealed significant alterations in the entire microbiome, including microeukaryotes, induced by the bacteriophages. These alterations led to a reduction in opportunistic microorganisms while preserving the beneficial ones. The genus *Proteobacteria* was found unaffected by three antibiotics in the bioreactor due to its antibiotic-resistant members, leading to its further growth, while the antibiotic-sensitive genus decreased in quantity. Furthermore, the phages notably influenced the metabolic pathway such as nitrogen, carbohydrate, and amino acid metabolisms by eliminating opportunistic microbes and providing improved growth conditions to bacterial species that are essential for effective reactor performance and wastewater treatment.

Keywords: bacteriophage; wastewater treatment; bioaugmentation; microbial community

check for updates

Citation: Reisoglu, Ş.; Aydin, S. Bacteriophage and Their Potential Use in Bioaugmentation of Biological Wastewater Treatment Processes. *Sustainability* **2023**, *15*, 12216. https://doi.org/10.3390/su151612216

Academic Editor: Andreas Angelakis

Received: 10 July 2023
Revised: 29 July 2023
Accepted: 3 August 2023
Published: 10 August 2023

1. Introduction

In recent years, water quality has notably decreased due to the increment in both population and industrial development. Wastewater treatment plants (WWTP) receive tons of wastewater coming from various sources such as hospitals, slaughterhouses, livestock farming, pharmaceutical facilities, and household sewage. In conditions where the treatment is not efficacious, effluent water may still have pollutants resulting in adverse effects on both human health and aquatic environments through being released from the treatment process. Therefore, the appropriate treatment of wastewater is of critical importance in terms of disposal. Various treatment methods such as land application, chemical treatment, and discharge in sewers are in use, however, there are constraints to their utilization since unfavorable odor and contamination of aquatic environments issue in the exploitation of the techniques of land application and discharge in sewers. In addition, it is not recommended to use a chemical treatment, as chemicals used in the process raise the treatment cost, and the challenge in the disposal of remaining sludge-containing chemicals makes the process cumbersome and costly [1]. On the contrary, the BWT approach provides a substantial remedy with its eco-friendly and cost-effective characteristic in removing pollutants from wastewater. In BWT, diverse functional populations of bacteria that are responsible for removing organic matter, nitrogen, and phosphorus from the wastewater play a critical role in the effective treatment of wastewater. Heterotrophic bacteria take a substantial role in organic matter removal, ammonia-oxidizing bacteria (AOB) in nitrification, denitrifying bacteria in nitrogen removal, and phosphate-accumulating organisms (PAOs) in phosphorus removal. Appropriate pollutant removal can be achieved through collaborative interactions among these functional groups [2]. In this process, bacterial strains adjust their population density and behaviors via a cell-to-cell communication pathway named quorum sensing (QS), which is also responsible for the formation of bacterial biofilms. In addition to the formation of biofilm, bacteria through QS also regulate the removal of pollutants, and pathogenesis control, and they determine the composition and behavior of the biofilm community. Therefore, the manipulation of the bacterial community in BWT is of critical interest, as it has a substantial role in obtaining effective treatment and optimal removal [3].

In recent years, alternative treatment systems have attracted a lot of attention in order to treat wastewater with a high quality, due to the limited area and strict discharge issues in conventional systems. Membrane bioreactor (MBR) is a promising strategy that has been frequently implemented in the treatment of both domestic and industrial wastewater in recent years, and it is an innovative technology for the treatment and reuse of treated water by combining the activated sludge process with the membrane filtration technique. In this system, pollutants with a molecular weight greater than the molecular weight cut-off of the membranes are trapped via the membrane's sieving effect. This brings pollutants into contact with the degrading microorganisms that exist inside the MBR, allowing for their full degradation [4]. The usage of MBR technology ensures augmented treatment efficacy, specifically if the wastewater contains effluents from diverse sources when compared to the conventional treatment approaches, including activated sludge system, and physical or chemical treatment. MBR requiring relatively low-maintenance needs offers several advantages that make it a more appealing technology. Firstly, it provides a consistent treatment quality for the wastewater, regardless of the sludge conditions in the bioreactor. MBR is also beneficial in a lower production of sludge while facilitating high removal rates of nutrients, organic compounds, and pollutants. Another advantage of MBR usage is its physical footprint, which occupies a smaller area compared to conventional treatment systems, resulting in space-saving. Additionally, the combination of BWT with membrane filtration ensures an advanced capability to retain microorganisms, which have recently turned into a major concern in WWTP [5].

Microorganisms have a pivotal role in the efficiency of the BWT process, however, characteristics of both bio-sludge and wastewater influence the effectiveness of microorganisms. Therefore, optimization of the interactions between microbial strains and modifications in the microbial community can have a substantial impact on a sustainable process. In this regard, bacteriophages or simply phages have the potential to enhance aerobic BWT processes by targeting specific bacterial strains. The control of the microbial community through selective targeting with phages can provide solutions to the challenges seen in aerobic BWT and can be helpful in the optimization of the microbial community dynamics, resulting in improved efficiency in wastewater treatment [6]. Since phage treatment with single species can face bacterial resistance development, the usage of specific phage mixtures, known as phage cocktails, offers improved treatment in various processes. In addition, single-phage species exhibit limited effectiveness in the mitigation of biofilm formation, while phage cocktails containing diverse phage species increase the host range and alleviate the phage resistance by bacterial pathogens [7]. The significance of phage treatment's adaptability lies in its achievement of effective implementation within both aerobic and anaerobic BWT process, as well as across various systems without necessitating alterations to the pre-existing operational dynamics of the systems [6].

Microbial community analyses have a substantial part to understand and regulate the internal dynamics of the microbial communities in BWT. Although 16S rRNA sequencing is a widely used approach in the identification and characterization of microbial communities, it has various limitations, especially in providing high-resolution data about genetic diversity and community dynamics. Therefore, researchers have focused on whole genome shotgun (WGS) metagenomic sequencing to obtain comprehensive data about microbial communities, and to eliminate the limitations of 16S rRNA sequencing [8]. Even though WGS is a relatively more expensive sequencing method, it provides a deeper understanding of microbial community, reaching beyond simple taxonomic identification. Through WGS, it is possible to explore the entire genetic makeup, metabolic pathways, and microbial interactions of the community [9].

In this study, shotgun metagenomic sequencing was used to investigate the impact of the bioaugmentation of phage cocktails on the functional microbiota, which is of great importance in the BWT process, with the addition of three common antibiotics (erythromycin, tetracycline, and sulfamethoxazole) that anthropogenically affect aquatic spots. The pyophage cocktail used in the current study was acquired from the Eliava Institute located in Tbilisi, Georgia. The pyophage cocktail has proven effective against five bacterial species, namely *Staphylococcus* sp., *E. coli*, *Streptococcus* sp., *Proteus* sp., and *Pseudomonas* sp. The results indicated that phages altered the whole microbiome including microeukaryotes, and substantially affected the metabolic pathways related to carbohydrates and amino acid metabolisms. This study reveals the potential of phages as promising tools to enhance the functional capabilities of microbial communities in WWTP systems.

2. Materials and Methods

2.1. Operation of Aerobic Membrane Bioreactor

The previous study which investigated the effect of pyophage cocktail on biofilm formation through 16S sequencing provided comprehensive details regarding the setup and operation of the bioreactors [10]. Three aerobic membrane bioreactors were used in a continuous feeding mode in the experimental setup to examine the bioaugmentation of wastewater treatment via phage cocktail. Pharmaceutical wastewater conditions were simulated, and erythromycin, tetracycline, and sulfamethoxazole antibiotics were included in these bioreactors, according to the previous research study [10]. A submerged hollow-fiber membrane module with a surface area of 0.125 m^2 was provided for each bioreactor. An air diffuser was used at the base of the bioreactors to ensure membrane surface-cleaning and aeration, and the volume of the aeration tank with dimensions of $R \times H = 7.25$ cm $\times$ 34 cm was 5 L. A hollow-fiber membrane module with an average pore size of 0.04 μm was used to perform ultrafiltration. To measure the pressure, the transmembrane pressure (TMP) sensor was attached to a vacuum line, and values were recorded on a computer automatically. Three different sets of bioreactors were prepared: the control (C1) reactor included only aerobic sludge, the control with aerobic sludge included antibiotics (C2) and the BP bioreactor included a pyophage cocktail in addition to aerobic sludge and antibiotics [10]. All bioreactors were run under the same environmental conditions to eliminate the alterations that could result from external factors. Samples were collected twice daily from the influent, effluent, and reactor sludge, and were analysed to define different parameters such as chemical oxygen demand (COD), total suspended solids (TSS), total solids (TS), antibiotic removal rate, and the occurrence of antibiotic transformation products. When the TMP value exceeded 25 kPa, the membrane module was removed from the reactors. Samples obtained from the bioreactor sludge were extracted and cleaned through OASIS HLB cartridges (6 mL, 200 mg, Waters, Milford, MA, USA) to prepare the samples for LC/MS (liquid chromatography/mass spectrometry) analysis.

2.2. Microbial Community Analysis Using Shotgun Metagenomic Sequencing

Shotgun metagenomic sequencing is an efficient approach to studying biodiversity and functional characteristics of microbial communities. To detect the alterations in aerobic microbial communities and the persistence of antibiotic resistance genes (ARGs) in the presence of antibiotics and pyophage cocktail, Illumina sequencing was employed and the preparation for the sequencing was realized at Gentera Biotechnology Co., Ltd. (Istanbul, Turkey). Sludge from the BP reactor was obtained as the sample for DNA extraction, and whole genome DNA extraction was realized through Genomic DNA Isolation Kit (Norgen Biotek. Corp., Thorol, ON, Canada) according to instructions of the manufacturer. Quality control was conducted at each phase during the library construction. An amount of 1% of agarose gel electrophoresis was employed in DNA purification and integration, and the measurement of DNA concentration was realized with Qubit 2.0 fluorometer quantitation with a Qubit dsDNA Assay Kit (Life Technologies, CA, USA). In library construction, each sample was prepared with 1 µg DNA, and NEBNext® Ultra™ DNA Library Prep Kit for Illumina (MA, USA) was used to construct libraries. After samples were tagged with specific index codes, Covaris™ S220 Sonicator (Covaris, Inc., Woburn, MA, USA) was employed in the physical fractionation of each sample to 350 bp. Following obtaining the fragments as A-tailed and end-polished, fragments were ligated with PCR-amplification for Illumina sequencing. The purification of samples was performed through the AMPure XP system, and the Agilent 2100 Bioanalyzer and Q-PCR were applied to provide sufficient enrichment. Afterwards, the cBot Cluster Genereation System was employed for the sequencing following the manufacturer's instructions. Subsequently, paired-end reads were generated from library preparations by using Illumina NextSeq 500/HiSeq platforms. Quality control and host filtering were applied to the obtained raw data to acquire clean data, which was later to be used in metagenome assembly. Bases with a value below 38 and reads surpassing the 40 bp threshold were subjected to trimming during the data pre-processing. In addition, reads with excessive ambiguous nucleotides (N) were discarded, while reads that overlapped significantly with adapters were trimmed. The Soap 2.21 software was employed to reduce potential contamination from the host DNA.

Following the quality control, MEGAHIT was used in the metagenome assembly as the early step [11]. Scaftigs, meaning uninterrupted sequences within scaffolds, were obtained by trimming at the time the "N" symbol occurred. Scaftigs with a length below 500 bp acquired from mixed reads and individual samples were cut. The unutilized reads formed by the combination of each sample were analyzed to identify the low-abundance species. MetaGeneMark was used to perform gene prediction, using scaftigs assembled from mixed and individual samples. Afterwards, the predicted genes were assembled for the dereplication, which subsequently resulted in the construction of an exhaustive gene catalog. The Clean Data obtained in the previous steps were subjected to ORF prediction through MetaGeneMark [12]. The SoapAligner tool that mapped the clean data to the catalog was used to perform the calculation of the gene abundance by using the total quantity of mapped reads and gene length [13]. This approach allowed for the determination of gene-abundance values (Figure 1).

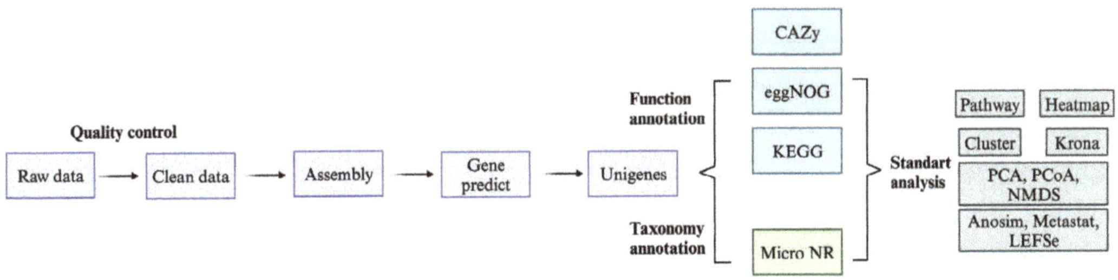

Figure 1. Workflow of metagenome analysis.

The MEGAN software (version 6.21.14) was used in the taxonomical annotation of metagenomic homologs and made a comparison with the NR database from NCBI (NCBI: version 2016-11-05) [14]. For the function determination of coding sequences and relative abundances of different types of functions in the metagenome, similarity research was conducted in databases such as KEGG, eggnog, and CAZy. Analysis involving NMDS, Anosim, and PCA clustering was performed based on taxonomic and functional abundance tables. In addition, LEfSe and Metastats were employed for statistical analysis when comparing the functional contents of different groups. In addition, the identification of ARGs abundance and distribution in species, the dataset was compared to the Comprehensive Antibiotic Resistance Database (CARD). In this comparison, BLASTp, which is a bioinformatics tool that compares protein sequences, was utilized and the e-value threshold in the comparison was $\leq 1 \times 10^{-5}$, providing the determination of the statistical significance of matches between the dataset and the CARD database [15].

3. Results and Discussion

3.1. Interactions between Bacteriophage and Indigenous Prokaryotic Microbial Diversity in Aerobic Membrane Bioreactor

In WWTPs, phages and bacteria are critical microbial groups in terms of microbial dynamics, the process of material circulation, and energy metabolism. Phages are viruses that target and infect specific bacterial strains, and they substitute a large abundance when compared to bacteria, with an approximate ratio of 10:1. Bacteriophage displays synergistic interactions with bacteria in communities, indicating their impact on bacterial communities. They can infect bacterial strains at a rate of approximately 10^{23} infections per second, and this ability has the potential to decrease target populations, especially antibiotic-resistant ones [16]. Previous studies have shown that pathogenic bacteria commonly found in municipal and industrial wastewater include *Staphylococcus* spp., *E. coli*, *Streptococcus* spp., *Proteus* spp., and *Pseudomonas* spp., as shown in [17,18]. As expected, with the addition of the cocktail containing the phages specific to these pathogenic strains to the reactor, these strains were hardly detected in the reactor. On the other hand, previous studies also reveal that *Proteobacteria* and *Bacteroidetes* are the most dominant bacterial phyla, while commonly found genera involve *Aeromonas*, *Acinetobacter*, *Pseudomonas*, *Bacillus*, etc., in different WWTPs [19]. Illumina sequencing provided comprehensive data about the microbial community structure, and indicated that the exposure to the combination of pyophage cocktail and antibiotics resulted in substantial alterations in the microbial community structure of the reactor sludge (Figure 2).

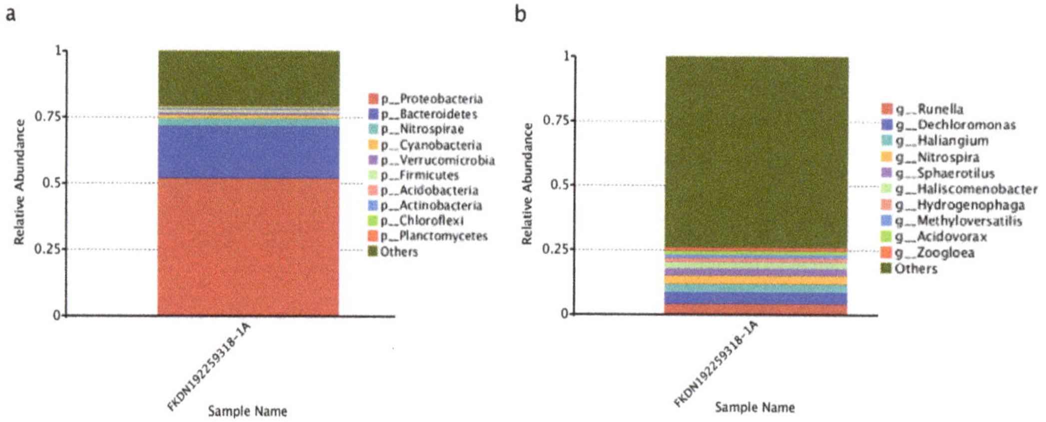

Figure 2. The abundance of the predominant ten taxa in the sludge of the BP reactor. (**a**) At the phylum level, *Proteobacteria* were found to be the most predominant phylum. (**b**) At the genus level, *Runella* and *Dechloromonas* were the two most dominant genera.

At the phylum level, *Proteobacteria* accounted for 62% of the bacterial population, showing an increase due to the tolerance, while the second most abundant phylum *Bacteroidetes* has a ratio of 23% in the BP reactor with antimicrobials, similar to the previous study [20]. On the other hand, *Firmicutes* is known as the third predominant phylum in WWTPs [21]. However, with the exposure of the pyophage cocktail along with antibiotics, the abundance of *Firmicutes* had a low rate, accounting for only 5% of the total bacterial population in the BP reactor. Previous studies have revealed that members belonging to the *Firmicutes* phylum can harbor antibiotic-resistance genes [22,23]. Considering this information, it is feasible to suggest that the reduction seen in the abundance of *Firmicutes* may be attributed to the effect of the pyophage cocktail.

Furthermore, *Nitrospirae* was found as the third predominant phylum in the bacterial population of the BP reactor sludge, constituting 4% of the entire bacterial population. *Nitrospirae* is a predominant phylum of nitrite-oxidizing bacteria (NOB) that has a substantial role in biological nutrient removal in wastewater systems, especially with low levels of dissolved oxygen and substrate availability [24]. Compared to the relative abundance of *Nitrospirae* in normally operating wastewater treatment systems, the rate of *Nitrospirae* detected in the BP reactor in the current study is higher, and it is appropriate to conclude that this increase is due to the pyophage cocktail [25,26]. Considering the lack of success in the nitrification process is generally linked to the low concentration of nitrifying bacteria in activated sludge systems [27], this increment in *Nitrospirae* demonstrates that the pyophage cocktail has the potential to improve the performance of MBR.

The development of the bacterial community within the BP reactor in the presence of the pyophage cocktail and antibiotics are also can be observed at the genus level. Previous studies have indicated that the genera of *Dechloromonas* and *Nitrospira* were one of the core microbial groups in most WWTPs and could show resistant characteristics when faced with antibacterial agents [28]. In the current study, *Runella* (5%), *Dechloromonas* (5%), *Nitrospira* (4%), *Haliangium* (4%), and *Haliscomenobacter* (3%) were identified as the most prevalent genera within the bacterial community. The importance of these five predominant genera, despite the existence of antimicrobial agents, points to possible resistance and their ability to develop in a community in which antibiotic-sensitive ones could have decreased, and this observation aligns with previous studies that have revealed similar findings [10,20]. Another genus found and notable in the BP reactor is *Zoogloea* (2%), and this group of bacteria is typically found in activated sludge and involved in the degradation of organic and inorganic compounds in wastewater [29].

3.2. Interactions between Bacteriophage and Indigenous Eukaryotic Microbial Diversity in Aerobic Membrane Bioreactor

We investigated eukaryotic microorganisms in bioreactors and found that these microorganisms constitute only a small percentage of the whole community. Phyla of *Ascomycota* (10%), *Chytridiomycota* (2%), and *Basidiomycota* (1%) were observed as the core fungal groups, while microalgal phylum *Chlorophyta* (2%) and diatom group *Thalassiosira* (1%) were the other notable eukaryotic microorganisms. A significant amount of microorganism assemblage was evaluated as unclassified, which could be the result of the lack of reference sequence data. *Ascomycota* and *Chytridiomycota* are able to secrete cellulase, thus, these fungal groups can be used in the treatment of cellulose-containing wastewater due to their cellulose-degrading abilities. However, the relative abundance of *Ascomycota* with a ratio of 10% in eukaryotes, is of notable importance (Figure 3). In a study conducted on aerobic sludge, the high abundance of *Ascomycota* detected in the sludge was correlated with the addition of rich cellulose in the membrane reactor [30]. Based on this study, the high abundance of *Ascomycota* phylum with a ratio of 10% in the current study may be cellulose-containing plants that account for as much as 32% of the eukaryotic organisms in the reactor. On the other hand, the genus *Aspergillus*, which constitutes 17% of *Ascomycota* in the BP reactor, is a filamentous fungus and is widely employed as a bioremediation agent for wastewater contaminated with heavy metal, petroleum, and other pollutants [31].

In the BP reactor antimicrobials, the existence of the *Chlorophyta* group (2%) in eukaryotic microorganisms, is noteworthy. The microalgal group *Chlorophyta* has been used in various streams, including industrial and municipal wastewater treatments with the purpose of nutrient removal. This utilization can be attributed to the substantial adaptation capability of microalgae in specific conditions [32]. Even though a significant challenge in the cultivation of microalgae in untreated wastewater is the presence of other microorganisms that restrict biomass productivity, *Chlorophyta* in the BP reactor has shown the ability to thrive in the existence of the combination of pyophage cocktail and antibiotics. Thus, if the BP reactor operates with long-term exposure to these antimicrobials, the microalgal group *Chlorophyta* may have improved growth opportunities. Therefore, the application of a pyophage cocktail with antibiotics would effectively hamper the development of microbes and this would result in the increased capacity and potential of the wastewater treatment systems for nutrient removal. These findings suggest that the pyophage cocktail positively influences the WWTP process by enhancing the development of microorganisms that are beneficial in bioremediation. By targeting specific pathogens or opportunistic microorganisms, the pyophage cocktail may help create a more suitable environment for these species to thrive, ultimately improving the treatment efficiency and overall performance of the system. Indeed, it is also vital to consider the potential synergistic relationship between antibiotics and the pyophage cocktail in the current study.

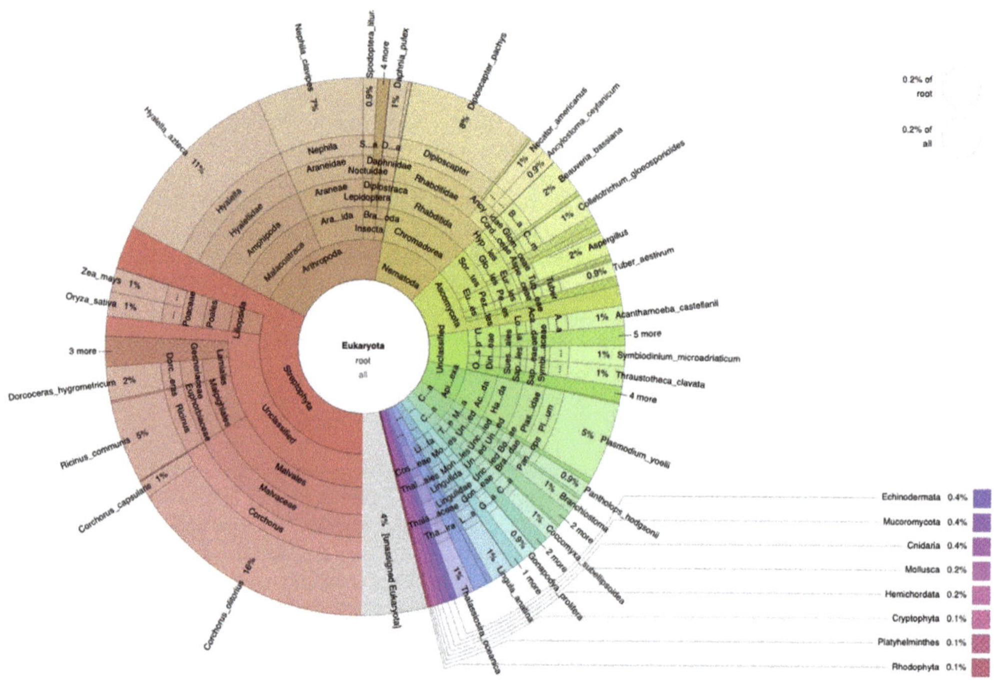

Figure 3. The relative abundance of *Ascomycota* in the BP reactor. *Ascomycota* accounts for a significant 10% of all eukaryotes in the BP reactor.

3.3. Interactions between Bacteriophage and the Metabolic Pathway

To gain further insights into the effects of bacteriophage on the metabolic pathway of the microbial community from the proteomic level, Kyoto Encyclopedia of Genes and Genomes (KEGG) pathway classification in the BP reactor was also carried out. Predominantly seen metabolic pathways are amino acid metabolism, carbohydrate metabolism, energy metabolism, cofactors and vitamins metabolism, nucleotide metabolism, lipid metabolism, xenobiotics biodegradation, and metabolism and biosynthesis of other seconder metabolites (Figure 4). The core genera in nitrogen metabolism were observed as *Nitrosomonas* (0.2%) and *Nitrospira* (4%), which are known as potential nitrifying bacterial groups, and they have been identified as predominant ammonia-oxidizing bacteria (AOB) in different activated sludge systems, indicating their tolerance to environmental alterations [33]. Therefore, the application of the combination of a pyophage cocktail and antibiotics did not appear to affect the existence of these genera. The group of these two is known for their contribution to the reactor performance by enhancing nitrification. Specifically, the genus of comammox *Nitrospira* has a substantial role in the nitrification process, which is the pivotal step in the nitrogen cycle. The process of nitrification briefly occurs with the conversion of ammonia (NH_3) to nitrate (NO_3^-) via two sequential oxidation reactions. While it was considered at first nitrification was realized by two different bacterial groups comprising ammonia-oxidizing bacteria (AOB) and nitrite-oxidizing bacteria (NOB). With the discovery of comammox *Nitrospira*, this understanding was totally changed. The unique feature of this genus is its capability in performing both ammonia oxidation and nitrite oxidation without the need for additional bacteria to realize the second step. In other words, the members of *Nitrospira* can perform the whole nitrification independently, increasing the efficiency of the process. Their abundance in diverse environments such as soil, fresh water and WWT systems, make them a key player in nitrogen metabolism [34]. In fact, the metabolic capabilities of the genus *Nitrospira* reach beyond nitrite oxidation, covering various abilities including the growth on formate and hydrogen under aerobic conditions, and the reduction of nitrate to nitrite when it is in anoxic condition [35].

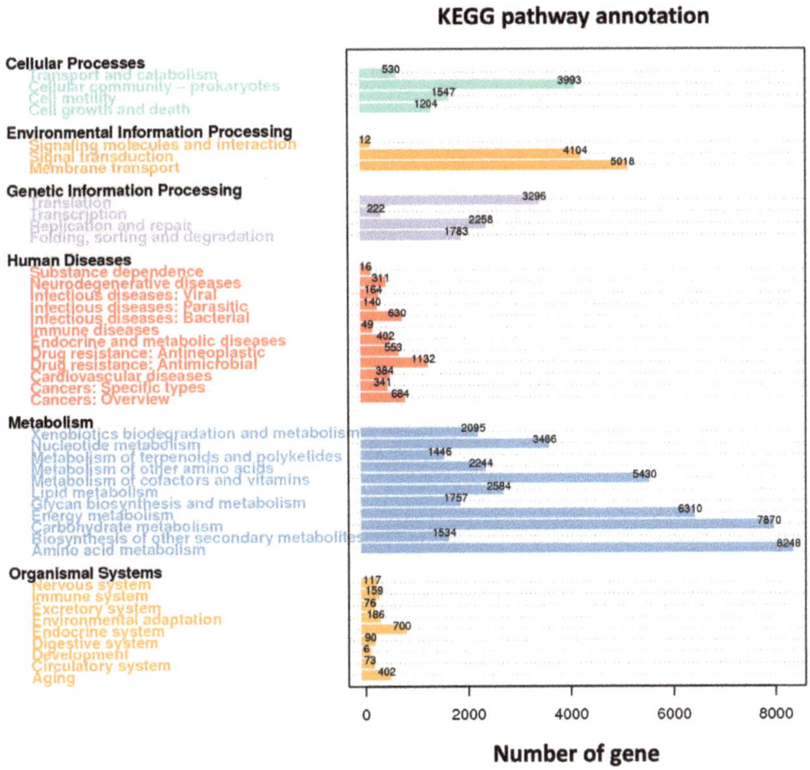

Figure 4. KEGG pathway classification in the BP reactor.

The BP reactor exhibited the existence of *Dechloromonas* (5%), *Zoogloea* (2%), and *Accumulibacter* (2%) genera, constituting a total of 9% of the bacterial population, well known for denitrification potentials. Especially, species in *Zoogloea* genus have been identified as denitrifying aerobic heterotrophic bacterial strains in traditionally operating activated sludge systems, providing nitrate reduction even in high levels of oxygen [36]. *Haliangium ochraceum* (1%) species have been associated with both carbohydrate and amino acid metabolisms. This group of bacteria displays genes encoding key enzymes that took part in the production of polyketide and peptide metabolites, including polyketide synthases and nonribosomal peptide synthases. They also have the ability to decompose macromolecules such as starch, DNA, casein, gelatine, and chitin [37].

Phosphorus removal of cofactor metabolism in activated sludge systems can be attributed to the activity of phosphorus-accumulating organisms (PAOs). The genus *Accumulibacter* is also recognized as a PAO genus [38], thus the existence of *Accumulibacter* (2%) in the BP reactor indicates this genus contribution to phosphorus removal as a PAO. Photosynthetic *Rhodobacterales* constituting 3% of the bacterial genome in the BP reactor have been previously described in energy and vitamin metabolisms [39]. In addition, the genus of *Sorangium* (0.9%) has a substantial role in the synthesis of approximately half of the secondary metabolites [40]. These metabolic pathways are widely encountered in wastewater treatment systems, and the genera and species above mentioned are one of the key players in the pathways. Despite the presence of a phage cocktail, the maintenance of these metabolic pathways suggests that the usage of phages for pathogen removal and the enrichment of WWTP does not have an adverse impact on bacterial metabolic pathways, which are crucial in reactor systems. Furthermore, the phage cocktail may have provided suitable environments for bacteria that are responsible for these pathways, by eliminating opportunistic pathogens from the BP reactor, resulting in the persistence of the pathways.

3.4. Interactions between Bacteriophage and Occurrence of Antibiotic Resistance Genes in Aerobic Membrane Bioreactor

WWTPs serve as prominent hotspots for the dissemination of resistance genes and the proliferation of resistant bacteria. This kind of contaminated environment harbors a diverse array of ARGs and a great amount of bacteria, fostering microbial interactions. Furthermore, WWTPs are subject to various stresses such as microplastic and heavy metal, which further enhance the selection and horizontal gene transfer (HGT). With the analysis of antibiotic resistance genes (ARGs) alongside microbial community structures in the bioreactor, we obtained insight into the correlations among ARGs. The presence of antibiotics led to an enrichment of gram-negative *Proteobacteria* in the BP reactor, which exhibited the highest levels of antibiotic resistance profiles. The considerable presence and wide range of *Proteobacteria* in the BP reactor could potentially be attributed to antibiotic-resistant species of *Proteobacteria* encompassing numerous significant human pathogens, including *Klebsiella* spp., *Salmonella* spp., *Acinetobacter* spp., and various others.

The increase in *Proteobacteria* was also observed in a previous study that indicated the addition of antibiotics into the influent caused a remarkable change in the microbial community composition, and an increase in the abundance of ARGs in the membrane reactor [41]. Different resistance genes occur through different mechanisms. Figure 5 illustrates different phyla containing ARGs in the BP reactor. Since *Proteobacteria* is the predominant genus, the most dominant resistant mechanism used by bacteria to acquire resistance genes is RND-type Drug Efflux Pumps, while mtrA has been identified as the common resistance gene (Figure 6b). RND antibiotic efflux pumps are a kind of efflux pump present in Gram-negative bacterial phylum such as *Proteobacteria*, and they significantly contribute to this group of bacteria developing resistance against a wide range of antibiotics. The mtrA gene, also known as mtrR, is a crucial resistant gene in various Gram-negative bacteria, and this gene encodes a transcriptional repressor protein that is responsible for the MtrCDE efflux pump system expression. This pump system plays a substantial role in bacterial resistance to different antimicrobials. The MtrA gene ensures that bacterial cells actively release antibiotics by upregulating the MtrCDE efflux pump expression, resulting in a decrease in intracellular concentrations of antibiotics and making them less efficient [42]. Although the existence of *Staphylococcus aureus* substitutes a very small amount in the whole bacterial genome (0.003%), the resistance gene (arlR) belonging to this species was observed as one of the most prominent resistance genes in the BP reactor. ArlR gene is part of a two-component regulatory system named arlRS and this system has been strongly associated with several gene regulations that play a role in antimicrobial resistance mechanisms [43]. Interestingly, lytic phage application did not appear to substantially improve the ARGs abundance; nevertheless, they were in high abundance in the BP reactor. In further investigations through phage bioaugmentation, it would be beneficiary to incorporate supplementary agents for the bioaugmentation, that can mitigate ARG formation, particularly by targeting *Proteobacteria*, in the aerobic membrane bioreactor.

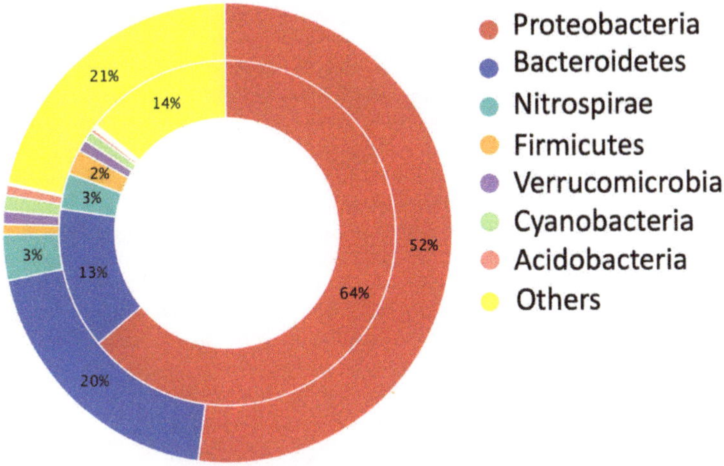

Figure 5. The inner circle indicates the abundance of diverse phyla containing ARGs, while the outer ring represents the rate of ARGs across diverse phyla within this group.

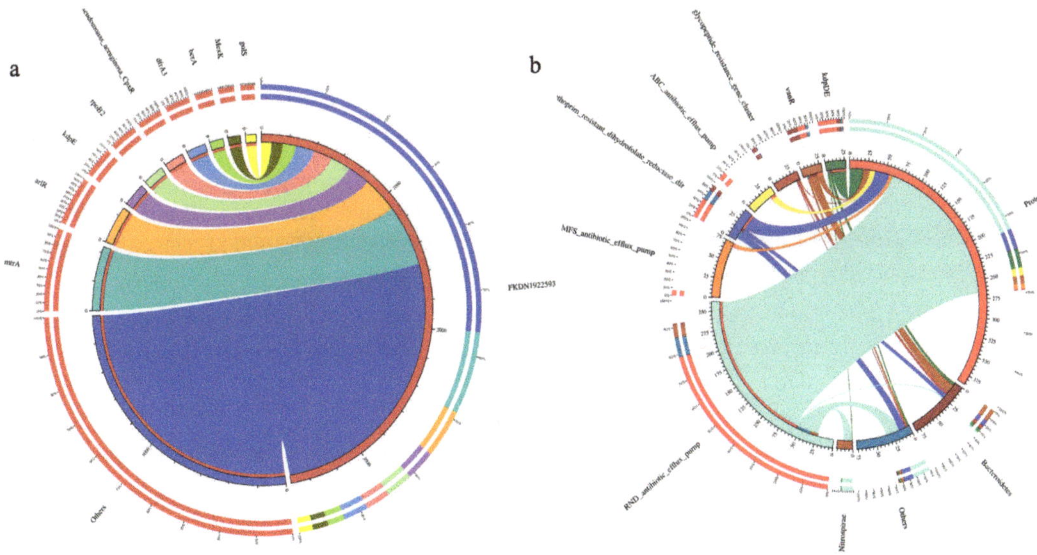

Figure 6. (**a**) Overview Circle chart of resistance genes, (**b**) the relationship between resistance mechanism and phyla.

3.5. The Potential Use of Bacteriophages in the Aerobic Membrane Reactor

Except for the function of phages in removing membrane contamination by targeting pathogens, phages display various capabilities in aerobic WWTPs in which microorganisms have a game-changer role and augment the treatment efficiency. Filamentous bacterial species such as *Gordonia* and *Nocordia* generally result in bulking and foaming issues in the activated sludge. Chemical approaches have been used to tackle this problem, however, this application can also have adverse results by hampering the development of bacteria that are key players in the process [44]. As an alternative approach, studies revealed the potential of phage treatment in controlling bulking and foaming in sludge systems. Four bacteriophages were isolated and mixed to reduce the *Gordonia* abundance, and the phage cocktail provided a successful reduction in comparison with the control [45]. Furthermore, the polyvalent phage isolated from a WWTP succeeded in the removal of different *Gordonia* and *Nocardia* species [46]. Excessive sludge-forming caused by filamentous bacteria and EPS secreted by *Zoogloea* and *Thauerea* genera is the other problem to tackle in activated sludge systems [47]. The high abundance of EPS prevents the dehydration of sludge through retaining water and subsequently results in the reduction of the sedimentation capacity and biofilm formation. The success of phages in disrupting these biofilm formations has been demonstrated by various studies [48].

The PN/AMX is a useful method in nitrogen removal and generation of low sludge in WWTP. Nevertheless, the competition between NOB (*Nitrobacter* and *Nitrospiraceae*) and anammox bacteria poses a difficulty. A study using *N. multiformis* indicated that two pyophages were induced under specific conditions, such as acidic pH, and the increment of these viral particles led to a decrease in nitrification activity through the reduction in *N. multiformis* abundance [49]. Phage-mediated disruption has also achieved successful reductions in nitrification through the removal of certain bacterial species in different research studies [10,50,51].

4. Conclusions

The current study investigated the impact of combining a phage cocktail with three antibiotics frequently present in wastewater on the microbial community, in a bioreactor using shotgun metagenomic analyses. The phage cocktail used in this study was specific to *Staphylococcus* sp., *E. coli*, *Streptococcus* sp., *Proteus* sp., and *Pseudomonas* sp. Pathogens, which are normally and frequently found in wastewater. The results indicate that the phage treatment was successful with no detection of these species in the BP reactor. This is of importance, since if untreated appropriately, these pathogens could potentially be released into aquatic environments. Antibiotics used in addition to the pyophage cocktail provided synergy with the phage treatment, being effective in removing opportunistic bacterial species that are outside the host range of the phage cocktail and generally found in wastewater but barely detected in the BP reactor. This combination supported bacteria such as NOB, which are responsible for effective reactor performance and wastewater treatment. The impact of the phage cocktail on metabolic pathways was also examined, and it was observed that the phage application did not have adverse effects on these metabolic pathways. In fact, the pyophage cocktail may have even supported bacterial groups involved in pathways, by removing opportunistic species from the reactor. However, more comprehensive knowledge about the effects of phages on microbial community dynamics will require further investigation under long-term exposure.

Author Contributions: Conceptualization, S.A.; methodology, S.A.; software, S.A.; validation, S.A.; resources, S.A.; writing—original draft preparation, Ş.R. & S.A.; writing—review and editing, Ş.R. & S.A.; project administration, S.A.; funding acquisition, S.A. All authors have read and agreed to the published version of the manuscript.

Funding: This study was funded by the Scientific Research Projects Coordination Unit of Istanbul University. Grant Number: FYL-2023-39728.

Institutional Review Board Statement: Not applicable.

Informed Consent Statement: Not applicable.

Data Availability Statement: Data will be available on request.

Conflicts of Interest: The authors declare no conflict of interest.

References

1. Aziz, A.; Basheer, F.; Sengar, A.; Irfanullah; Khan, S.U.; Farooqi, I.H. Biological wastewater treatment (anaerobic-aerobic) technologies for safe discharge of treated slaughterhouse and meat processing wastewater. *Sci. Total Environ.* **2019**, *686*, 681–708. [CrossRef]
2. Liu, R.; Li, Z.; Han, G.; Cun, S.; Yang, M.; Liu, X. Bacteriophage ecology in biological wastewater treatment systems. *Appl. Microbiol. Biotechnol.* **2021**, *105*, 5299–5307. [CrossRef] [PubMed]
3. Maddela, N.R.; Sheng, B.; Yuan, S.; Zhou, Z.; Villamar-Torres, R.; Meng, F. Roles of quorum sensing in biological wastewater treatment: A critical review. *Chemosphere* **2019**, *221*, 616–629. [CrossRef] [PubMed]
4. Goswami, L.; Kumar, R.V.; Borah, S.N.; Manikandan, N.A.; Pakshirajan, K.; Pugazhenthi, G. Membrane bioreactor and integrated membrane bioreactor systems for micropollutant removal from wastewater: A review. *J. Water Process. Eng.* **2018**, *26*, 314–328. [CrossRef]
5. Lebleu, N.; Roques, C.; Aimar, P.; Causserand, C. Effects of membrane alterations on bacterial retention. *J. Membr. Sci.* **2009**, *348*, 56–65. [CrossRef]
6. Reisoglu, Ş.; Aydin, S. Bacteriophages as a promising approach for the biocontrol of antibiotic resistant pathogens and the reconstruction of microbial interaction networks in wastewater treatment systems: A review. *Sci. Total Environ.* **2023**, *890*, 164291. [CrossRef]
7. Zhang, B.; Yu, P.; Wang, Z.; Alvarez, P.J.J. Hormetic Promotion of Biofilm Growth by Polyvalent Bacteriophages at Low Concentrations. *Environ. Sci. Technol.* **2020**, *54*, 12358–12365. [CrossRef]
8. Lazarevic, V.; Whiteson, K.; Gaïa, N.; Gizard, Y.; Hernandez, D.; Farinelli, L.; Østerås, M.; François, P.; Schrenzel, J. Analysis of the salivary microbiome using culture-independent techniques. *J. Clin. Bioinform.* **2012**, *2*, 4. [CrossRef]
9. Hasan, N.A.; Young, B.A.; Minard-Smith, A.T.; Saeed, K.; Li, H. Microbial Community Profiling of Human Saliva Using Shotgun Metagenomic Sequencing. *PLoS ONE* **2014**, *9*, 97699. [CrossRef]
10. Aydin, S.; Can, K. Pyophage cocktail for the biocontrol of membrane fouling and its effect in aerobic microbial biofilm community during the treatment of antibiotics. *Bioresour. Technol.* **2020**, *318*, 123965. [CrossRef]
11. Li, D.; Liu, C.-M.; Luo, R.; Sadakane, K.; Lam, T.-W. MEGAHIT: An ultra-fast single-node solution for large and complex metagenomics assembly via succinct de Bruijn graph. *Bioinformatics* **2015**, *31*, 1674–1676. [CrossRef] [PubMed]
12. Zhu, W.; Lomsadze, A.; Borodovsky, M. Ab initio gene identification in metagenomic sequences. *Nucleic Acids Res.* **2010**, *38*, e132. [CrossRef] [PubMed]
13. Gu, S.; Fang, L.; Xu, X. Using SOAPaligner for Short Reads Alignment. *Curr. Protoc. Bioinform.* **2013**, *44*, 11. [CrossRef] [PubMed]
14. Huson, D.H.; Beier, S.; Flade, I.; Górska, A.; El-Hadidi, M.; Mitra, S.; Ruscheweyh, H.J.; Tappu, R. MEGAN Community Edition-Interactive Exploration and Analysis of Large-Scale Microbiome Sequencing Data. *PLoS Comput. Biol.* **2016**, *12*, e1004957. [CrossRef]
15. Alcock, B.P.; Raphenya, A.R.; Lau, T.T.Y.; Tsang, K.K.; Bouchard, M.; Edalatmand, A.; Huynh, W.; Nguyen, A.-L.V.; Cheng, A.A.; Liu, S.; et al. CARD 2020: Antibiotic resistome surveillance with the comprehensive antibiotic resistance database. *Nucleic Acids Res.* **2019**, *48*, D517–D525. [CrossRef]
16. Worley-Morse, T.O.; Gunsch, C.K. Modeling phage induced bacterial disinfection rates and the resulting design implications. *Water Res.* **2015**, *68*, 627–636. [CrossRef]
17. Kumaraswamy, R.; Amha, Y.M.; Anwar, M.Z.; Henschel, A.; Rodríguez, J.; Ahmad, F. Molecular Analysis for Screening Human Bacterial Pathogens in Municipal Wastewater Treatment and Reuse. *Environ. Sci. Technol.* **2014**, *48*, 11610–11619. [CrossRef]
18. Yasir, M. Analysis of Microbial Communities and Pathogen Detection in Domestic Sewage Using Metagenomic Sequencing. *Diversity* **2020**, *13*, 6. [CrossRef]
19. Zhao, R.; Feng, J.; Liu, J.; Fu, W.; Li, X.; Li, B. Deciphering of microbial community and antibiotic resistance genes in activated sludge reactors under high selective pressure of different antibiotics. *Water Res.* **2018**, *151*, 388–402. [CrossRef]
20. Du, B.; Wang, Q.; Yang, Q.; Wang, R.; Yuan, W.; Yan, L. Responses of bacterial and bacteriophage communities to long-term exposure to antimicrobial agents in wastewater treatment systems. *J. Hazard. Mater.* **2021**, *414*, 125486. [CrossRef]
21. Aydin, S.; Can, K.; Çalışkan, M.; Balcazar, J.L. Bacteriophage cocktail as a promising bio-enhancer for methanogenic activities in anaerobic membrane bioreactors. *Sci. Total Environ.* **2022**, *832*, 154716. [CrossRef] [PubMed]
22. Wang, G.; Deng, D.; Hu, C.; Lou, L.; Luo, L.; He, J.; Tian, D.; Xiao, Y.; He, Y.; Zhang, S.; et al. More effective removal of antibiotic resistance genes from excess sludge by microwave integrated fenton treatment. *Int. Biodeterior. Biodegrad.* **2020**, *149*, 104920. [CrossRef]
23. Schwarz, S.; Shen, J.; Wendlandt, S.; Feßler, A.T.; Wang, Y.; Kadlec, K.; Wu, C.M. Plasmid-Mediated Antimicrobial Resistance in Staphylococci and Other Firmicutes. *Microbiol. Spectr.* **2014**, 421–444. [CrossRef] [PubMed]
24. Geng, M.; You, S.; Guo, H.; Ma, F.; Xiao, X.; Zhang, J.; Ma, X. Response of aerobic granular sludge to loading shock: Performance and proteomic study. *Chem. Eng. J.* **2022**, *444*, 136458. [CrossRef]
25. Yang, Y.; Wang, L.; Xiang, F.; Zhao, L.; Qiao, Z. Activated sludge microbial community and treatment performance of wastewater treatment plants in industrial and municipal zones. *Int. J. Environ. Res. Public Health* **2020**, *17*, 436. [CrossRef]
26. Xiao, X.; Guo, H.; Ma, F.; Zhang, J.; Ma, X.; You, S. New insights into mycelial pellets for aerobic sludge granulation in membrane bioreactor: Bio-functional interactions among metazoans, microbial communities and protein expression. *Water Res.* **2023**, *228*, 119361. [CrossRef]

27. Wanner, J.; Ruzickova, I.; Krhutkova, O.; Pribyl, M. Activated sludge population dynamics and wastewater treatment plant design and operation. *Water Sci. Technol.* **2000**, *41*, 217–225. [CrossRef]
28. Wu, L.; Ning, D.; Zhang, B.; Li, Y.; Zhang, P.; Shan, X.; Zhang, Q.; Brown, M.R.; Li, Z.; Van Nostrand, J.D.; et al. Global diversity and biogeography of bacterial communities in wastewater treatment plants. *Nat. Microbiol.* **2019**, *4*, 1183–1195. [CrossRef]
29. Wang, X.; Hu, M.; Xia, Y.; Wen, X.; Ding, K. Pyrosequencing Analysis of Bacterial Diversity in 14 Wastewater Treatment Systems in China. *Appl. Environ. Microbiol.* **2012**, *78*, 7042–7047. [CrossRef]
30. Chen, C.; Gan, Z.; Xu, R.; Meng, F. Cellulose-induced shifts in microbial communities and microbial interactions in an anoxic/aerobic membrane bioreactor. *J. Water Process. Eng.* **2021**, *42*, 102106. [CrossRef]
31. El-Bondkly, A.M.A.; El-Gendy, M.M.A.A. Bioremoval of some heavy metals from aqueous solutions by two different indigenous fungi Aspergillus sp. AHM69 and Penicillium sp. AHM96 isolated from petroleum refining wastewater. *Heliyon* **2022**, *8*, e09854. [CrossRef]
32. Abinandan, S.; Shanthakumar, S. Challenges and opportunities in application of microalgae (Chlorophyta) for wastewater treatment: A review. *Renew. Sustain. Energy Rev.* **2015**, *52*, 123–132. [CrossRef]
33. Geets, J.; Boon, N.; Verstraete, W. Strategies of aerobic ammonia-oxidizing bacteria for coping with nutrient and oxygen fluctuations. *FEMS Microbiol. Ecol.* **2006**, *58*, 1–13. [CrossRef] [PubMed]
34. Latocheski, E.C.; da Rocha, M.C.V.; Braga, M.C.B. Nitrospira in wastewater treatment: Applications, opportunities and research gaps Graphical abstract. *Rev. Environ. Sci. Biotechnol.* **2022**, *21*, 905–930. [CrossRef]
35. Koch, H.; van Kessel, M.A.H.J.; Lücker, S. Complete nitrification: Insights into the ecophysiology of comammox Nitrospira. *Appl. Microbiol. Biotechnol.* **2019**, *103*, 177–189. [CrossRef] [PubMed]
36. Straub, K.L.; Schönhuber, W.A.; Buchholz-Cleven, B.E.E.; Schink, B. Diversity of Ferrous Iron-Oxidizing, Nitrate-Reducing Bacteria and their Involvement in Oxygen-Independent Iron Cycling. *Geomicrobiol. J.* **2004**, *21*, 371–378. [CrossRef]
37. Ivanova, N.; Daum, C.; Lang, E.; Abt, B.; Kopitz, M.; Saunders, E.; Lapidus, A.; Lucas, S.; Del Rio, T.G.; Nolan, M.; et al. Complete genome sequence of Haliangium ochraceum type strain (SMP-2T). *Stand. Genom. Sci.* **2010**, *2*, 96–106. [CrossRef] [PubMed]
38. Mielczarek, A.T.; Nguyen, H.T.T.; Nielsen, J.L.; Nielsen, P.H. Population dynamics of bacteria involved in enhanced biological phosphorus removal in Danish wastewater treatment plants. *Water Res.* **2013**, *47*, 1529–1544. [CrossRef]
39. Sasaki, K.; Watanabe, M.; Suda, Y.; Ishizuka, A.; Noparatnaraporn, N. Applications of photosynthetic bacteria for medical fields. *J. Biosci. Bioeng.* **2005**, *100*, 481–488. [CrossRef]
40. Chen, H.; Wang, M.; Chang, S. Disentangling Community Structure of Ecological System in Activated Sludge: Core Communities, Functionality, and Functional Redundancy. *Microb. Ecol.* **2020**, *80*, 296–308. [CrossRef]
41. Wen, Q.; Yang, L.; Zhao, Y.; Huang, L.; Chen, Z. Insight into effects of antibiotics on reactor performance and evolutions of antibiotic resistance genes and microbial community in a membrane reactor. *Chemosphere* **2018**, *197*, 420–429. [CrossRef]
42. Rouquette, C.; Harmon, J.B.; Shafer, W.M. Induction of the mtrCDE-encoded ef¯ux pump system of Neisseria gonorrhoeae requires MtrA, an AraC-like protein. *Mol. Microbiol.* **1999**, *33*, 651–658. [CrossRef] [PubMed]
43. Otarigho, B.; O Falade, M. Open access Analysis of antibiotics resistant genes in different strains of *Staphylococcus aureus*. *Bioinformation* **2018**, *14*, 113–122. [CrossRef]
44. Collivignarelli, M.C.; Baldi, M.; Abbà, A.; Caccamo, F.M.; Miino, M.C.; Rada, E.C.; Torretta, V. Foams in Wastewater Treatment Plants: From Causes to Control Methods. *Appl. Sci.* **2020**, *10*, 2716. [CrossRef]
45. Liu, M.; Gill, J.J.; Young, R.; Summer, E.J. Bacteriophages of wastewater foaming-associated filamentous Gordonia reduce host levels in raw activated sludge. *Sci. Rep.* **2015**, *5*, 13754. [CrossRef] [PubMed]
46. Petrovski, S.; Seviour, R.J.; Tillett, D. Prevention of Gordonia and Nocardia Stabilized Foam Formation by Using Bacteriophage GTE7. *Appl. Environ. Microbiol.* **2011**, *77*, 7864–7867. [CrossRef] [PubMed]
47. Luo, L.; Ge, Y.; Yuan, S.; Yu, Y.; Shi, Z.; Zhou, S.; Deng, J. Enhanced dewaterability of waste activated sludge by a combined use of permanganate and peroxymonosulfate. *RSC Adv.* **2019**, *9*, 27593–27601. [CrossRef]
48. Sheng, G.-P.; Yu, H.-Q.; Li, X.-Y. Extracellular polymeric substances (EPS) of microbial aggregates in biological wastewater treatment systems: A review. *Biotechnol. Adv.* **2010**, *28*, 882–894. [CrossRef]
49. Choi, J.; Kotay, S.M.; Goel, R. Various physico-chemical stress factors cause prophage induction in Nitrosospira multiformis 25196- an ammonia oxidizing bacteria. *Water Res.* **2010**, *44*, 4550–4558. [CrossRef]
50. Liao, Y.; Zhang, J.; Wang, M.; Wu, Y.; Wang, S.; Pan, Y.; Cao, G. Nitrogen removal from wastewater for heterotrophic nitrification-aerobic denitrification bacterium with the combination of bacteriophage DEY7 and Fe nanoparticles. *Biochem. Eng. J.* **2023**, *191*, 108805. [CrossRef]
51. Liu, R.; Qi, R.; Wang, J.; Zhang, Y.; Liu, X.; Rossetti, S.; Tandoi, V.; Yang, M. Phage-host associations in a full-scale activated sludge plant during sludge bulking. *Appl. Microbiol. Biotechnol.* **2017**, *101*, 6495–6504. [CrossRef] [PubMed]

sustainability

Article

Role of Vermicomposting Microorganisms in the Conversion of Biomass Ash to Bio-Based Fertilizers

Güldane Aslı Turp [1,*], Saim Ozdemir [1], Kaan Yetilmezsoy [2], Nurtac Oz [1] and Ali Elkamel [3,4]

1 Department of Environmental Engineering, Faculty of Engineering, Sakarya University, Esentepe, Sakarya 54187, Turkey; saimo@sakarya.edu.tr (S.O.); nuroz@sakarya.edu.tr (N.O.)
2 Department of Environmental Engineering, Faculty of Civil Engineering, Yildiz Technical University, Davutpasa, Esenler, Istanbul 34220, Turkey; yetilmez@yildiz.edu.tr
3 Department of Chemical Engineering, University of Waterloo, Waterloo, ON N2L 3G1, Canada; aelkamel@uwaterloo.ca
4 Department of Chemical Engineering, Khalifa University, Abu Dhabi P.O. Box 127788, United Arab Emirates
* Correspondence: g.asliturp@gmail.com

Abstract: A high pH, low solubility of bound plant nutrients, and negative impacts on microbial communities are common drawbacks of biomass ash (BA) vermicomposting. In this study, nutrient-rich BA mixed with cow manure was tested at three different application rates to obtain final nitrogen (N), phosphorus (P), and potassium (K) contents of 3.5%, 7.0%, and 10.0% for bio-based fertilizers via vermicomposting. The results showed that all BA blends made with cow manure increased fermentation temperatures and allowed successful worm activity during the subsequent vermicomposting phase. The order of indicator enzyme activities in all vermicomposting samples was urease (220 μg NH_4 g^{-1} h^{-1}) > β-glucosidase (95 μg PNP g^{-1} h^{-1}) > alkaline phosphatase (91 μg PNP g^{-1} h^{-1}) > arylsulfatase (83 μg PNP g^{-1} h^{-1}) > acid phosphatase (60 μg PNP g^{-1} h^{-1}). As an indicator of nutrient bioavailability, high correlations were observed between enzyme activities and microbial diversity in vermicompost samples. Determination coefficients (R^2) obtained from multiple linear regressions between enzyme activities and bacterial population for T_0, T_1, T_2, and T_3 were determined as 0.90, 0.65, 0.73, and 0.90, respectively. According to a novel metagenome-based approach proposed within the scope of the present study, the stimulatory effects of *Flavobacteriales*, *Burkholderiales*, *Saccharimonadales*, and *Pseudomonadales* on enzyme activities for the nutrient solubility were found to be significant and positive. The findings of this study demonstrated that worm composting could be a sustainable bio-based technology for the production of slow-release fertilizer from nutrient-rich waste material.

Keywords: biomass ash recycling; vermicomposting; nutrient availability; microbial community; enzyme activity

check for updates

Citation: Turp, G.A.; Ozdemir, S.; Yetilmezsoy, K.; Oz, N.; Elkamel, A. Role of Vermicomposting Microorganisms in the Conversion of Biomass Ash to Bio-Based Fertilizers. *Sustainability* **2023**, *15*, 8984. https://doi.org/10.3390/su15118984

Academic Editor: Helvi Heinonen-Tanski

Received: 24 April 2023
Revised: 26 May 2023
Accepted: 30 May 2023
Published: 2 June 2023

1. Introduction

The recovery of bio-based plant nutrients from a variety of waste resources to replace synthetic fertilizers is considered a key strategy for more sustainable agriculture [1] and waste management practices [2]. In terms of a circular bioeconomy and sustainable environment, composting is considered to be a unique and cost-effective process that provides high-value-added and useful products from various organic materials [3,4]. In this context, vermicomposting is recognized as an environment-friendly and sustainable biowaste treatment technology since it allows the recovery of nutrients from biowaste [5], the proliferation of beneficial microorganisms [6], and the recovery of plant growth-stimulating substances [7]. However, depending on the feedstock, the nutrient content is generally low, so nutrient enrichment is required to increase end-user acceptance [5,8]. In this regard, biomass power plant ash has the potential to enrich plant nutrients in vermicompost, allowing it to be used as a bio-based fertilizer [9]. On the other hand, the availability of nutrients in biomass ash to plants is limited as they are typically bound within the crystalline. Therefore, the further processing of BA should be explored in order to improve the nutrient efficiency and enrichment potential of worm compost by BA nutrients.

With the increase in the global demand for renewable biomass for energy production, biomass ash (BA) derived from power plants has been increasing for years. From the perspective of agricultural use, BA contains many plant nutrients such as potassium (K), magnesium (Mg), and phosphorus (P), which are important nutrients for plant growth [10]. Although the limited use of bio-ash in agriculture is attributed to its high pH and heavy metals, which are thought to affect plant growth, returning it to the field as a fertilizer for crop production is one of the most sustainable ways to minimize nutrient depletion in agro-ecosystems [11,12]. Biomass ash, which consists of crop residues, agricultural wastes, and poultry litter, contains significant amounts of fertilizer nutrients, especially phosphorus. In recent years, research on P recovery from waste streams has been triggered by an increased awareness of P scarcity and the need to close the P cycle. However, during the incineration process, 80% of P is converted to the less plant-available apatite form [13].

There are several chemical, physical, and biological ways to solubilize and recover plant nutrients from BA [14]. However, no approach or procedure can effectively recover plant nutrients since there are so many variations in BA materials [15]. In light of this problem, one goal of the present study is to assess the established biological waste treatment methods in order to increase the plant availability of nutrients from BA and release them back into the environment as bio-based fertilizer. The consecutive pre-fermentation and vermicomposting of BA with biowaste could be used as a feasible method to solubilize and recover recalcitrant nutrients from BA [16]. The use of inorganic amendments in the vermicomposting process has become significant regarding the enhancement of composting processes [17] and the nutritional values of the final product [8]. On the other hand, pH is reported to be key factor in the availability of nutrients in incineration ash [15]. In addition, BA composts are characterized by an alkaline pH, which can have adverse effects on the microbial community and their decomposition activities [18]. To create vermicompost of the highest nutrient content, it is essential to understand the changes in the microbial community, their enzymatic activities, and the nutrient availabilities during the vermicomposting process [12].

Vermicomposting is a worm-mediated biodegradation process using live epigeic worms such as *Eisenia fetida*. Worms play an important role in the process of nutrient dissolution and bioremediation by enhancing the population of beneficial bacteria [19]. The digestive system of worms contains many beneficial microorganisms, nitrogen-fixing bacteria, and enzymes [7]. Although microorganisms are actually responsible for the biochemical degradation of organic matter, worms decompose and condition the substrate, increasing the surface area for microorganism activity [20]. Worms mineralize organic matter through intestinal transit, digest it in the foregut and midgut, and then excrete it through the hindgut, interacting directly with microorganisms [6]. A diversified bacterial community in the gut accelerates the breakdown and mineralization of organic matter and even the accumulation of P and K [21]. Microorganisms involved in the vermicomposting process degrade complex organic compounds (e.g., lignin, cellulose, hemicellulose) into simpler forms and produce a variety of extracellular hydrolytic enzymes such as cellulase, protease, urease, phosphatase, lipase, and β-glucosidase [22], which are responsible for the transformation of nutrients. Therefore, the present study also aimed to clarify the bacterial diversity and enzyme activities responsible for BA nutrients in combination with cow dung.

Several studies have investigated the effectiveness of the aforementioned treatments in improving the leaching and bioavailability of biomass ash nutrients for use as a bio-based nutrient source and their effects on plant growth [10,23]. However, few studies have addressed the specific effects of BA on vermicompost microbial communities and their enzymatic activities related to nutrient availability. Therefore, in order to addressthe gaps in the literature, a novel metagenome-based approach was used in the present study to understand the functional bacterial groups and enzymatic function potential of microbial populations in BA-enriched vermicompost samples. Considering the multidimensional effect of earth worms on organic matter, the vermicomposting process was proposed as an effective process by which to improve the bioavailability of BA nutrients (Appendix A). In order to achieve this goal, the following specific objectives were selected for the current investigation: (1) to determine the effects of vermicomposting on the solubility and bioavailability of major plant nutrients; (2) to investigate the effect of BA on bacterial composition and abundance; and (3) to investigate the effect of BA on the bacterial enzyme activities responsible for nutrient dissolution.

2. Materials and Methods

2.1. Vermicomposting Process

In this study, vermicompost samples were prepared by a local vermicomposting company (Anatolian worm compost producers cooperative) using dairy cattle manure and BA from the 15 MW-biomass power plant located in Sakarya province, Turkey. A premixed combination of sawdust, forestry residues, hazelnut shells (20%), and poultry litter (80%) were used as biomass feedstock materials in the plant, which generates approximately 20,000 kg of ash daily. A dewatered form of cattle manure, consisting of a mixture of feces, urine and bedding straw litter, was obtained from a dairy cattle farm, in Sakarya, Turkey. A maximum of 7.0% was determined as the limit value for the sum of N, P, and K in the vermicompost of the Turkish compost regulation [24]. Considering this limit value, the amounts of BA were calculated to obtain the final NPK contents of 3.5% (T_1), 7.0% (T_2) and 10.0% (T_3). It is noted that T_0 without BA was also included in the study for comparison. The calculated amount of BA was uniformly mixed into the raw cattle manure, and then the mixtures were initially pre-composted for 21 days to remove toxic substances and potential pH shock that could harm the worms [25]. The mixtures were then poured into rectangular open plastic container boxes (length × width × depth = 50 × 34 × 30 cm) and epigeic (surface-dwelling) worms were added to each box (10 worms per kg of material) to allow for vermicompost production. The boxes were started with 3 kg of feed (on dry weight basis) from each application, and the same amount of material was added to the surface at weekly intervals. For each treatment, vermicompost-forming materials were kept at a 60 ± 2% moisture level by spraying surface water in a dark place at room temperature (25 ± 2 °C). Composting was monitored using daily temperature and humidity measurements for 60 days until a mature compost was obtained [25].

2.2. Enzyme Activity Analysis

In the present study, enzyme activity analysis was performed on matured 1 g wet vermicompost samples. Fresh samples of β-glucosidase, acid and alkaline phosphatase, and arylsulfatase were evaluated via incubation with p-nitrophenyl with colorless substrates of each enzyme. The acid phosphatase and alkaline phosphatase activities were analyzed using a colorimetric method and the enzyme activities were expressed as p-nitrophenol g^{-1} vermicompost h^{-1} [26].

Β-glucosidase activity was measured via the quantification of p-nitrophenol obtained after the incubation of samples with the nitrophenyl-β-D-glucopyranoside substrate. Free p-nitrophenols were measured using a spectrophotometer at 400 nm. The urease activity was analyzed using 1 g of vermicompost with urea solution as a substrate, and the NH_4 released was measured using a UV spectrometer at 578 nm and expressed as μg ammonium g^{-1} vermicompost h^{-1} on a moisture-free basis [8].

2.3. 16S Metagenome Analysis

The vermicomposted samples collected at the end of the vermicomposting experiments from each of the increasing treatments of BA, from T_0 (0.0%), T_1 (3.5%), T_2 (7.0%) to T_3 (10.0%), were used for microbial community analysis compared to the pre-composted treatment (K). 16S rRNA gene amplicon analysis was performed using QIIME 2. To create the library, purification was performed after amplification of the 16S rRNA gene with specific primers [7]. In the index PCR step, illumina binary indexes and adapters were added using the Nextera XT index kit, and then purification was performed. The concentration of the libraries created using real-time PCR (Kyratec, SC-300G, SuperCycler, Caboolture, QLD, Australia) was measured and diluted to 4 nM and normalized. Normalized samples were combined using the pooling method. After the library was prepared, each time a new dNTP (deoxyribose nucleotide triphosphate) was added to the sequencing using the synthesis method, the fluorescence of the added base was optically observed and recorded. The data generated after sequencing were converted to raw sequencing data (FASTA format) for the analysis (Figure 1). It is noted that the FASTA format is a text-based format used to encode nucleotide or amino acid sequences, where nucleotides or amino acids (proteins) are represented by single-letter codes (e.g., A, C, T, G, and N). In summary, raw sequence data are a FASTA or FASTQ text-based data format obtained directly from a sequencing device and used in a next-generation microbiome bioinformatics platform such as QIME.

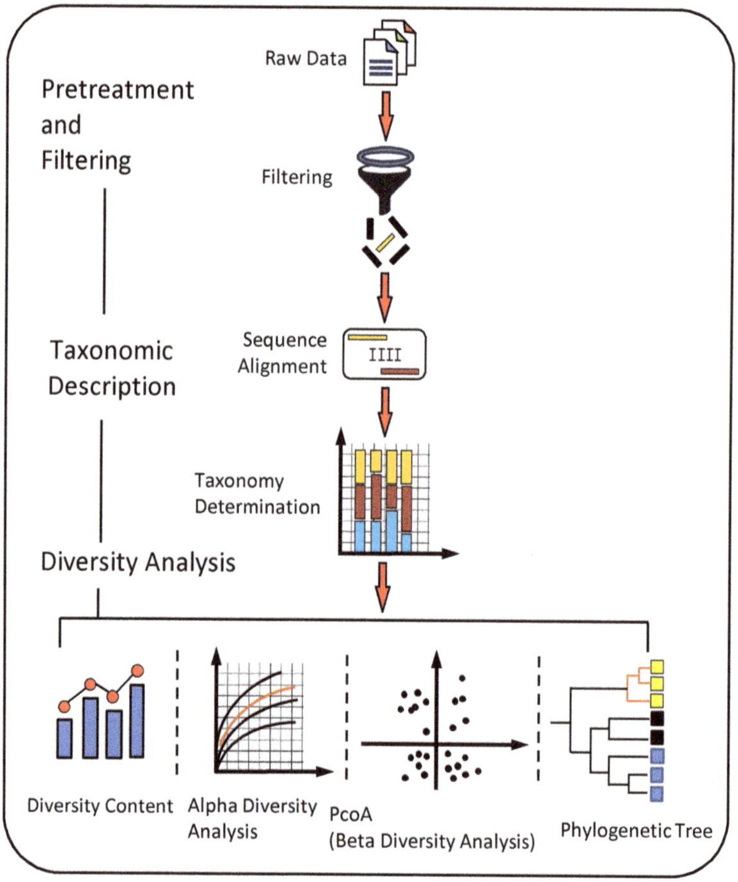

Figure 1. 16S rRNA amplicon-based microbiome analysis steps of pre-composted (K) and vermicompost (T_0, T_1, T_2, and T_3). Raw sequencing data were processed using Quantitative Insights into Microbial Ecology (QIIME 2).

2.4. Analytical Processing of Biomass Ash and Vermicompost Samples

The complementary X-ray diffraction (XRD) method was employed to identify the crystalline structure of the ash powder samples. A current of 40 mA was used to perform the XRD measurements on powder samples. The analysis was performed by using the Rigaku diffractometer (Rigaku, Tokyo, Japan) and using Cu K radiation at 40 kV and 130 mA in the scanning angle of 5–50° and at a scanning speed of 0.5° min^{-1} [5]. The Instrument software was used "o an'lyze the obtained diffraction patterns.

Vermicompost samples were oven-dried at 78 °C until a stable weight and then sieved using 2 mm mesh pieces before being subjected to further chemical analyses. The pH and electrical conductivity (EC) of both the experimental soil and vermicompost samples were measured based on a 1:5 (*w*/*v*) soil-to-water suspension using a pH meter (CG 840, Schott Geräte GmbH, Hofheim, Germany) and EC electrode (HQ14D, HACH, Loveland, CO, USA). In addition, organic matter (OM) was determined using an ignition loss technique with an ashing temperature of 550 °C for 4 h. The N in samples were determined according to the Kjeldahl procedure [3]. A colorimetrical determination procedure was used to measure the total content and plant available fraction of phosphorus after the extraction of 0.5 M of $NaHCO_3$ [5].

For total element analysis, vermicompost samples (250 mg) were digested in a microwave digester (ETHOS™, Milestone Srl., Sorisole (BG), Italy) by adding an acid mixture of 6 mL of HNO_3 (65%) and 1 mL of H_2O_2 (30%). According to the ammonium bicarbonate-diethylenetriaminepentaacetic acid (AB-DTPA) method, the available nutrients of the plant were determined after the extraction using 0.005 M DTPA. In the next step of this process, the digested samples were cooled and filtered through Whatman filter paper (0.45 μm) and diluted to 50 mL using ultrapure water (Merck Millipore, Molsheim, France). The total and plant available nutrients were measured using Inductively Coupled Plasma Optical Emission Spectrometry (ICP-OES) (Spectro Arcos, Kleve, Germany) [5].

2.5. Statistical Analysis

Experimental data were subjected to analysis of variance (full block design randomization) using Statgraphics Centurion version XVI (Statpoint Technologies Inc., Warrenton, VA, USA). Means that differed significantly were separated using Tukey's Honestly Significant Difference (HSD) test using an alpha (α) level of 0.005.

Alpha and beta diversity analyzes were performed according to the data obtained from the sequence results. Beta diversity shows the variation in the distribution of species between five different groups (K, T_0, T_1, T_2, and T_3), or the similarity or difference of the species content detected in the five different groups. Bray–Curtis difference principal coordinate analysis (pCoA) was used for beta diversity analysis. Alpha diversity is the number of different species detected in each sample. The phylogenetic diversity (PD) metrics were used for alpha diversity analysis [27]. To determine the relationship between the bacterial community by genus, the UPGMA unweighted double-group method with arithmetic mean was constructed from its matrix using a phylogenetic tree and Euclidean metrics.

3. Results and Discussion

3.1. Biomass Ash Properties and Composition

Numerous studies have shown that ash particles have a crystalline, amorphous, opaque, round, and porous structure with nanoscale voids [28]. Among these elements, P is important due to the its increasing scarcity [5]. However, most biomass combustion retains the P in the ash, which means that it can potentially be recovered. As seen in Figure 2, the main mineralogical composition of BA is as follows [5]: hydroxyapatite (15.6%), feldspars (15.6%), anhydrite (13.8%), calcite (12.5%), akermanite (9.7%), quartz (9.4%), potassium chloride (9.3%), brown millerite (6.2%), aphthalite (4.8%), hematite (2.1%), and diopsite (1.1%). The dominant peak, hydroxyapatite ($Ca_{10}(PO_4)_6(OH)_2$), is consistent with its high Ca and P contents. The combustion of biomass containing the crystal forms of Ca and P at temperatures above 800 °C is defined as the main mineral phase, and the solubility of apatite may vary depending on other contents [29].

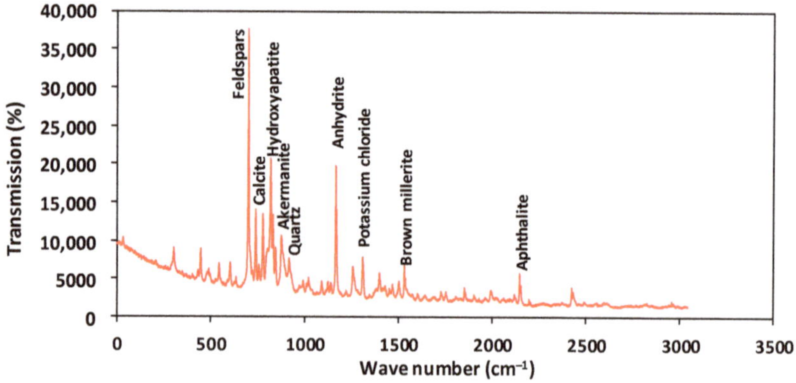

Figure 2. XRD pattern of experimental biomass ash and mineralogical main components [5].

The retention of elements in ash during the combustion of biomass fuels is important for the reutilization because major plant nutrients, such as K, P, Ca, Mg, and micro nutrients, such as Fe, Mn, Zn, and Cu, accumulate in ash [30]. The pH is influenced by these alkali metals and earth elements, and most biomass ash exhibits a noticeably high pH, as in this study (pH = 13.04). Similarly, the electrical conductivity values for BA were also recorded in a high range of 7.43 mS cm^{-1}. An excessively high pH and EC induced by BA in compost may be harmful for the microbial activity, since they cause a liming effect and osmotic problems [31]. On the other hand, some authors reported that alkaline additives may passivate heavy metals in amended composts [32,33]. Therefore, if successfully established, BA can enrich both the nutrient content of vermicompost and further reduce environmental harm.

It is obvious from Figure 3 that BA enrichment in the worm compost samples from T_0 to T_3 significantly reduced the bioavailable fraction of nutrients in the final product. This result is in agreement with previous findings that indicate that composting or vermicomposting by alkaline additives can have significant effects on nutrient availability [31]. In addition, liming is a common strategy used to passivate the heavy metals in organic waste compost applications [32]. In this regard, BA also helps to reduce the extractable contents and bioavailability of investigated heavy metals or micro plant nutrients in worm compost samples. The improvement of heavy metal passivation by BA is very clear in this vermicomposting system (Figure 3). However, the bioavailability or immobility of plant nutrients primarily depend on specific elements, complexation between metals and the dose of BA itself [34]. The immobilization effect of BA on some nutrients such as K was weak due to the ongoing microbial degradations or solubilizing effect of their metabolism products [21]. Overall, the addition of BA significantly enhanced plant nutrients, the passivation of heavy metals and the solubilization of major plant nutrients during the vermicomposting process. It is important to note that the total concentration of toxic heavy metals, such as Cd or Pb, was either below the detection limits or far below the national compost standards [24].

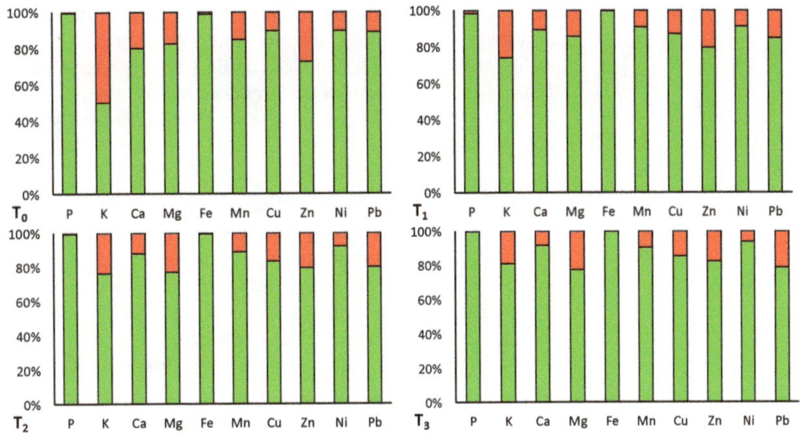

Figure 3. DTPA extractable percentages (red) of major plant nutrients and some heavy metals from worm compost samples containing increasing doses of biomass ash from T_0 to T_3 (The green bars show non-extractable fractions).

3.2. Enzyme Activities in Vermicompost

There are a number of criteria and parameters based on different characteristics that can be used to evaluate the nutrient transformations in vermicompost. Enzyme activities, which is one of them, are highly effective in the decomposition stages, impact the quality of the chemical components of the raw materials, and increase the diversity of the microbial communities in the soil where it is applied as a fertilizer [35]. Therefore, the enzyme activities are used as an indicator of organic matter conversion and nutrient transformation in vermicompost. Regarding the effects of BA addition on different hydrolytic enzyme activities, the increase in BA significantly reduced all of the enzyme activities studied compared to the T_0 control. All enzymes in the vermicompost showed higher activity in the T_0 that did not contain BA. The order of mean enzyme activities in all vermicompost samples was as follows: urease > β-glucosidase > alkaline phosphatase > arylsulfatase > acid phosphatase (Table 1).

Table 1. Enzymatic activities in vermicompost treatments with increasing dose of biomass ash (values are represented as mean ± standard deviation in three replicates).

Treatments	Acid Phosphatase (μg PNP g^{-1} h^{-1})	Alkaline Phosphatase (μg PNP g^{-1} h^{-1})	Arylsulphatase (μg PNP g^{-1} h^{-1})	β-Glucosidase (μg PNP g^{-1} h^{-1})	Urease (μg NH_4 g^{-1} h^{-1})
K	58.3 ± 2.3 c	90.6 ± 3.2 c	77.3 ± 3.3 bc	98.7 ± 2.1 c	235.0 ± 6.6 c
T_0	72.3 ± 2.4 a	101.6 ± 7.6 a	100.6 ± 7.2 a	115.7 ± 3.8 a	292.5 ± 6.2 a
T_1	65.5 ± 2.6 b	97.4 ± 5.6 b	89.4 ± 9.3 b	103.3 ± 3.4 b	263.0 ± 3.6 b
T_2	57.8 ± 3.1 c	85.2 ± 4.2 d	78.4 ± 3.1 c	88.3 ± 1.8 d	214.6 ± 3.8 d
T_3	46.3 ± 3.8 d	79.7 ± 4.6 d	61.6 ± 5.4 d	69.4 ± 2.5 e	101.4 ± 5.2 e
Significance	0.01 **	0.01 **	0.01 **	0.01 **	0.05 *

Significance: * $p < 0.05$, ** $p < 0.01$. Means ± standard deviations for each parameter followed by the same letter are not significantly different according to Tukey's HSD post hoc test at $p \leq 0.05$. PNP: p-nitrophenyl phosphate.

Some bacterial groups (e.g., *Proteobacteria* and *Firmicutes*) have been used effectively to dissolve the phosphorus found in apatite form in BA into the form that plants can take. In particular, hydrolytic bacterial enzymes such as phosphatase and arylsulfatase can make the P in apatite form soluble and usable [8]. In the present study, phosphatase activities were used to estimate the P mineralization status in worm compost samples. Phosphatase activities tended to decrease as the BA content increased, most probably because of the liming effect on the microbial activity. In addition, the reason for the high urease enzyme activity in all samples compared to other enzyme activities was attributed to the rapid proliferation of the microbial population, which is capable of producing the urease enzyme. It has been suggested that the decrease in all enzymatic activities due to the increase in the BA may be caused by the decrease in the available substrate ratio and the liming effect of the BA [12,17].

3.3. Microbial Community Structures

Metagenome analysis gives more comprehensive explanations of the genetic complexity of the microbial communities responsible for metabolic activities in the vermicomposting process [36]. According to the metagenomic results of the vermicompost samples prepared using cattle manure and BA, the alpha diversity analysis performed to determine the species richness in a single sample is shown in Figure 4. The fact that the lines become parallel to the right indicates that the number of reads is sufficient for analysis. In beta diversity analysis, the diversity within the microbial community between the experimental samples was calculated. According to Figure 5, the microbial biomass increases in T_0 and T_1, which had undergone worm digestion compared to K without worms. T_0 is the sample with the highest microbial diversity. Samples T_2 and T_3 also contain a similar diversity of microbial communities. Worms played an important role in the microbial diversity of the unmixed T_0 sample compared to the other samples. The increase in BA content negatively affected the number of microbial communities in samples T_2 and T_3 [37].

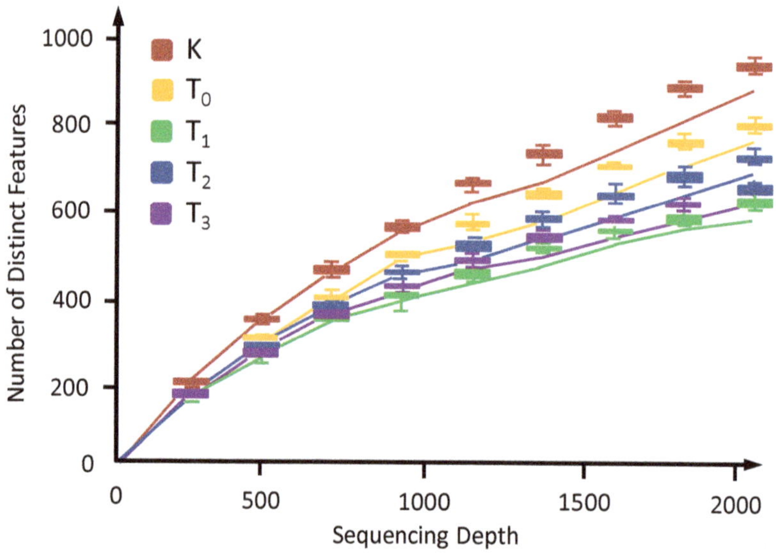

Figure 4. Rarefaction curves of the five pooled samples at percent (%) sequence similarity illustrate the species richness of vermicomposting microbiota. *X*-axis represents sequencing depth/samples. *Y*-axis shows the measures of species richness, determined as the number of distinct features.

After vermicomposting, the bacterial communities of samples T_0 (0%), T_1 (3.5%), T_2 (7.0%) and T_3 (10.0%) clustered, as shown in Figure 5. The Bacterial communities of samples T_2 and T_3 showed a similar distribution. The results revealed that the BA rates could affect bacterial communities with regard to their dispersal patterns in the vermicompost application.

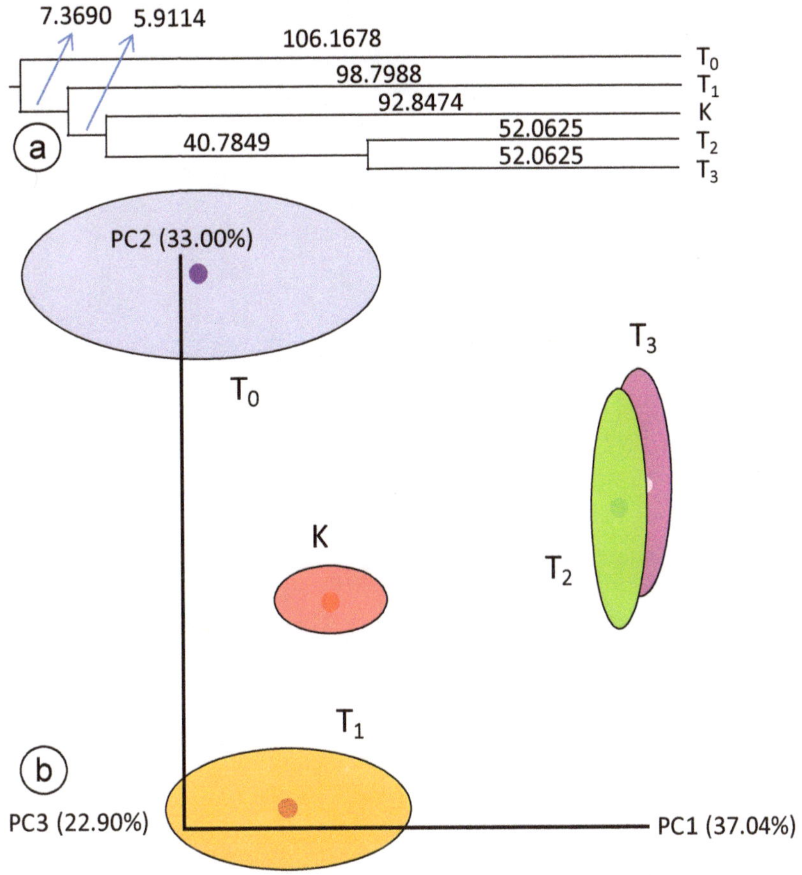

Figure 5. (**a**) Phylogenetic tree and (**b**) bacterial pCoA plots of beta-diversity estimates (Bray–Curtis) for treatment (K, T_0, T_1, T_2, and T_3) groups for 16S metagenomic strategies.

3.4. Bacterial Communities of Samples

Microorganisms are recognized as the main force of nutrient conversion, which takes place during the vermicomposting process. During the formation of vermicompost, the surface area that microorganisms can affect increases with the activity of worms, and accordingly, it facilitates the production of fertilizers with a high plant nutritional value [6,38]. As a result of the analysis performed to provide information about the diversity and functional potential of vermicompost bacteria, the bacterial community structure at the phylum level is shown in Figure 6. *Proteobacteria* (K-23.77%, T_0-18.44%, T_1-24.34%, T_2-23.5%, T_3-27.4%) *Bacterioids* (K-30.5%, T_0-15.7%, T_1-29%, T_2-11.49%, T_3-13.94%) and *Planctomycetota* (K-9.44%, T_0-15.32%, T_1-12.5%, T_2-20.3%, T_3-19.85%) were identified as the dominant bacterial phyla in all samples. This was in line with previous work on vermicompost applications [7,39].

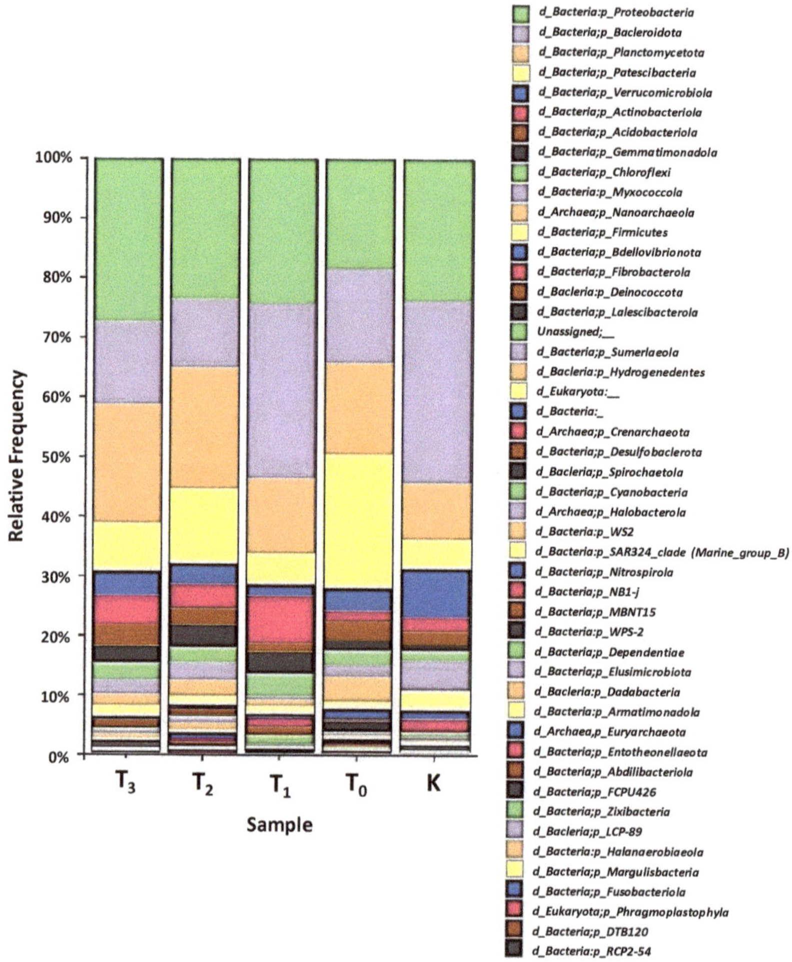

Figure 6. The stacked bar chart shows the taxonomic abundance at the phylum level of vermicompost microbes in the species of the samples. *X*-axis represents biological replicates of samples for each species, which are K, T_0, T_1, T_2, and T_3. *Y*-axis represents the taxon abundance.

Some studies have suggested that worms may form a microbial bacterial subpopulation that reflects the composition of ingested food [26,40,41]. The dominant bacterial phylum *Bacterioids*, *Proteobacteria* and *Planctomycetota* in samples T_0, T_1, T_2 and T_3 were similar in non-worm-digested sample K. A decreasing trend was found in the proportion of *Bacterioids* when comparing the pre-fermented treatment K and sample T_1 with T_0, T_2 and T_3. Most of the *Bacterioids* commonly seen in animal intestines and skins were represented by cattle manure used as raw material [42]. The fact that the genus *Flavomacterium*, which is called an opportunistic pathogen in the *Bacteriodata* family, was prominent in treatment K and in sample T_1 among all samples was throught to be a result of the high temperature during composting and the inability to resist the alkaline pH value of BA for the pre-fermented treatment K. It can also be concluded that the added BA ratio for T_1 is ideal to ensure P transformation during vermicomposting. It has been reported that *Chryseolinea* bacteria play significant roles in nitrogen fixation as important remineralizers that convert organic materials into micronutrients due to the relative abundance of *Chyophagales*, which is determined as the dominant genus belonging to the *Bacterioid* phyllium, with a significant positive correlation in T_0, T_1, T_2, and T_3 applications [43]. Gram-negative bacterial communities include most of the phyla, including *Proteobacteria*. Considering the effects of worms on microorganisms, it has been shown that gram-negative bacteria can survive in the worm gut at higher rates than gram-positive bacteria [39,43].

A hierarchical heat map was created by comparing the relative abundances of the 20 most common phyla from five samples (pre-fermented K, and worm-composted T_0, T_1, T_2, and T_3) (Table 2). At the phylum level, 47 bacterial phyla were estimated among the samples, and most of the samples shared similar phyla.

Table 2. Heatmap of the top 20 classified bacteria genera from samples (K, T_0, T_1, T_2, and T_3). The color gradient from blue to red indicates increasing species abundance. Numbers within the figure mean the level of species abundance (%).

Bacterium	K (%)	T_0 (%)	T_1 (%)	T_2 (%)	T_3 (%)
Bacteroidota	30.58	15.78	29.30	11.50	13.94
Planctomycetota	9.45	15.33	12.58	20.31	19.85
Patescibacteria	4.94	22.52	5.26	12.61	8.08
Proteobacteria	23.78	18.44	24.34	23.55	27.41
Verrucomicrobiota	8.10	3.79	1.87	3.46	4.09
Gemmatimonadota	1.13	1.82	3.71	3.82	2.91
Actinobacteriota	2.33	1.71	8.01	4.10	4.89
Nanoarchaeota	0.03	4.16	1.02	2.64	1.87
Fibrobacterota	2.16	0.26	1.53	0.32	0.24
Latescibacterota	-	1.75	0.05	0.20	0.27
Myxococcota	4.97	1.98	0.55	2.97	2.41
Crenarchaeota	-	0.44	0.02	0.62	0.22
Acidobacteriota	2.21	3.11	1.35	2.67	3.55
Deinococcota	0.10	0.28	1.38	0.92	1.07
Hydrogenedentes	0.26	0.37	0.06	1.18	0.77
Chloroflexi	1.64	2.42	3.81	2.55	2.76
Halobacterota	0.26	0.21	0.10	0.06	0.19
Spirochaetota	0.36	0.15	0.38	0.31	0.24
Firmicutes	3.3105	1.2473	1.4015	1.5638	1.9464
Bdellovibrionota	1.4373	1.5168	0.6069	0.3561	0.2749

3.5. Bacterial Communities and Enzyme Activity Relationships

Worms have digestive enzymes in their guts that help digest and break down macromolecules. These enzymes play an important role in the dissolution of organic matter and are considered an indicator for microbial activity [12]. By increasing the availability of P in BA, the phosphatase enzymes produced by many microorganisms make P soluble. The production of hydrolytic enzymes increases the release of P from fly ash [26]. In the vermicomposting process, the phosphatase enzyme activity dissolves the P in apatite form and makes it available to the the plant. Phosphatases can be divided into different classes based on their substrate range (mixed or specific) and pH optimum (acid or alkaline). The phyla of *Proteobacteria*, *Bacteroidota*, *Firmicutes*, and *Actinobacteriota*, and the genus *Pseudomonas*, *Bacillus*, *Rhizobium*, and *Burkholderia*, were identified as phosphorus-solubilizing bacteria [38,44].

The population of these bacterial classes in the samples are shown in Figure 7. The abundance of *Proteobacteria* increased to 18.44%, 24.34%, 23.55%, and 27.41% for T_0, T_1, T_2, and T_3, respectively, with the increase in the amount of BA in the compost samples. *Proteobacteria* include a variety of bacteria associated with the carbon, nitrogen, and sulfur cycle. The variation in enzyme activity may be due to the organic matter substrate that is available to the distinctive bacterial population [45]. Studies have shown that some microorganisms that play a vital role in the nitrogen cycle are *Proteabacteria*. In light of this information, *Proteabacteria* caused an increase in urease enzyme activity in all samples [31,46]. *Bacteriodata* and *Planctomycetota* were the other dominant phyla. The abundance of *Bacteriodata* was 17.78%, 29.30%, 11.50%, and 13.94% for T_0, T_1, T_2, and

T_3, respectively. The phyla of *Planctomycetota* accounted for 15.33%, 12.58%, 20.31%, and 19.85% in T_0, T_1, T_2, and T_3. Previous studies have reported that *Planctomycetota* are well adapted to different habitats, maintain high numbers in microbial communities and have the remarkable ability to participate in nitrogen metabolism in natural environments [47,48]. The significant concentration of the *Bacteriodata* phylum in the T_1 sample also indicated that it had an impact on the alkaline phosphatase enzyme activity value. The dominant population of *Flavobacterium*, belonging to the *Bacteriodata* phylum, showed higher activity than T_0, T_2, and T_3 by activating P availability in T_1, with 11.50% [49]. The T_0, T_2, and T_3 values for the dominant strain of *Flavobacterium* were 3.43%, 2.31%, and 3.23%, respectively. Considering these values, although worm digestion had a negative effect on *Flavobacterium*, the addition of BA to cattle manure showed a positive effect on phosphatase enzyme activity.

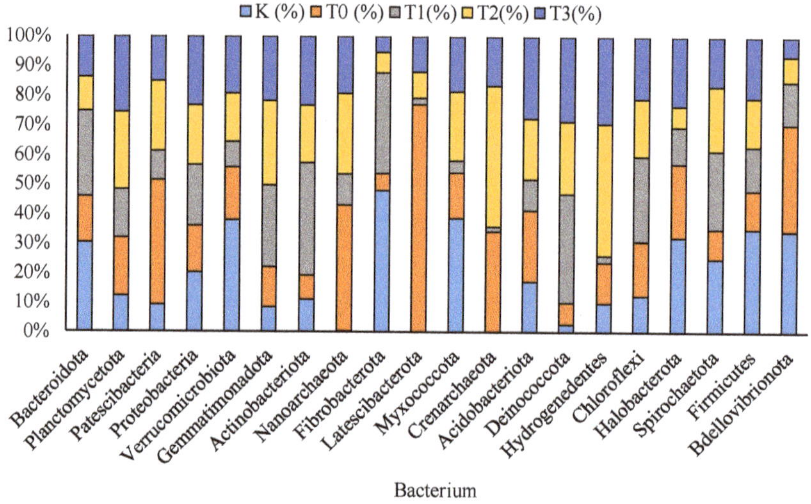

Figure 7. Bacterial abundance at phylum level of vermicompost samples collected at the end of vermicomposting experiments from each of the increasing BA applications from K (pre-composted), T_0 (0.0%), T_1 (3.5%), T_2 (7.0%) to T_3 (10.0%).

The urease enzyme activity showed the highest activity values in all samples. The urease enzyme plays an important role in converting complex organic nitrogenous matter into simpler forms. The population of *Cytophagales* producing hydrolytic enzymes in order to convert nitrogen forms, which are one of the most important organic substances of plant growth, to urea forms accounted for 8.44%, 6.14%, 5.97% and 8.09% in T_0, T_1, T_2, and T_3, respectively. The urease enzyme activity (Table 1) was determined as $T_0 > T_1 > T_2 > T_3$. The gradual decrease in the urease enzyme activity in the samples showed a parallel decrease for the bacterial population. The reason for this may be related to the diversity of nitrogen-solving bacteria, the abundance of bacteria, the increase in BA content, and the decrease in nutritional value.

The results of the enzyme activity and metagenomic analysis performed on the samples that were vermicomposted by adding cattle manure and BA, as well as the the results obtained using the multiple linear regression application, are presented in Table 3. This table shows that the above bacterial species stimulate the enzymes. It is evident from the results that during the vermicomposting process, worms indirectly stimulate microbial populations via the degradation and homogenization of organic matter, which increase the surface area that is available for microbial colonization and decomposition [20,22]. In this study, it was observed that the enzyme activity in vermicomposts was related to the bacterial population, and that the bacterial phylum listed in Table 2 significantly supported acid phosphatase, alkaline phosphatase, arylsulphatase, β-glucosidase and urease activities. Positive correlations between enzyme activities and the microbial community can increase the yield of recovered phosphorus during the vermicompost process using dominant bacteria (Figure 7).

Table 3. Results of enzyme and metagonomic analyses using multiple linear regression.

Enzymatic Activities	Treatments	Bacteria Population (%)	R^2
Acid phosphatase (μg PNP g^{-1} h^{-1}), Alkaline phosphatase (μg PNP g^{-1} h^{-1}), Arylsulphatase (μg PNP g^{-1} h^{-1}), β-glucosidase (μg PNP g^{-1} h^{-1}), Urease (μg NH_4 g^{-1} h^{-1})	T_0	*Candidatus_Kaiserbacteria* (4.55%) *Candidatus_Magasanikbacteria* (4.24%) *Woesearchaeales* (4.16%) *Saccharimonadales* (4.04%) *Flavobacteriales* (3.43%)	0.90
	T_1	*Rhizobiales* (3.63%) *Planctomycetales* (3.60%) *Pseudomonadales* (2.88%) *Alphaproteobacteria* (2.67%) *Actinomarinales* (2.41%)	0.65
	T_2	*Saccharimonadales* (5.11%) *Flavobacteriales* (2.31%) *Actinomarinales* (2.01%) *Planctomycetales* (2.43%) *Burkholderiales* (3.98%)	0.73
	T_3	*Burkholderiales* (3.51%) *Flavobacteriales* (3.23%) *Pseudomonadales* (3.16%) *Rhizobiales* (2.31%) *Saccharimonadales* (2.91%)	0.90

3.6. Strengths and Limitations of the Present Application

This study proved that during the vermicomposting process, plant nutrients that are strictly bound to the biomass ash are transformed by worms, due to their gut microbial composition and their digestive enzymes, into a more bioavailable form for plant growth. Besides composting bacterial flora, the worm compost itself increased the bacterial composition of the BA vermicompost samples. As a biotechnological system, several physical, chemical and biological reactions play a critical role in nutrient transformation. Although biological activities are prone to nutrient dissolution from biomass ash, the results of the present study do not distinguish the mechanism responsible for nutrient bioavailability. Nonetheless, this study suggests that vermicomposting BA is one of the best options, as it offers a low-cost and environmentally sound strategy by which to obtain bioactive compounds enriched with plant fertilizer nutrients.

3.7. Future Recommendations for Current Research Topic

It is recommended that some future research in this area is still required to provide new process-related insights and to understand the relevant methodology with more certainty. Thus, issues that are not included in the present analysis, but that are important to include in a future study, are summarized below.

(1) A future investigation is needed to evaluate and focus on the entire microbial community, including yeasts and fungi genera, next to the bacterial community that is responsible for the transformation of plant nutrients. It may provide a good insight into the mechanism involved in the reaction to dissolution and help to describe the system's behavior in terms of adverse alkalinity and electrical conductivity during earthworms' ingestion of BA-enriched bio-waste substrates.

(2) Future research is encouraged to provide data on the possible dissolution of phosphorus after the enrichment of the vermicomposting process by acidifying bacteria groups.

This is important in order to gain an understanding of the outcome of the transformation and plant availability of less labile Fe- and Al-bound P fractions that specifically need acid dissolution.

(3) It will be interesting for future research to look into the subject of nutrient use efficiency and the response of crops to the nutrient-rich worm compost samples reported in the present study. This will provide information regarding the justification of the product in terms of the crop response, its economic applicability and the acceptance of the final user.

(4) The study also needs to be extended to other domains, such as food wastes, that specifically contain high levels of moisture and generate acidic conditions during the composting process. It may provide different findings (e.g., the optimization of pile moisture and the determination of what happens to major plant nutrients, including P, Ca, Mg, and Fe, during the BA composting process) from various perspectives.

4. Conclusions

The vermicomposting of cattle manure by adding different proportions of biomass ash had a strong effect on worm activity and the enzyme activities of the microbial community. This was seen in the high correlations between enzyme activities and the bacterial population. The results of this study shed light on the relationship between microbial diversity and the enzyme activities that dissolve phosphorus in the form of apatite. According to the bacteria populations (%), the stimulating effect of *Flavobacteriales* (T_0-3.43%, T_2-2.31%, T_3-3.23%), *Burkholderiales* (T_2-3.98%, T_3-3.51%), *Saccharimonadales* (T_0-4.04%, T_2-5.11%, T_3-2.91%) and *Pseudomonadales* (T_1-2.88%, T_3-3.16%) showed a positive effect on enzyme activities. The determination coefficients (R^2) obtained from multiple linear regressions between the enzyme activities and bacterial population for T_0, T_1, T_2, and T_3 were determined as 0.90, 0.65, 0.73, and 0.90, respectively. As a result of the experiments supporting each other with regard to the samples containing biomass ash after vermicompost application, the importance of microorganisms in the solubilization of phosphorus in the apatite form in the biomass was demonstrated. In addition, the presence of microorganisms proved that P was dissolved and that its dynamics continued.

Author Contributions: Conceptualization, S.O. and K.Y.; methodology, G.A.T., K.Y. and N.O.; software, G.A.T. and K.Y.; formal analysis, G.A.T., K.Y. and N.O.; resources, G.A.T. and S.O.; data curation, G.A.T., S.O., K.Y., N.O. and A.E.; writing—original draft, G.A.T., S.O. and K.Y.; writing—review and editing, G.A.T., S.O., K.Y., N.O. and A.E.; visualization, G.A.T., S.O., K.Y. and A.E.; investigation, G.A.T.; supervision, S.O. and K.Y.; project administration, S.O.; funding acquisition, S.O. All authors have read and agreed to the published version of the manuscript.

Funding: This study was supported by the project code 2022-7-25-143 of Sakarya University Scientific Research Projects Coordinatorship.

Institutional Review Board Statement: Not applicable.

Informed Consent Statement: Informed consent was obtained from all subjects involved in the study.

Data Availability Statement: Not applicable.

Acknowledgments: This work was carried out as a part of the Ph.D. thesis by Güldane Aslı Turp, completed under the supervision of Saim Ozdemir. The authors also highly acknowledge the valuable insights of the reviewers and academic editor who have contributed extensively to improving the article's quality.

Conflicts of Interest: The authors declare no conflict of interest.

Appendix A. Conversion of Biomass Ash (BA) to Bio-Based Fertilizer

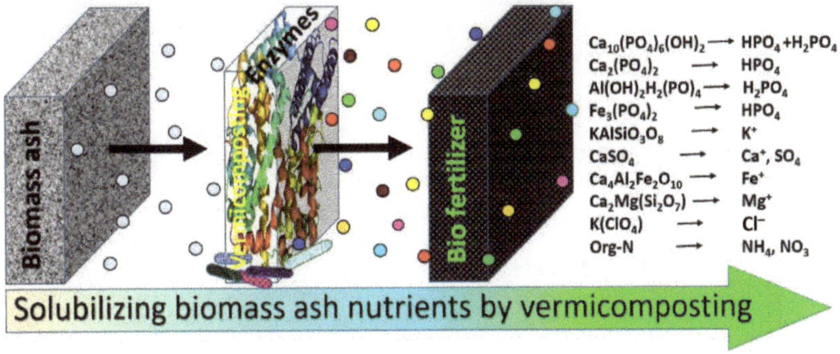

Figure A1. An illustrated presentation showing the solubilization of BA nutrients during the vermicomposting process.

References

1. Morey, L.; Fernández, B.; Tey, L.; Biel, C.; Robles-Aguilar, A.; Meers, E.; Riau, V. Acidification and solar drying of manure-based digestate to produce improved fertilizing products. *J. Environ. Manag.* **2023**, *336*, 117664. [CrossRef] [PubMed]
2. Yetilmezsoy, K.; Ilhan, F.; Kiyan, E.; Bahramian, M. A comprehensive techno-economic analysis of income-generating sources on the conversion of real sheep slaughterhouse waste stream into valorized by-products. *J. Environ. Manag.* **2022**, *306*, 114464. [CrossRef] [PubMed]
3. Czekała, W.; Nowak, M.; Piechota, G. Sustainable management and recycling of anaerobic digestate solid fraction by composting: A review. *Bioresour. Technol.* **2023**, *375*, 128813. [CrossRef]
4. Piechota, G.; Unpaprom, Y.; Dong, C.D.; Kumar, G. Recent advances in biowaste management towards sustainable environment. *Bioresour. Technol.* **2022**, *368*, 128326. [CrossRef] [PubMed]
5. Turp, G.A.; Turp, S.M.; Ozdemir, S.; Yetilmezsoy, K. Vermicomposting of biomass ash with bio-waste for solubilizing nutrients and its effect on nitrogen fixation in common beans. *Environ. Technol. Innov.* **2021**, *23*, 101691. [CrossRef]
6. Dume, B.; Hanc, A.; Svehla, P.; Michal, P.; Chane, A.D.; Nigussie, A. Vermicomposting technology as a process able to reduce the content of potentially toxic elements in sewage sludge. *Agronomy* **2022**, *12*, 2049. [CrossRef]
7. Wang, F.; Zhang, W.; Miao, L.; Ji, T.; Wang, Y.; Zhang, H.; Ding, Y.; Zhu, W. The effects of vermicompost and shell powder addition on Cd bioavailability, enzyme activity and bacterial community in Cd-contaminated soil: A field study. *Ecotoxicol. Environ. Saf.* **2021**, *215*, 112163. [CrossRef] [PubMed]
8. Ozdemir, S.; Turp, G.A. The impact of the pyroligneous acid-assisted biomass ash vermicompost on dry beans through climatic and agroecosystem changes. *J. Mater. Cycles Waste Manag.* **2023**, *25*, 490–500. [CrossRef]
9. Gupta, S.; Sharma, S.; Aich, A.; Verma, A.K.; Bhuyar, P.; Nadda, A.K.; Mulla, S.I.; Kalia, S. Chicken feather waste hydrolysate as a potential biofertilizer for environmental sustainability in organic agriculture management. *Waste Biomass Valor.* **2023**, 1–17. [CrossRef]
10. Zhang, Z.; He, F.; Zhang, Y.; Yu, R.; Li, Y.; Zheng, Z.; Gao, Z. Experiments and modelling of potassium release behavior from tablet biomass ash for better recycling of ash as eco-friendly fertilizer. *J. Clean Prod.* **2018**, *170*, 379–387. [CrossRef]
11. Liang, F.; Feng, L.; Liu, N.; He, Q.; Ji, L.; De Vrieze, J.; Yan, S. An improved carbon fixation management strategy into the crop–soil ecosystem by using biomass ash as the medium. *Environ. Technol. Innov.* **2022**, *28*, 102839. [CrossRef]
12. Karwal, M.; Kaushik, A. Co-composting and vermicomposting of coal fly-ash with press mud: Changes in nutrients, micronutrients and enzyme activities. *Environ. Technol. Innov.* **2020**, *18*, 100708. [CrossRef]
13. Tesfaye, F.; Lindberg, D.; Moroz, M.; Hupa, L. Investigation of the K-Mg-Ca sulfate system as part of monitoring problematic phase formations in renewable-energy power plants. *Energies* **2020**, *13*, 5366. [CrossRef]
14. Pesonen, J.; Kuokkanen, T.; Rautio, P.; Lassi, U. Bioavailability of nutrients and harmful elements in ash fertilizers: Effect of granulation. *Biomass. Bioenergy.* **2017**, *100*, 92–97. [CrossRef]
15. Semerci, N.; Kunt, B.; Calli, B. Phosphorus recovery from sewage sludge ash with bioleaching and electrodialysis. *Int. Biodeterior. Biodegrad.* **2019**, *144*, 104739. [CrossRef]
16. Quintela-Sabarís, C.; Mendes, L.A.; Domínguez, J. Vermicomposting as a sustainable option for managing biomass of the invasive tree *Acacia dealbata* link. *Sustainability* **2022**, *14*, 13828. [CrossRef]
17. Dede, C.; Ozer, H.; Dede, O.H.; Celebi, A.; Ozdemir, S. Recycling nutrient-rich municipal wastes into ready-to-use potting soil: An approach for the sustainable resource circularity with inorganic porous materials. *Horticulturae* **2023**, *9*, 203. [CrossRef]
18. Fu, X.; Huang, K.; Cui, G.; Chen, X.; Li, F.; Zhang, X.; Li, F. Dynamics of bacterial and eukaryotic community associated with stability during vermicomposting of pelletized dewatered sludge. *Int. Biodeter. Biodegrad.* **2015**, *104*, 452–459. [CrossRef]

19. Goswami, L.; Gorai, P.S.; Mandal, N.C. Microbial fortification during vermicomposting: A brief review. *Recent Adv. Microb. Biotechnol.* **2021**, 99–122. [CrossRef]
20. Aira, M.; Domínguez, J. Microbial and nutrient stabilization of two animal manures after the transit through the gut of the earthworm *Eisenia fetida* (Savigny, 1826). *J. Hazard. Mater.* **2009**, *161*, 1234–1238. [CrossRef]
21. Yürürdurmaz, C. Impact of different fertilizer forms on yield components and macro–micronutrient contents of cowpea (*Vigna unguiculata* L.). *Sustainability* **2022**, *14*, 12753. [CrossRef]
22. da Silva, L.F.; da Silva, E.F.; Morais, F.M.S.; Portela, J.C.; de Oliveira, F.H.T.; de Freitas, D.F.; de Almeida Ferreira, E.; Gurgel, M.T.; Pinheiro, A.M.; Lima, R.B.; et al. Potential of vermicomposting with mixtures of animal manure and vegetable leaves in the development of *Eisenia foetida*, microbial biomass, and enzymatic activity under semi-arid conditions. *J. Environ. Manag.* **2023**, *330*, 117169. [CrossRef] [PubMed]
23. Asquer, C.; Cappai, G.; De Gioannis, G.; Muntoni, A.; Piredda, M.; Spiga, D. Biomass ash reutilisation as an additive in the composting process of organic fraction of municipal solid waste. *Waste Manag.* **2017**, *69*, 127–135. [CrossRef] [PubMed]
24. Official Gazette. *Regulation on Organic, Mineral and Microbial Sources of Fertilizer Used in Agriculture*, 23rd ed.; No: 30341; Presidency General Directorate of Law and Legislation: Ankara, Turkey, 2018. (In Turkish)
25. Ozdemir, S.; Dede, G.; Dede, O.H.; Turp, S.M. Composting of sewage sludge with mole cricket: Stability, maturity and sanitation aspects. *Int. J. Environ. Sci. Technol.* **2019**, *16*, 5827–5834. [CrossRef]
26. Busato, J.G.; Lima, L.S.; Aguiar, N.O.; Canellas, L.P.; Olivares, F.L. Changes in labile phosphorus forms during maturation of vermicompost enriched with phosphorus-solubilizing and diazotrophic bacteria. *Bioresour. Technol.* **2012**, *110*, 390–395. [CrossRef] [PubMed]
27. Kolbe, A.R.; Aira, M.; Gómez-Brandón, M.; Pérez-Losada, M.; Domínguez, J. Bacterial succession and functional diversity during vermicomposting of the white grape marc *Vitis vinifera* v. Albariño. *Sci. Rep.* **2019**, *9*, 7472. [CrossRef] [PubMed]
28. Ondrasek, G.; Kovačić, M.B.; Carević, I.; Štirmer, N.; Stipičević, S.; Udiković-Kolić, N.; Filipović, V.; Romić, D.; Rengel, Z. Bioashes and their potential for reuse to sustain ecosystem services and underpin circular economy. *Renew. Sustain. Energy Rev.* **2021**, *151*, 111540. [CrossRef]
29. Leng, L.; Bogush, A.A.; Roy, A.; Stegemann, J.A. Characterisation of ashes from waste biomass power plants and phosphorus recovery. *Sci. Total Environ.* **2019**, *690*, 573–583. [CrossRef]
30. Ozdemir, S.; Demir, M.S. Biofuel characteristics and combustion emissions of poultry litter and lignocellulosic biomass. *Environ. Prog. Sustain. Energy* **2021**, *40*, e13555. [CrossRef]
31. Xu, Z.; Qi, C.; Zhang, L.; Ma, Y.; Li, G.; Nghiem, L.D.; Luo, W. Regulating bacterial dynamics by lime addition to enhance kitchen waste composting. *Bioresour. Technol.* **2021**, *341*, 125749. [CrossRef]
32. Singh, J.; Kalamdhad, A.S. Effects of lime on bioavailability and leachability of heavy metals during agitated pile composting of water hyacinth. *Bioresour. Technol.* **2013**, *138*, 148–155. [CrossRef] [PubMed]
33. Ozdemir, S.; Turp, S.M.; Oz, N. Simultaneous dry-sorption of heavy metals by porous adsorbents during sludge composting. *Environ. Eng. Res.* **2020**, *25*, 258–265. [CrossRef]
34. Dědina, M.; Jarošíková, A.; Plíva, P.; Dubský, M. The effect of ash admixture on compost quality and availability of nutrients. *Sustainability* **2022**, *14*, 1640. [CrossRef]
35. Alidadi, H.; Hosseinzadeh, A.; Najafpoor, A.A.; Esmaili, H.; Zanganeh, J.; Takabi, M.D.; Piranloo, F.G. Waste recycling by vermicomposting: Maturity and quality assessment via dehydrogenase enzyme activity, lignin, water soluble carbon, nitrogen, phosphorous and other indicators. *J. Environ. Manag.* **2016**, *182*, 134–140. [CrossRef] [PubMed]
36. Borker, S.S.; Thakur, A.; Khatri, A.; Kumar, R. Quality assessment, safety evaluation, and microbiome analysis of night-soil compost from Lahaul valley of northwestern Himalaya. *Waste Manag.* **2022**, *149*, 42–52. [CrossRef]
37. Belyaeva, O.N.; Haynes, R.J. Chemical, microbial and physical properties of manufactured soils produced by co-composting municipal green waste with coal fly ash. *Bioresour. Technol.* **2009**, *100*, 5203–5209. [CrossRef]
38. Khan, H.; Akbar, W.A.; Shah, Z.; Rahim, H.U.; Taj, A.; Alatalo, J.M. Coupling phosphate-solubilizing bacteria (PSB) with inorganic phosphorus fertilizer improves mungbean (*Vigna radiata*) phosphorus acquisition, nitrogen fixation, and yield in alkaline-calcareous soil. *Heliyon* **2022**, *8*, e09081. [CrossRef]
39. Aira, M.; Olcina, J.; Pérez-Losada, M.; Domínguez, J. Characterization of the bacterial communities of casts from *Eisenia andrei* fed with different substrates. *Appl. Soil. Ecol.* **2016**, *98*, 103–111. [CrossRef]
40. Dominguez, J.; Aira, M.; Kolbe, A.R.; Gómez-Brandón, M.; Pérez-Losada, M. Changes in the composition and function of bacterial communities during vermicomposting may explain beneficial properties of vermicompost. *Sci. Rep.* **2019**, *9*, 9657. [CrossRef]
41. Pérez-Losada, M.; Narayanan, D.B.; Kolbe, A.R.; Ramos-Tapia, I.; Castro-Nallar, E.; Crandall, K.A.; Domínguez, J. Comparative analysis of metagenomics and metataxonomics for the characterization of vermicompost microbiomes. *Front. Microbiol.* **2022**, *13*, 854423. [CrossRef]
42. Tang, M.; Wu, Z.; Li, W.; Shoaib, M.; Aqib, A.I.; Shang, R.; Yang, Z.; Pu, W. Effects of different composting methods on antibiotic-resistant bacteria, antibiotic resistance genes, and microbial diversity in dairy cattle manures. *J. Dairy Sci.* **2023**, *106*, 257–273. [CrossRef] [PubMed]
43. Tao, K.; Zhang, X.; Chen, X.; Liu, X.; Hu, X.; Yuan, X. Response of soil bacterial community to bioaugmentation with a plant residue-immobilized bacterial consortium for crude oil removal. *Chemosphere* **2019**, *222*, 831–838. [CrossRef] [PubMed]

44. Wu, Q.; Wan, W. Insight into application of phosphate-solubilizing bacteria promoting phosphorus availability during chicken manure composting. *Bioresour. Technol.* **2023**, *373*, 128707. [CrossRef]
45. Hou, T.; Zhou, Y.; Cao, X.; Li, W.; Zhang, S.; Zhao, Y.; Chen, L.; An, Q.; Meng, L. Effects of microbial inoculum on microbial community and enzyme activity involved in nitrogen-sulfur metabolism during sewage sludge composting. *Sci. Total Environ.* **2023**, *858*, 159954. [CrossRef] [PubMed]
46. Bao, J.; Lv, Y.; Qv, M.; Li, Z.; Li, T.; Li, S.; Zhu, L. Evaluation of key microbial community succession and enzyme activities of nitrogen transformation in pig manure composting process through multi angle analysis. *Bioresour. Technol.* **2022**, *362*, 127797. [CrossRef]
47. Gao, W.; Liu, P.; Ye, Z.; Zhou, J.; Wang, X.; Huang, X.; Deng, X.; Ma, L. Divergent prokaryotic microbial assembly, co-existence patterns and functions in surrounding river sediments of a Cu-polymetallic deposit in Tibet. *Sci. Total Environ.* **2022**, *851*, 158192. [CrossRef]
48. Suarez, C.; Dalcin Martins, P.; Jetten, M.S.; Karačić, S.; Wilén, B.M.; Modin, O.; Persson, F. Metagenomic evidence of a novel family of anammox bacteria in a subsea environment. *Environ. Microbiol.* **2022**, *24*, 2348–2360. [CrossRef]
49. Lidbury, I.D.; Scanlan, D.J.; Murphy, A.R.; Christie-Oleza, J.A.; Aguilo-Ferretjans, M.M.; Hitchcock, A.; Daniell, T.J. A widely distributed phosphate-insensitive phosphatase presents a route for rapid organophosphorus remineralization in the biosphere. *Proc. Natl. Acad Sci. USA* **2022**, *119*, e2118122119. [CrossRef]

sustainability MDPI

Article

Application of Artificial Intelligence to Predict CO_2 Emissions: Critical Step towards Sustainable Environment

Ahmed M. Nassef [1,2,*], Abdul Ghani Olabi [3,4], Hegazy Rezk [1,5] and Mohammad Ali Abdelkareem [3,6]

1 Department of Electrical Engineering, College of Engineering in Wadi Alddawasir, Prince Sattam bin Abdulaziz University, Wadi Alddawasir 11991, Saudi Arabia
2 Computers and Automatic Control Engineering Department, Faculty of Engineering, Tanta University, Tanta 31733, Egypt
3 Sustainable Energy & Power Systems Research Centre, RISE, University of Sharjah, Sharjah P.O. Box 27272, United Arab Emirates
4 School of Engineering and Applied Science, Aston University, Aston Triangle, Birmingham B4 7ET, UK
5 Department of Electrical Engineering, Faculty of Engineering, Minia University, Minia 61111, Egypt
6 Department of Chemical Engineering, Faculty of Engineering, Minia University, Minia 61111, Egypt
* Correspondence: a.nasef@psau.edu.sa

Abstract: Prediction of carbon dioxide (CO_2) emissions is a critical step towards a sustainable environment. In any country, increasing the amount of CO_2 emissions is an indicator of the increase in environmental pollution. In this regard, the current study applied three powerful and effective artificial intelligence tools, namely, a feed-forward neural network (FFNN), an adaptive network-based fuzzy inference system (ANFIS) and long short-term memory (LSTM), to forecast the yearly amount of CO_2 emissions in Saudi Arabia up to the year 2030. The data were collected from the "Our World in Data" website, which offers the measurements of the CO_2 emissions from the years 1936 to 2020 for every country on the globe. However, this study is only concerned with the data related to Saudi Arabia. Due to some missing data, this study considered only the measurements in the years from 1954 to 2020. The 67 data samples were divided into 2 subsets for training and testing with the optimal ratio of 70:30, respectively. The effect of different input combinations on prediction accuracy was also studied. The inputs were combined to form six different groups to predict the next value of the CO_2 emissions from the past values. The group of inputs that contained the past value in addition to the year as a temporal index was found to be the best one. For all the models, the performance accuracies were assessed using the root mean squared errors (RMSEs) and the coefficient of determination (R^2). Every model was trained until the smallest RMSE of the testing data was reached throughout the entire training run. For the FFNN, ANFIS and LSTM, the averages of the RMSEs were 19.78, 20.89505 and 15.42295, respectively, while the averages of the R^2 were found to be 0.990985, 0.98875 and 0.9945, respectively. Every model was applied individually to forecast the next value of the CO_2 emission. To benefit from the powers of the three artificial intelligence (AI) tools, the final forecasted value was considered the average (ensemble) value of the three models' outputs. To assess the forecasting accuracy, the ensemble was validated with a new measurement for the year 2021, and the calculated percentage error was found to be 6.8675% with an accuracy of 93.1325%, which implies that the model is highly accurate. Moreover, the resulting forecasting curve of the ensembled models showed that the rate of CO_2 emissions in Saudi Arabia is expected to decrease from 9.4976 million tonnes per year based on the period 1954–2020 to 6.1707 million tonnes per year in the period 2020–2030. Therefore, the finding of this work could possibly help the policymakers in Saudi Arabia to take the correct and wise decisions regarding this issue not only for the near future but also for the far future.

Keywords: sustainable environment; CO_2 emissions; artificial intelligence; data modelling

check for updates

Citation: Nassef, A.M.; Olabi, A.G.; Rezk, H.; Abdelkareem, M.A. Application of Artificial Intelligence to Predict CO_2 Emissions: Critical Step towards Sustainable Environment. *Sustainability* **2023**, *15*, 7648. https://doi.org/10.3390/su15097648

Academic Editors: Kaan Yetilmezsoy, Pallav Purohit and Steve W. Lyon

Received: 2 March 2023
Revised: 9 April 2023
Accepted: 25 April 2023
Published: 6 May 2023

1. Introduction

Saudi Arabia is the biggest exporter of fossil fuels in the world. Controlling CO_2 emissions will help in improving the quality of life all over the world. Therefore, every country has to keep an eye on the CO_2 emissions within its region. To do this, the data have to be recorded from sensors, and a qualitative prediction model has to be built. Higher prediction values from this model are alarm signs for expected harmful situations that could inform and help the authorities to take suitable actions.

Despite the fact that the increase in CO_2 emissions is harmful to human beings and the environment, an adequate amount of CO_2 is an essential requirement for plants and agriculture. According to records, the United Nations announced in the COP27 conference that was held in Sharm Elsheikh, Egypt in November 2022 that CO_2 emissions have increased by 1% in 2022 to reach 37.5 billion tonnes [1]. This big number is an alarm to the whole world that the global warming problem will continue, and hence, facing it is a must. Moreover, the promises have to be transferred to real actions on the ground [2]. To be on the right track to reducing CO_2 emissions, two strategies have to be considered seriously. The first is to decrease the utility of CO_2 emission fuels by increasing the use of other alternatives, such as renewable and green energy sources. The energy obtained from photovoltaic (PV) cells is an example of a renewable energy source. In fact, research is growing rapidly in this direction [3,4]. The second is to increase green agricultural areas. Fortunately, Saudi Arabia is currently adopting two routes to positively participate in decreasing the global warming catastrophe. To help authorities in taking the proper decision, a trustable forecasting tool is required.

The need for data forecasting can be encountered in many engineering applications. Therefore, it obtained great interest from many researchers of different disciplines and directions. To build a forecasting model, there are many tools and techniques to do so. However, building a robust model that can correctly predict future values from past values is a challenging task. In this regard, artificial intelligence (AI) and machine learning (ML) techniques occupy a substantial and competitive position among other tools. They are considered the best-nominated approaches that can perfectly handle data modelling and forecasting tasks [5–9].

The applications of artificial intelligence (AI) and deep learning (DL) tools are growing from day to day. These applications can be categorized into two main fields: classification and prediction. The applications of classifying tools include the fingerprint classifier, voice recognition, image, etc. However, from the prediction tool perspective, applications might include modelling and forecasting of time-series data, traffic, stock, weather, etc. The reliability of a predicting model comes from the appropriate selection of the model's type, and then a successful training phase is implemented. There are many AI techniques to be used as predictors, such as artificial neural networks (ANNs) [10], adaptive network-based fuzzy inference systems (ANFISs) [11], long short-term memory (LSTM) [12], regression support vector machines (RSVMs) [13], autoregressive integrated moving averages (ARIMAs) [14], etc.

Similarly, AI techniques can be applied to forecast CO_2 emissions based on the measured time-series data. Vlachas et al. used LSTM as a forecasting model to forecast time-series of a high-dimensional, chaotic system. The results have been compared with those obtained by the Gaussian processes (GPs) and showed that LSTM performed better than the GP [15]. Amarpuri et. al. applied a deep learning hybrid model based on a convolution neural network (CNN) and long short-term memory network (LSTM), namely, (CNN-LSTM), to predict CO_2 levels in India in the year 2020. This study aimed to provide an estimated value of the CO_2 emissions in India, as the country had an agreement to reduce the carbon dioxide levels to 30–35% of the level in the year 2005 according to the Paris Agreement [16]. Kumari and Singh used three statistical models, two linear regression models and a deep learning model to predict CO_2 emissions in India for the next ten years based on the data from 1980 to 2019 [14]. The statistical models included the autoregressive integrated moving average (ARIMA) model, the seasonal autoregressive integrated moving average with exogenous factors (SARIMAX) model, and the Holt–Winters model.

The machine learning models included linear regression and random forest techniques, while the deep learning-based model included long short-term memory (LSTM). They concluded that LSTM outperformed the other five models, which produced an RMSE value of 60.635. Meng and Noman applied four machine learning (ML) SARIMA models, namely, SARIMAX, to predict total CO_2 emissions, taking into account the effect of COVID-19 pandemic circumstances in the reduction of the emission value [17]. The study included a comparison between the four prediction models to suggest the effective one based on the lowest value of mean absolute percentage error (MAPE). The authors considered the forecasting periods as near future prediction, future prediction and far future prediction for the years 2022 to 2027, 2022 to 2054 and 2022 to 2072, respectively. They showed that the post-COVID-19 model forecasts CO_2 emission in a reasonable behaviour. Birjandi et. al. applied an artificial neural network (ANN) model of eleven neurons in the hidden layer with two different activation functions, *normalized RBF* and *tansig*, to predict CO_2 in four Southeastern countries: Malaysia, Indonesia, Singapore and Vietnam [18]. They concluded that the ANN model with the normalized RBF produced the highest correlation value with the measured data.

Researchers are promoted to apply AI techniques in the forecasting of CO_2 emissions applications. Examples of reputative AI tools are the FFNN, ANFIS and LSTM. Many recent studies used the FFNN [19,20], ANFIS [21,22] and LSTM [14,23,24] techniques to build a forecasting model of CO_2 emissions. For the FFNN modelling approach, Mutascu used the single-layer, 20-neuron feed-forward artificial neural network approach to predict CO_2 emissions in the United States of America [25]. The model was built based on the annual data collected during the years ranging from 1984 to 2020. The simulation results showed that the forecasting accuracy was 92.42%, and the forecasting error percentage was 7.58%. The RMSE and the R^2 were used as the statistical assessment of the modelling accuracy. Moreover, Wang et. al. combined several machine learning tools in a hybrid procedure to forecast CO_2 emissions in China [26]. They applied four two-stage procedures. Support vector regression (SVR) with an ANN was one of the used procedures. Regarding the ANFIS modelling tool, Cansiz et. al. applied some AI techniques that included an ANN and ANFIS to predict the CO_2 emissions from the transportation sector [21]. Their results showed that the ANN model was the best. For the LSTM technique, Kumari and Singh conducted a comparison study between some ML tools, including LSTM, to predict CO_2 emissions in India [14]. They concluded that the LSTM model was the best to predict CO_2 emissions, as it produced the lowest RMSE of 60.635 among the other techniques.

Particularly, some researchers applied several AI tools to forecast the CO_2 emissions in Saudi Arabia. Hamieh et. al. conducted a study to quantify the CO_2 emissions from industrial facilities in Saudi Arabia [27]. They studied the CO_2 emissions from six industrial entities, such as electricity generation, desalination, oil refining, cement, petrochemicals and iron and steel, which are responsible for more than 70% of the country's total emissions. This was to register the details about the emission source locations, rates and characteristics. Alam and AlArjani compared CO_2 emission forecasting in Gulf countries based on an annual basis [28]. They used statistical tools such as autoregressive integrated moving averages (ARIMAs), artificial neural networks (ANNs) and Holt–Winters exponential smoothing (HWES) as the forecasting models. Their conclusion was to use the ANN model for forecasting the CO_2 in Gulf countries, as it produced the best results relative to the ARIMA and the HWES models. Alajmi studied the modelling of carbon emissions and electricity generation in SA [29]. This study used a structural time-series model (STSM) and logarithmic mean Divisia index (LMDI) for estimating the long-run elasticities. The results showed that three controlling variables such as gross domestic product (GDP), electricity generation and population are significantly affecting carbon dioxide (CO_2) emissions. Furthermore, the findings from utilizing the logarithmic mean division index (LMDI) analysis showed that there was an increase of 1377.56 million tonnes in CO_2 emissions from the three controlling factors between 1980 and 2017 in Saudi Arabia. Habadi and Tsokos built a statistical-based time-series model to forecast carbon dioxide in the atmosphere in the

Middle East and the atmospheric temperature in Saudi Arabia. The model is of a seasonal autoregressive integrated moving average (seasonal-ARIMA) type, and the data are based on the monthly temperature from 1970–2015 [30]. Althobaiti and Shabri performed a comparison between the nonlinear grey Bernoulli model NGBM (1,1) and the GM (1,1) model for predicting the yearly CO_2 emissions in Saudi Arabia from 2017 to 2021 based on the data from 1970 to 2016 [31]. The study concluded that the NGBM (1,1) produced better results based on the mean absolute percentage error (MAPE), which was below 10%. Alkhathlan and Javid studied and analyzed the underlying energy demand trend (UEDT) for the total carbon emissions and carbon emissions from the domestic transport sector in Saudi Arabia using a structural time-series technique [32]. The considered data was in the period from 1971 to 2013. The study demonstrated that there is a positive correlation between the growth in real income and CO_2 emissions.

From the aforementioned literature review, it can be noticed that some researchers have been interested in CO_2 emissions either in Saudi Arabia or the Middle East area. In the following paragraph, a brief description of each study is presented but from a critical viewpoint.

In [27], the work presents a database for the volumetric concentrations of CO_2 emissions from all major stationary industrial sources in Saudi Arabia. However, the paper does not show any modelling or forecasting approach. In [28], the authors developed a comparative study between three prediction models, namely, ARIMA, ANN and Holt–Winters exponential smoothing. They used 55 data samples that represent the recordings during the years 1960:2014. Their goal was to build the best model for forecasting CO_2 emissions in the annual range of 2015:2025. The paper showed that the ANN was the best, but the paper does not show the value of the training/testing ratio to assess the model's reliability. In [29], the authors built a structural time series model (STSM) to study the effect of gross domestic product (GDP), electricity generation and population on CO_2 emissions in Saudi Arabia between 1980 and 2017. The paper does not show any data-forecasting strategy. In [30], the authors used a statistical forecasting technique, namely, the autoregressive integrated moving average (ARIMA), to predict the atmospheric CO_2 in the entire Middle East between 1996–2015. The forecasting range was applied monthly through only the years 2014:2015. In [31], the study used a statistical method, namely, the nonlinear grey Bernoulli model, to forecast CO_2 in Saudi Arabia. The model was built using a dataset in the range of 1970:2016, and the forecasting period represents only the annual range between 2017:2021.

Despite all the abovementioned studies being related to CO_2 emissions, none of them has used machine learning or artificial intelligence tools to model CO_2 emissions in Saudi Arabia. Moreover, none of them has applied the obtained model to forecast the annual emissions between 2020 to 2030. Almost all of these studies have used statistical methods in their work. Furthermore, all of them have used an entire set of data to build a model. However, in our proposed work, three ensembled machine learning tools, such as FFNN, ANFIS and LSTM, have been used to perform the forecasting task. In addition, during the modelling procedure, the optimal training/testing ratio of 70/30 has been applied [33].

In conclusion, up to the knowledge of the authors, this is the first use of an ensemble composed of three AI and ML tools, such as FFNN, ANFIS and LSTM, to forecast CO_2 emissions in Saudi Arabia for the period from 2020 to 2030. Accordingly, the authors believe that the proposed strategy is considered novel.

The work contribution can be highlighted as follows:

- Forecasting CO_2 emissions in Saudi Arabia for the period 2020–2030, which has not been studied before.
- Forecasting models have been built using three artificial intelligence tools, namely, FFNN, ANFIS and LSTM.
- The dataset was obtained from a verified repository to ensure the obtained models are trustable [34].
- The optimal train/test ratio of 70/30 has been applied to avoid overfitting [33].

Table 1 summarizes the key comparisons between the previous studies and our proposed work. From the table, it can be concluded that the work proposed in this research paper is novel and unique.

Table 1. Key comparisons between previous studies and our proposed work.

Reference	Publication Year	Model	Model Type	Modelling Range (Years)	No. of Samples	Forecasting Range (Years)	Train/Test Ratio
[27]	2022	-	-	-		-	-
[28]	2021	ARIMA, ANN and Holt–Winters Exponential Smoothing	Statistical and ANN	1960:2014	55	2015:2025	100/0
[29]	2022	STSM	Statistical	1980:2017	38	-	100/0
[30]	2017	ARIMA	Statistical	1996–2015	20	2014:2015	100/0
[31]	2022	Nonlinear Grey Bernoulli Model	Statistical	1970:2016	47	2017:2021	100/0
Proposed methodology		**FFNN, ANFIS and LSTM**	**AI and ML**	**1954:2020**	**67**	**2020:2030**	**70/30**

Fortunately, the CO_2 emission resulting from all contributors in Saudi Arabia is only 674.9 million tonnes. However, the emissions resulting from the entire world reached a value of 37.5 billion tonnes. This implies that Saudi Arabia has contributed only 1.8%. Despite this small value of contribution, the country is adopting a strategy to reduce these emissions gradually to reach net zero emissions by 2060. From this perspective, this paper presents a study to forecast CO_2 emissions using three powerful artificial intelligence tools, such as FFNN, ANFIS and LSTM, for the coming years up to 2030.

In conclusion, FFNN, ANFIS and LSTM are three machine learning tools that have proven their efficiency in data modelling and forecasting. Therefore, they are adopted in this study to build the forecasting model of CO_2 emissions in Saudi Arabia up to 2030.

The rest of the paper is organized as follows: The structure, function and mechanism of the three AI models are presented in Section 2. In Section 3, the data preparation is illustrated. However, in Section 4, the modelling phase is introduced in the results and discussion. In Section 5, the proposed ensemble model is presented. Finally, this work's conclusions are summarized in Section 6.

2. Artificial Intelligence Models

In this study, three competitive AI modelling tools have been applied to predict CO_2 emissions in Saudi Arabia. In the following paragraphs, a general idea of the FFNN, ANFIS and LSTM AI tools will be presented in the following subsections.

2.1. FFNN Model

In 1943, McCulloch and Pitts proposed the mathematical model of the artificial neuron called the perceptron, which is a single arithmetic unit. This perceptron tries to exactly mimic the signal processing procedure of the natural neuron. Particularly, the human brain is composed of billions of these tiny neural cells. These cells are linked together to compose a network to be able to execute a certain complex task. The interconnection between neurons is accomplished through the cell's dendrites and synapses. The dendrite is a receiving input terminal, while the synapse is a transmitting output terminal. For a neuron to receive signals from its neighbouring cells, the dendrites are connected to the synapses of these cells to receive a signal from every individual preceding it. The strength of each received signal is mainly based on the synaptic strength between the neuron and its exciting cell. This strength is mathematically modelled by a weight value. All the weighted inputs are summed together in the cell's soma, and the output is triggered to the cell's synapse through the axion as soon as it reaches or exceeds a certain critical value. Therefore, the output is generated by feeding the summed weighted inputs through an activation function. This output is propagated to the next cell to process the input signals with the same procedure to end up with a final output to the network. Modelling these actions is a little bit simple even it if can be used to model very complicated problems. The primary perceptron unit (artificial neuron) is shown in Figure 1a [35]. However, the mathematical representation of signal processing in the perceptron from receiving the inputs to transmitting its output is illustrated in Equations (1) and (2) [36]:

$$u_k = \sum_{j=1}^{m} \omega_{k_j} x_j \quad (1)$$

$$y_k = \varnothing(u_k + b_k) \quad (2)$$

where

u_k: the sum of the weighted inputs;
y_k: the output of the kth perceptron;
ω_{k_j}: the jth synaptic weight sent to the kth preceding neuron;
x_j: the jth input from the preceding neuron;
b_k: the bias of the kth neuron which is a constant;
$\varnothing(.)$: the activation function.

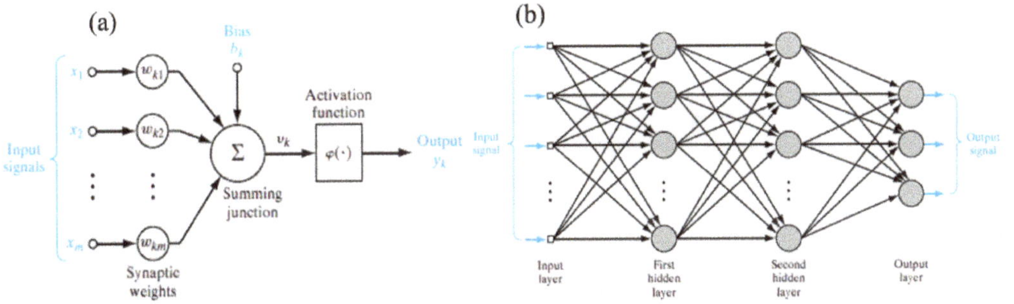

Figure 1. (**a**) Perceptron model, labelled k [35], and (**b**) two hidden-layer MLPANN [37].

Despite the simplicity of the mathematical model of the perceptron, which is the primitive model of the artificial neuron, it can efficiently model highly nonlinear engineering problems. This satisfaction comes from the principle of "Divide and Conquer". This principle deals with a complex problem by dividing it into several minor problems and attacks each problem separately. By adopting this strategy, combining simple processing units in a network can definitely deal with highly complicated problems efficiently. In particular, a multi-layer perceptron artificial neural network (MLPANN), known in the AI field as the ANN, is an example of this strategy. In an MLPANN, every layer is composed of many single units, and the outputs of every layer are transmitted to the next layer and so on until the final output is calculated at the last layer. The first and last layers are assigned to the inputs and outputs, respectively, while the in-between layers are hidden. It is worth mentioning that the number of hidden layers and the number of their units are problem-specific, and they can be determined by a trial-and-error technique to identify the optimal numbers. The feed-forward neural network (FFNN) is one of the ANN architectures, and it is considered in this study. The FFNN layout is shown in Figure 1b [37]. Generally speaking, the most important issue in dealing with the ANN is the determination of the best set of synaptic weights. Usually, this issue is considered an optimization problem. Therefore, the network is trained using a set of input–output experimental data to obtain the required weights' vector that satisfies the minimum mean squared errors (MSE) between the experimental output and the network's output. This learning procedure is called a supervised-learning technique. The most common analytical algorithm that is used to train the FFNN is back-propagation (BP). However, the ANN can also be trained by integrating other recent and modern optimization algorithms, such as GA, PSO, ACO, etc.

2.2. ANFIS Model

The concept of fuzzy sets first appeared in the mid-sixties of the last century by Lotfi Zadeh [38]. He extended and generalized the classical binary set that is composed of only two values {0, 1} to accept elements with multi-degree memberships in the range [0 1]. This new concept needed new types of logic and algebra similar to Boolean logic and Boolean algebra. Therefore, fuzzy logic (FL) and fuzzy mathematics emerged. Nevertheless, this concept is extended to fuzzy numbers, fuzzy inference, fuzzy rules, etc. The work with the fuzzy set is analogical to the real-life way of thinking. In a binary set, the sharp threshold is applied to the membership degree of its instances. With this threshold, any element either belongs (takes a value of 1) or does not belong (takes a value of 0) to the set. However, in fuzzy sets, partial membership is possible. This partial degree of membership is very natural to match our real-life style. In this regard, the representation of an event might take a value of 0.3 or 0.8 depending on its degree of belonging to such a set through predefined membership functions (MFs). Therefore, partial belonging to multiple sets at the same time is valid in the concept of fuzzy sets.

The development of fuzzy systems first emerged by applying fuzzy logic in engineering applications in the mid-seventies of the last century by Ebrahim Mamdani. Professor Mamdani proposed the Mamdani-type fuzzy rule that takes the form as shown in Equation (3) [39]:

$$\text{IF a is A and b is B THEN c is C} \tag{3}$$

where a and b are the input variables; c is the output variable and A, B and C are the MFs of a, b and c, respectively.

The second keystone in the development of fuzzy systems is the Sugeno-type fuzzy rule which was proposed by Takagi, Sugeno and Kang (TSK) and takes the form as in Equation (4) [39]:

$$\text{IF a is A and b is B THEN c} = f(\text{a, b}) \tag{4}$$

where a and b are the input variables; c is the output variable; A and B are the MFs of inputs a and b, respectively, and f(a, b) is a mathematical function of the inputs.

In 1993, Kang proposed the adaptive network-based fuzzy inference system, ANFIS [40]. The ANFIS is a product of the collaboration between the FL and the ANN to mimic the fuzzification, inference engine and defuzzification processes. Consequently, the ANFIS is emulated in a networked structure as the same as the ANN, as shown in Figure 2. The fuzzification and the defuzzification processes are responsible for converting the crisp values of the variables to fuzzy values and vice versa, respectively. The inference engine is responsible for producing the final output by inferring the output of each individual rule through a Min operator and then aggregating these outputs to deliver the final single output through a Max operator. This inference technique is called the Min–Max procedure. The optimal parameters of the MFs and the weights of the connected network are determined through a training algorithm. The ANFIS is usually trained through a supervised learning procedure that uses a hybrid algorithm including the gradient descent (GD) and the least squares estimate (LSE). However, other modern optimization techniques can also be applied to ANFIS training.

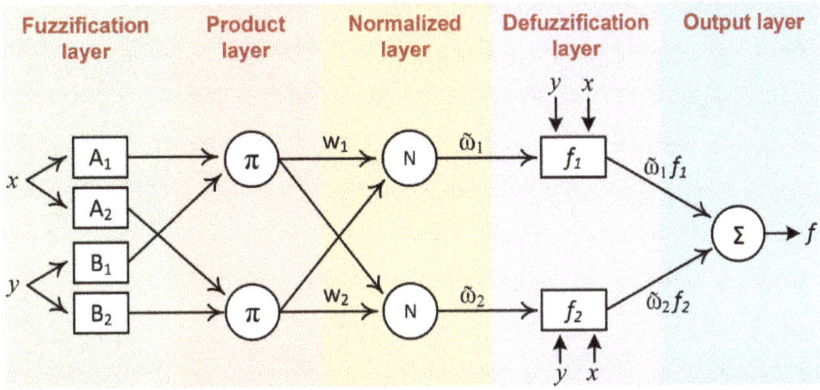

Figure 2. The networked structure of the ANFIS.

2.3. LSTM Model

Long short-term memory (LSTM) is one of the artificial intelligence and deep learning tools that proved its efficiency in prediction models. It was first proposed by Hochreiter and Schmidhuber in 1997 [41]. It can be considered a higher version of the classical recurrent neural network (RNN). However, LSTM overcomes one of the main drawbacks of using the RNN, which is known as the vanishing gradient problem [42]. The RNN is a variant of neural networks that takes the past value of the output to be truncated with the present inputs vector to produce the future value of the output. It suffers from the premature convergence problem when the optimizer gets stuck in the local optima during the training process. This occurs when the gradient is getting smaller and smaller and hence produces an unchanged weight vector. LSTM sorted out this problem by taking into consideration the significant data samples even if it has been seen for a long time. Therefore, the working mechanism of the LSTM network is composed of three gates: the forget gate, the input gate and the output gate. Despite the first gate being called the forget gate, it is actually responsible for memorising significant data from the past. The structure of the LSTM network (cell) with its operational gates is shown in Figure 3.

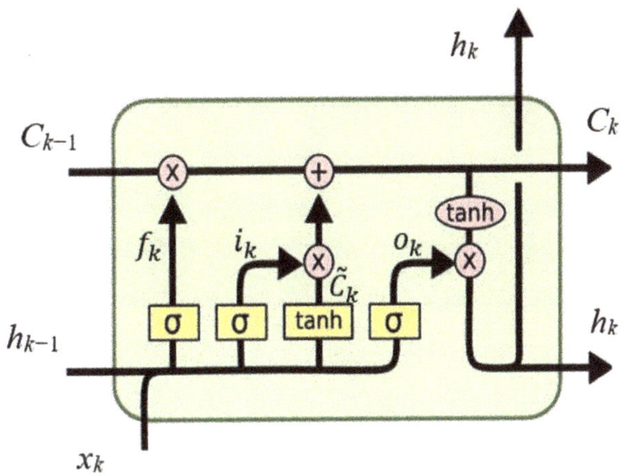

Figure 3. Structure of the LSTM network with its operational gates.

Equations (5)–(10) describe the working operations of the LSTM cell as follows [43]:

$$f_k = \sigma\left(x_k U^f + h_{k-1} W^f\right) \tag{5}$$

$$i_k = \sigma\left(x_k U^i + h_{k-1} W^i\right) \tag{6}$$

$$o_k = \sigma(x_k U^o + h_{k-1} W^o) \tag{7}$$

$$\tilde{C}_k = \tanh(x_k U^g + h_{k-1} W^g) \tag{8}$$

$$C_k = f_k C_{k-1} + i_k \tilde{C}_k \tag{9}$$

$$h_k = \tanh(C_k) * o_k \tag{10}$$

where f_k, i_k and o_k are the responses of the forget, input and output gates, respectively; k is a time index; x_k and y_k are the network's input and output vectors, respectively; U^n and W^n are the weight vectors associated with the input and output vectors of the gate n, respectively; C_k denotes the network's state; σ is the sigma function that maps its inputs to the range [0 1], as in Equation (11) and tanh is the hyperbolic tangent function that maps its inputs to the range [−1 1], as in Equation (12).

$$\sigma(x) = \frac{1}{1 + e^{-x}} \tag{11}$$

$$tanh(x) = \frac{e^x - e^{-x}}{e^x + e^{-x}} \tag{12}$$

3. Data Preparation

In this study, the data were downloaded from the "Our World in Data" website [34]. This data repository recorded many data with different categories for every country in the world. However, this study is only interested in the data for the CO_2 emissions in Saudi Arabia. The available data are in the range from 1936 to 2020, but the two years 1951 and 1952 are missing. Therefore, this study considered only the data in the range from 1953 to 2020, which ends up with 67 data samples. Figure 4 shows the plot of the collected CO_2 emissions in Saudi Arabia within the considered yearly range. However, without effective monitoring and controlling of the tremendous increase, it will negatively contribute to the global warming problem not only in the Middle East region but also all over the world. It is clear from the figure that since 1966, the CO_2 emissions in Saudi Arabia were increasing rapidly, and this matches with the third industrial revolution which started in 1950 by emerging the digital world, internet, computer programming and computer networks, etc.

This study addressed the selection of the data samples randomly for training and testing. In other words, to select random samples from the entire range for the training phase, the remaining samples were reserved for the testing phase. The ratio between the training and testing samples was set to 70:30 for all models. On the other hand, this study investigated the best input combination that produced the best performance. Table 2 illustrates the six groups of the input combination that have been examined in this study to predict the next value of the CO_2 emission (CO_2(t + 1)) in Saudi Arabia, where t is an index to the year.

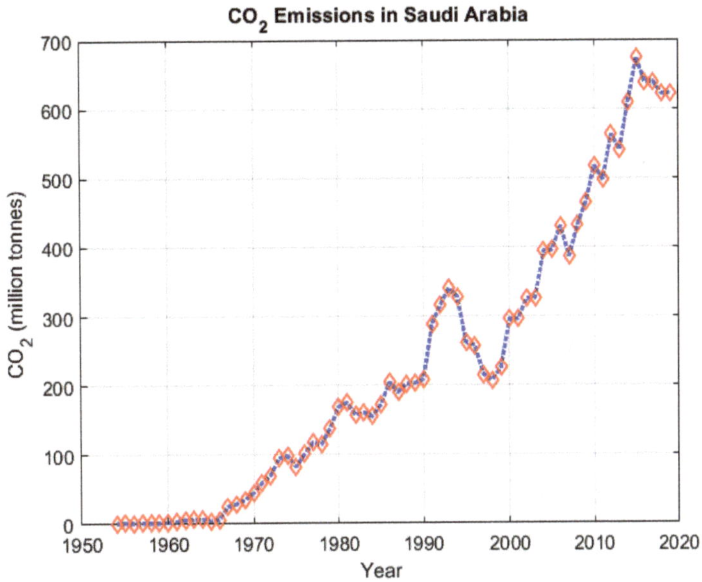

Figure 4. The plot of yearly CO_2 emissions in Saudi Arabia through 1953–2020.

Table 2. The description of input combinations examined in this study.

Gr #	Input Combination	Description
1	Present	$CO_2(t)$
2	Present and change	$CO_2(t) + \Delta CO_2(t)$
3	Present and two past values	$CO_2(t) + CO_2(t-1) + CO_2(t-2)$
4	Year and present	$t + CO_2(t)$
5	Year and present and change	$t + CO_2(t) + \Delta CO_2(t)$
6	Year and present and two past values	$t + CO_2(t) + CO_2(t-1) + CO_2(t-2)$

The data are now ready for the modelling procedure. Therefore, in the following section, the methodologies of building the considered FFNN, ANFIS and LSTM models to forecast CO_2 emissions in Saudi Arabia are described.

4. Results and Discussion

The available 67 data points were reformulated to be in an input–output data matrix form. For predicting the next sample from the present sample, the data matrix was prepared to have 66 data records. These records were divided into 46 samples for training, and the remaining 20 samples were held for testing with the optimal ratio of 70:30, respectively. The code was built using MATLAB R2022 in an intel CORE i7, 8 MB RAM computer, and the findings of the current work are presented in the following subsections. The obtained results from the addressed AI models will be presented in the following subsections.

4.1. FFNN

The study of predicting CO_2 in Saudi Arabia included the feed-forward neural network (FFNN) as one of the artificial intelligence tools that proved its efficiency in many engineering applications [44]. Despite the FFNN's success in pattern recognition applications in the field of data classification, it also produces a completive result in case of data regression problems. Therefore, the FFNN was nominated in this study as the first AI tool. Before starting the modelling phase, the input–output data were normalized to be in the range [0 1] by dividing the whole data set by the maximum value of the input and output vectors. This normalization procedure is beneficial for the learning phase because it unifies the inputs' ranges to the learning algorithm while looking for the optimal weights' vector. For a fair comparison, the FFNN model was trained with the same optimal training-to-testing ratio of 70:30. The structure of the network included n-input and single output architecture with one hidden layer that contained ten perceptrons (units). The number of inputs, n, was dependent on the input combination (number of inputs in the group) to be studied. The network was trained with the backpropagation (BP) algorithm for 10 epochs until the lowest RMSE value for the testing data was reached. Table 3 presents the resulting RMSE and R^2 as comparison statistical measures for the training and testing phases.

Table 3. The RMSE and the coefficient of determination for the training and testing data of the FFNN model.

		RMSE			R^2		
Gr #	Inputs	Training	Testing	Average	Training	Testing	Average
1	**Present**	27.033	19.404	**23.2185**	0.98394	0.98956	**0.98675**
2	Present & Change	28.798	18.855	23.8265	0.97928	0.99208	0.98568
3	Two past values & Present	38.331	18.303	28.317	0.96089	0.99524	0.978065
4	**Year & Present**	25.015	14.545	**19.78**	0.98494	0.99703	**0.990985**
5	Year & Present & Change	31.858	16.869	24.3635	0.97764	0.97581	0.976725
6	**Year & two past values & Present**	22.947	19.481	**21.214**	0.98574	0.99289	**0.989315**

The table illustrates that the input combinations in groups (Gr. #1, Gr. #4 and Gr. #6) are the best three groups that produced the minimum average RMSE and the maximum R^2 values. Figure 5 presents the plots of the resulting outputs of the FFNN.

The prediction accuracy plot shown in Figure 5b demonstrates the good modelling phase where the model's outputs are very close to the measured data samples for both training and testing subsets. These prediction accuracies are reinforced by the regression curve shown in Figure 5c, which shows that the FFNN has been trained correctly, and its output has more than a 99% correlation value with the measurements. Figure 5d illustrates also the convergence curves during the training phase, and it shows that the network is well-trained after 10 epochs. To show the entire performance of the FFNN model, the model predictions of the whole data set were plotted versus the experimental data points, as shown in Figure 6. The curves in Figure 6 show that the predicted output of the FFNN is very close to the measured data and tracks it well, which indicates the success of the training phase.

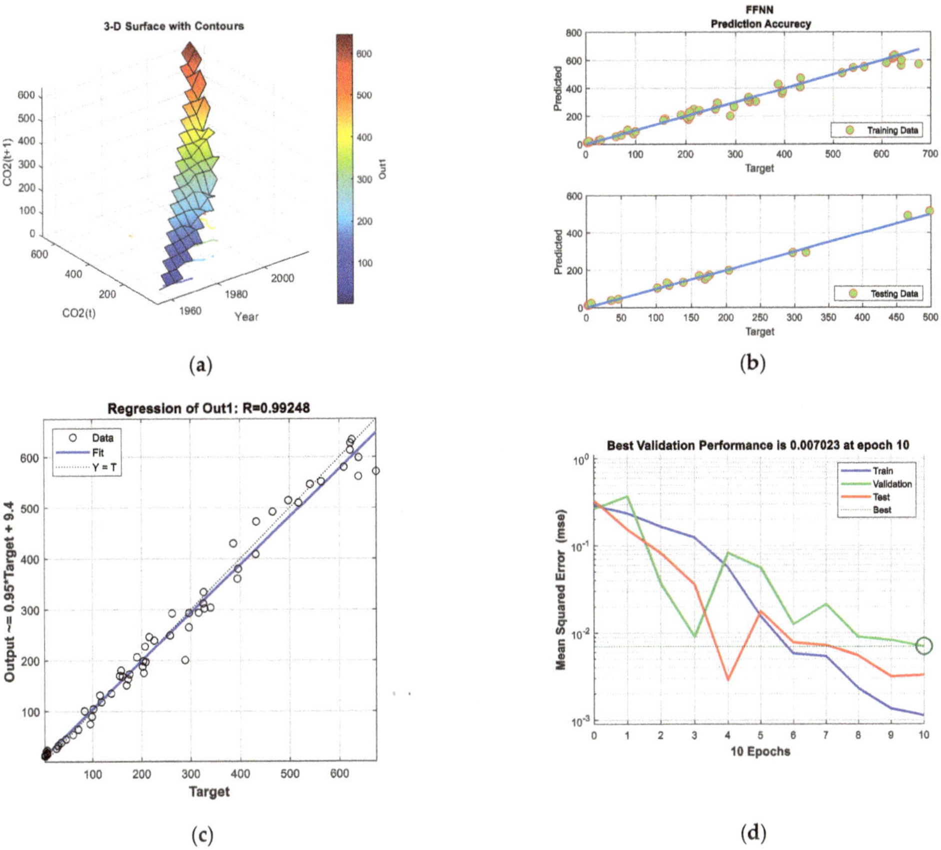

Figure 5. The modelling results of the FFNN: (**a**) 3D surface, (**b**) training and testing prediction accuracies, (**c**) regression curve and (**d**) MSE of performance curves.

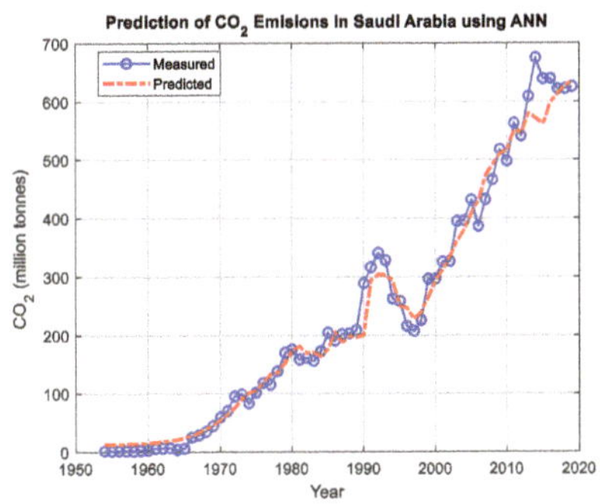

Figure 6. The predictions of CO_2 emissions in Saudi Arabia using the FFNN model from 1954–2020.

4.2. ANFIS

The ANFIS model parameters were assigned as the Gaussian-shape, the subtractive clustering (SC) and the weighted average (Wavg) for the fuzzification, rules generation and defuzzification processes, respectively. For obtaining a smoother prediction curve, the

Gaussian-shape is a very suitable MF, as it gives a fine transition from one predicted point to the next. On the contrary, the other triangular or trapezoidal shapes produce jumps in the predictions. On the other hand, the SC technique produced the optimal as well as the smallest number of rules that can effectively handle the trend of the dataset. The ANFIS structure used in this work is the Sugeno-type fuzzy model. In this type, each rule considers the consequence (output) as a linear function of the antecedent (inputs). Accordingly, the defuzzification method aggregates the outputs by obtaining the weighted average of each rule to end up with a final single-valued output. To build a robust model, the process of continuous training was conducted for the ANFIS model until a stopping criterion was met. In this study, the training process stopped when the testing data's root mean squared error (RMSE) reached the minimum value. The statistical markers were calculated in both phases to assess the performance of the resulting model in each phase. Table 4 illustrates the values of the root mean squared error (RMSE), coefficient of determination (R^2) and number of generated fuzzy rules (#Rules). The first metric (RMSE) measures the accuracy of the model's predictions, while the latter (R^2) measures the tracking capability. The model is said to be better when its predictions give the lowest RMSE and the biggest R^2 values for both groups of datasets (training and testing). R^2 is defined as the squared of the correlation coefficient, which implies that its value is in the range [0 1] where 0 and 1 indicate no correlation and full correlation between the predicted and actual signals, respectively. By referring to Table 4 and according to the list of the lowest three average values and the list of the highest three coefficients of determination, it can be noticed that there are three input combinations shared between the two lists. The first combination is the year and the present value (Gr. #4), the second combination is the year and present value and its rate of change (Gr. #5), while the third combination is the year with the present value and the two past values (Gr. #6).

Table 4. The RMSE and the coefficient of determination for the training and testing data of the fuzzy model.

Gr #	Inputs	RMSE			R^2			# Rules
		Training	Testing	Average	Training	Testing	Average	
1	Present	30.1601	13.1190	21.63955	0.9768	0.9962	0.9865	3
2	Present and change	30.6822	14.9410	22.8116	0.9773	0.9934	0.98535	4
3	Present and two past values	28.1576	14.9611	21.55935	0.9787	0.9955	0.9871	4
4	**Year and present**	28.0677	13.7224	**20.89505**	0.9821	0.9954	**0.98875**	3
5	**Year and present and change**	23.4659	16.7586	**20.11225**	0.9868	0.9939	**0.99035**	5
6	**Year and present and two past values**	21.3019	15.4643	**18.3831**	0.9894	0.9954	**0.9924**	4

In the randomly selected sets of training and testing samples, the fuzzy modelling succeeded in obtaining the better model. This is because the subtractive clustering algorithm used the entire range of the input samples. On the contrary, when using sequential data sets for training and testing, the fuzzy model builds its rules based on a part of the entire range, and hence, it is difficult for it to predict data far from this trained range. It can be noticed from Table 4 that in fuzzy modelling, adding the time index to a group of inputs keeps the temporal feature to the data and hence enhances the values of average prediction and the R^2. This is clear when comparing the average and R^2 values obtained from (Gr. #1, Gr. #4) and (Gr. #2, Gr. #5) with those obtained from (Gr. #3, Gr. #6). The fuzzy model structure and the membership functions of the input combination Gr. #4 are shown in Figure 7a,b, respectively. The training and testing predictions of the ANFIS model are plotted against the first and second inputs and are shown in Figure 8a,b, respectively. However, the training and testing prediction errors plotted against the first and second inputs are shown in Figure 8c,d, respectively. The model predictions are noticed to be well-matched with the measured samples for both training and testing subsets. This is supported by the prediction accuracy curve plotted in Figure 9a. The 3D surface shown in Figure 9b is also generated to illustrate the dependency and behaviour of the output on the variation of the inputs. For more inspection of the ANFIS model predictions, the predictions in both training and testing subsets were plotted in 3D view, as shown in Figure 10a,b, respectively.

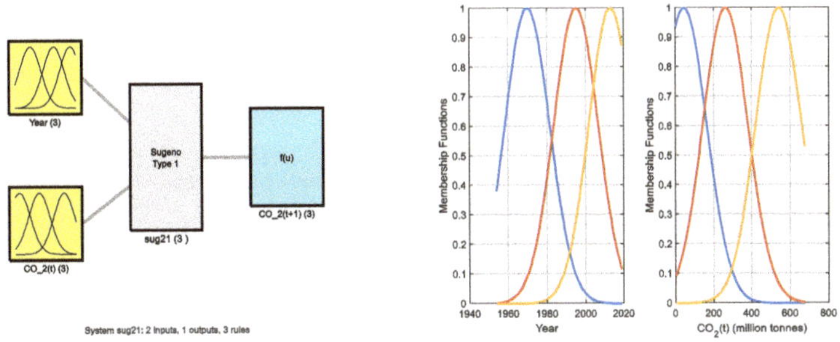

(**a**) (**b**)

Figure 7. (**a**) The structure of the fuzzy model and (**b**) the membership functions of the input combination in the case of Gr. #4.

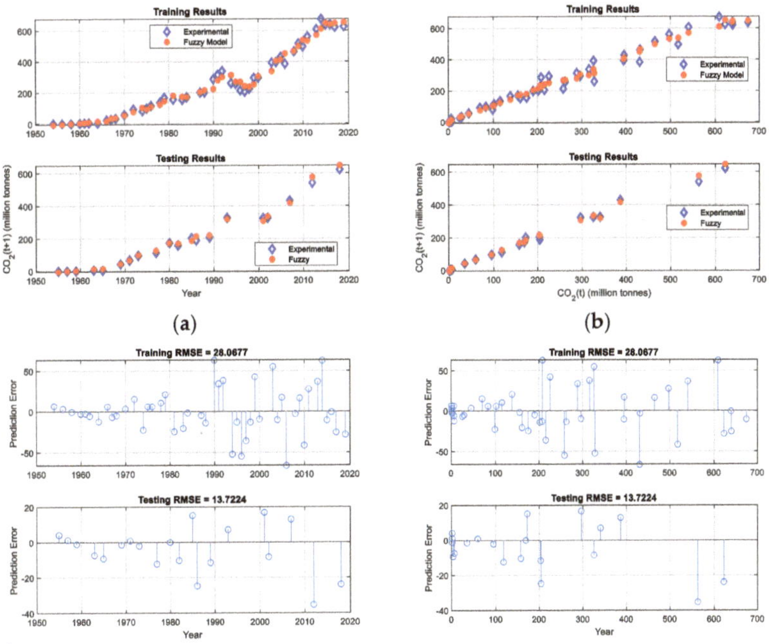

(**a**) (**b**)

(**c**) (**d**)

Figure 8. The plots of the fuzzy model performance vs. the experimental data in case of Gr. #4 (**a**) training and testing performances against the first input, (**b**) training and testing performances against the second input, (**c**) training and testing prediction errors against the first input, (**d**) training and testing prediction errors against the second input.

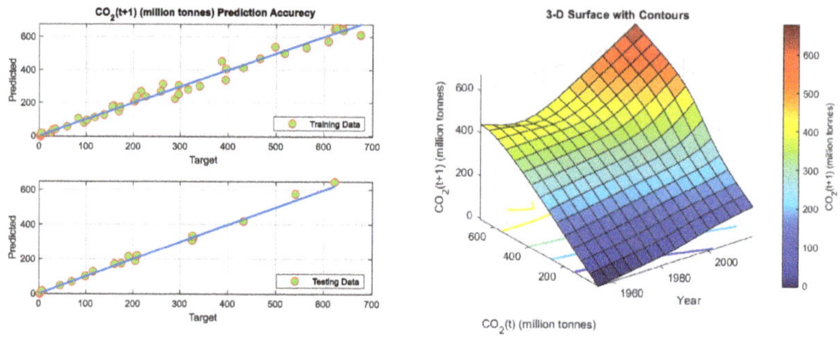

(**a**) (**b**)

Figure 9. (**a**) The fuzzy model's prediction accuracy of training and testing data. (**b**) The fuzzy model 3D surface.

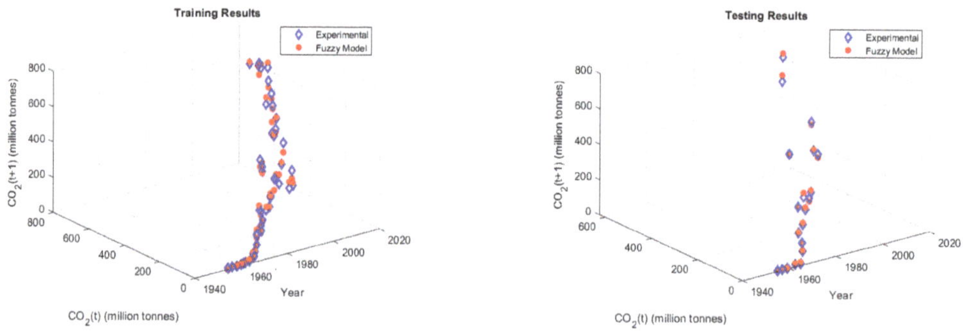

(**a**) (**b**)

Figure 10. The fuzzy model's predictions vs. experimental measurements in 3D view: (**a**) training and (**b**) testing.

To show the entire performance of the ANFIS model, the model predictions of the whole data set were plotted versus the experimental data points, as shown in Figure 11.

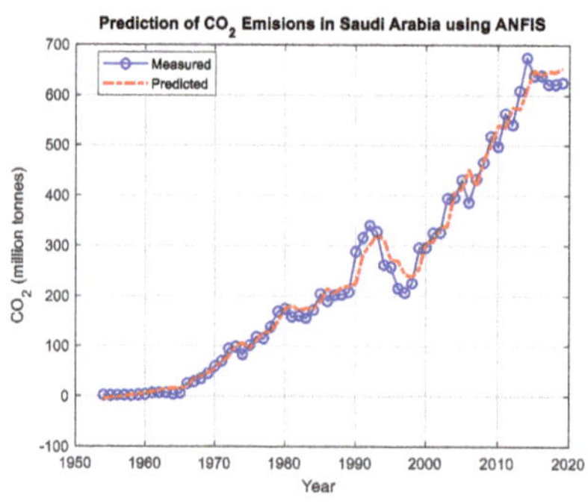

Figure 11. The predictions of CO_2 emissions in Saudi Arabia using the ANFIS model from 1954–2020.

The curves in Figure 11 show that the predicted output of the ANFIS is very close to the measured data and tracks it efficiently, which indicates a successful training phase.

4.3. LSTM

Before the training procedure starts, the data is standardised with its mean and standard deviation as follows:

$$y(t) = \frac{x(t) - \mu}{\sigma} \quad (13)$$

where $x(t)$ and $y(t)$ are the data sample and its standardized value at time t, respectively. μ and σ are the mean and standard deviation values of the training dataset.

Table 5 introduces the resulting RMSE and the R^2 values obtained through the training and testing phases for the LSTM model. By referring to Table 5 and according to the list of the lowest three average values and the list of the highest three coefficients of determination, the three best groups are Gr. #2, Gr. #4 and Gr. #6.

Table 5. The RMSE and the coefficient of determination for the training and testing data of the LSTM model.

		RMSE			R^2		
Gr #	Inputs	Training	Testing	Average	Training	Testing	Average
1	Present	22.5454	25.1811	23.86325	0.9888	0.9856	0.9872
2	**Present and change**	15.376	21.7467	**18.56135**	0.9948	0.9911	**0.99295**
3	Present and two past values	18.4012	22.5109	20.45605	0.992	0.9914	**0.9917**
4	**Year and present**	17.2924	13.5535	**15.42295**	0.9919	0.9971	**0.9945**
5	Year and present and change	14.712	19.6651	**17.18855**	0.9957	0.9688	0.98225
6	Year and present and two past values	18.6523	23.0988	20.87555	0.9922	0.9859	0.98905

Figure 12 illustrates the plots of the predicted samples using LSTM against the observed (real) data for both training and testing samples of Gr. #4. Figure 13 presents the errors between the LSTM predictions and the real data of the randomly selected subsets for the same input group. The lowest training and testing RMSE values resulting from the modelling phase were found to be 17.2924 and 13.5535, respectively, which occurred at Gr. #4.

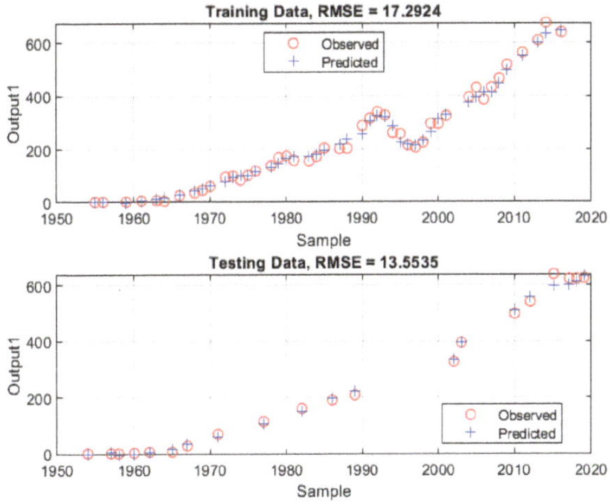

Figure 12. The plots of the LSTM predictions against the real data for training and testing samples when the inputs are the year and the present value.

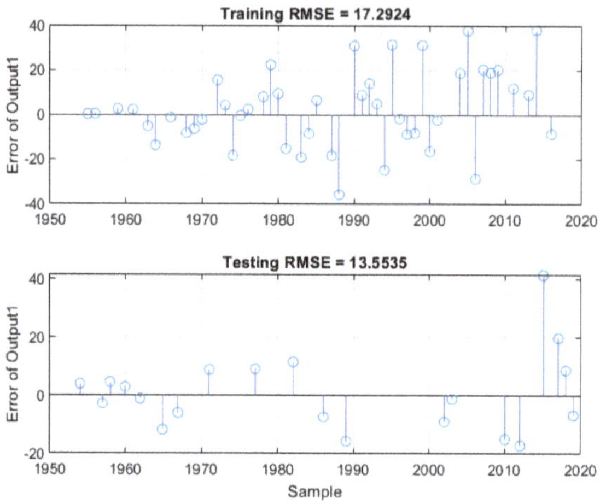

Figure 13. The errors between the LSTM predictions and the real data in the case of random selection strategy when the inputs are the year and the present value.

To show the entire performance of the LSTM model, the model predictions of the whole data set were plotted versus the experimental data points, as shown in Figure 14. The curves in Figure 14 show that the predicted output of the LSTM model is very close to the measured data and tracks it well, which implies that the training phase was successful.

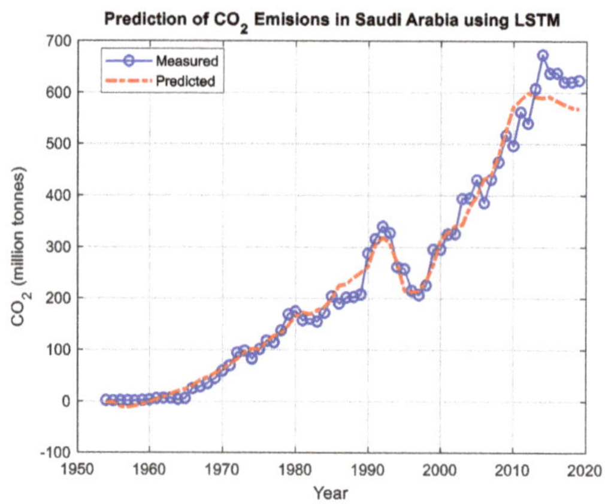

Figure 14. The predictions of CO_2 emissions in Saudi Arabia using the LSTM model from 1954–2020.

Referring to Tables 3–5, it can be concluded that input combination #4 is common in the best results list for the three LSTM, ANFIS and FFNN models. Therefore, the investigation study will be focused only on Gr. #4. Figure 15 shows the predictions of every individual model of the three considered AI models and their forecasting curves until the year 2030. Based on the FFNN forecasting curve, the CO_2 emissions will remain constant with a slight change for the next ten years. However, the ANFIS forecasting model predicts that the emissions will increase for the next ten years but at a smaller rate than in the previous period. On the other hand, the LSTM expects that the curve will decrease for a while and then start to increase again.

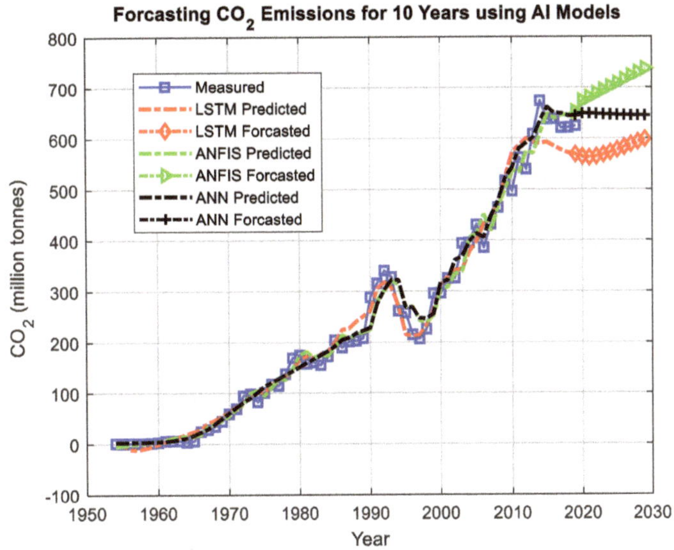

Figure 15. The forecasting curves of CO_2 emissions in Saudi Arabia using FFNN, ANFIS and LSTM.

For Gr. #4, the percentage of the mean absolute error (MAPE) to the average of the testing data for the FFNN, ANFIS and LSTM models were calculated, and their values were found to be 2.939%, 2.82% and 10.6881%, respectively.

5. Ensembled Model

Based on the "no free lunch" principle, a single model cannot have the power of two united models. In this study, the overall decision was considered with the ensemble of the three AI models' responses to benefit from the collaboration of these models. This ensemble is represented by the average value of their outputs to end up with the final single decision, as shown in Figure 16. For the entire data set, the RMSE and the R^2 values were calculated for each individual model as well as their averages. Table 6 presents the statistical markers of the RMSE and the R^2 for the three considered AI models and their average values. It can be noticed that the RMSE of the ensemble model is the lowest among any individual model. Furthermore, the resulting output of the ensembled model is more correlated with the measured data than the output of any individual model based on the highest R^2 value. The experimental data and the prediction values of the three models and their ensemble from 1954–2030 are tabulated in Table A1 in Appendix A.

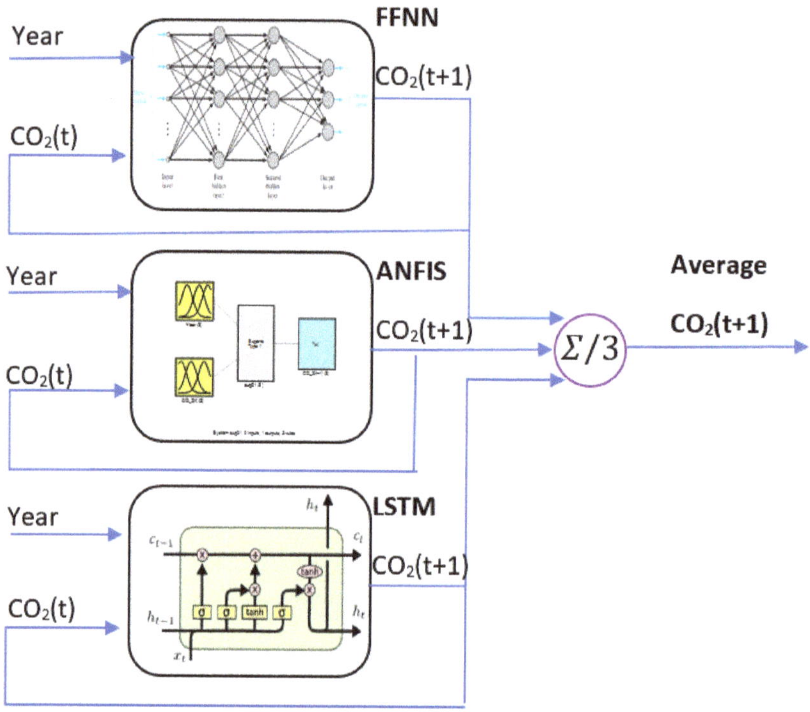

Figure 16. The diagram of the forecasting ensemble of the three AI models.

Table 6. The RMSE and R^2 of the considered models' responses and their average.

RMSE				R^2			
FFNN	ANFIS	LSTM	AVG	FFNN	ANFIS	LSTM	AVG
26.275	24.62	32.834	**22.298**	0.98502	0.98544	0.97458	**0.98932**

Figure 17 shows the predictions of the three AI models and their forecasting curves until the year 2030. The figure shows the average value of the ensembled model. From the average value curve, it can be noticed that with the current strategy, the CO_2 emissions are expected to be increased but at a slight rate. From the available measurements, the increment in CO_2 emissions was 9.4976 million tonnes/year in the period from 1953 to 2020. However, according to results from the AI models, it is expected that the increment will be only 6.1707 million tonnes/year in the period from 2020 to 2030. This implies that the increment rate decreases by 35.03%. This supports the strategy that the Saudi Arabian authorities adopted in order to decrease CO_2 emissions to reach net zero by 2060. Eventually, the findings of this study and the predictions of the artificial intelligence models suggest that keeping the rate decrement within 35% every 10 years would help in reaching net zero by 2060 or before.

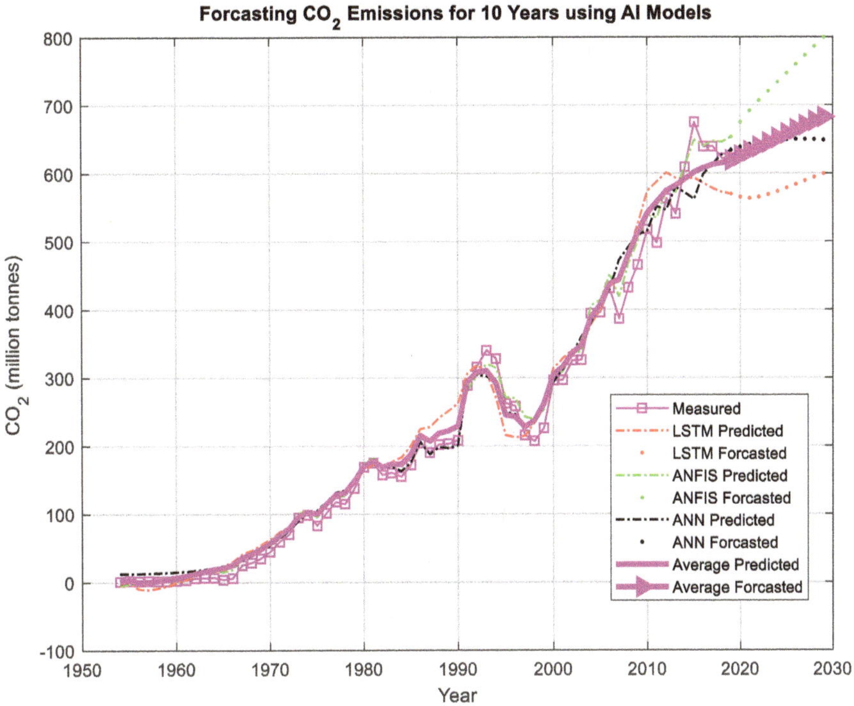

Figure 17. Predictions of the AI models and their forecasting responses up to the year 2030.

To obtain a robust model, the possibility of the occurrence of overfitting has to be avoided. Thus, two strategies have been adopted in our research. The first one is to use an appropriate ratio between the training and testing samples to perform the training phase properly. In system modelling and machine learning fields, the ratio of 80:20 is commonly used [45]. In this regard, Joseph proposed a criterion for calculating the optimal training/testing splitting ratio, R_s, based on the number of parameters, n. Accordingly, this ratio is calculated as shown in Equation (14) [33]:

$$R_s = \sqrt{n} : 1 \quad (14)$$

Therefore, by following this criterion and with the number of input parameters $n = 2$, the optimal training/testing ratio should be 70/30, which was adopted in our study. The second strategy is to assess the model predictions during the testing phase via the values of the statistical markers of RMSE and R^2. Small and big RMSE and R^2 values, respectively, indicate a successful training phase. Subsequently, the obtained model is reliable and can produce trustable predictions.

The models' predictions have been validated with an unseen record of CO_2 emissions in Saudi Arabia. This measurement has recently been added to the database repository to express emissions in the year 2021, which has a value of 672.38 tonnes. The measured, predicted and average values of this validation record when applying the FFNN, ANFIS and LSTM models are illustrated in Table 7. Moreover, the prediction error percentages obtained by these models, including their ensemble, are introduced in Table 8. From Table 8, it can be noticed that the ANFIS model produced the best prediction accuracy with the lowest error value of 0.3704%. On the other hand, the LSTM model had the highest prediction error value that reached 15.9288%. However, the FFNN presented a moderate error value of 638.4638%. Consequently, the percentage error value of the model's ensemble was only 6.8675%, which implies that a forecasting accuracy of 93.1325% has been gained. These obtained high-accuracy predictions proved the reliability of the adopted forecasting strategy of using artificial intelligence tools as a forecasting ensemble. By comparing our findings with the other previous works, it is noticed that the results obtained by our suggested ensemble are better than those obtained by a recent study of forecasting CO_2 emissions in the USA. This study produced a forecasting accuracy and prediction error of 92.42% and 7.58%, respectively [25]. However, our proposed methodology produced accuracy and error values of 93.1325% and 6.8675%, respectively. Furthermore, our results are compared with the one obtained in [14]. The LSTM model in [14] produced prediction RMSE and R^2 values of 60.635 and 0.990, respectively, while our LSTM model delivers values of 15.42295 and 0.9945, respectively. This indicates that our suggested strategy is outstanding and promising. Usually, a forecasting model is classified according to its resulting percentage error as inaccurate, reasonable, good or highly accurate when the percentage error is greater than 50%, 20–50%, 10–20% and less than 10%, respectively [46]. Therefore, the proposed FFNN and ANFIS can be classified as highly accurate models, while the LSTM model can be considered good. In addition, as our proposed ensemble produced a validating error of 6.8675%, it can certainly be classified as a highly accurate forecasting model.

Table 7. The measured, predicted and average values of the validating data obtained by the considered AI models.

Original Data (Measured)			Predicted CO_2(t + 1)			
Inputs		Output	AI Models' Predictions			
t(Year)	CO_2(t)	CO_2(t + 1)	FFNN	ANFIS	LSTM	Ensemble
2020	625.508	672.38	**638.4638**	**674.8708**	**565.2777**	**626.204**

Table 8. The prediction errors of the validation record obtained by the considered AI models.

Measured	Prediction Error (%)			
CO_2(t + 1)	FFNN	ANFIS	LSTM	Ensemble
672.38	**5.0442**	**0.3704**	**15.9288**	**6.8675**

In conclusion, the main contribution of this work can be summarized in the following points, which makes it novel:

- An ensemble of three AI models, namely, FFNN, ANFIS and LSTM, has been built to forecast CO_2 emissions in Saudi Arabia during the period from 2020 to 2030.
- The optimal train/test ratio of 70:30 was used throughout the training phase [33].
- The ensemble produced a forecasting error percentage of 6.8675% for the validating sample.
- The forecasting models are classified as highly accurate based on the formula in [46].

6. Conclusions

Tracking carbon dioxide (CO_2) emissions obtained major interest all over the world in order to effectively monitor global warming. Therefore, this work presents a study of using artificial intelligence (AI) tools to forecast CO_2 emissions in Saudi Arabia, the biggest fossil fuels exporter in the world, through the period from the year 2020 to 2030. Three powerful AI prediction models have been applied in this study. These models include the feed-forward neural network (FFNN), the adaptive network-based fuzzy inference system (ANFIS) and long short-term memory (LSTM). The annual measurements of CO_2 emissions cover the period from 1936 to 2020 but with some missing values. The dataset was filtered to cover only the period from 1954 to 2020. For all models, the available records were divided randomly with the optimal ratio of 70:30 into two groups for training and testing, respectively. Every model was trained until the lowest root mean squared error (RMSE) of the testing data was obtained. The study also investigated the best input combination that produces the best performance. The predicting models' performances have been assessed based on the smallest RMSE and the biggest coefficient of determination (R^2) values. The group containing the past value and the temporal annual index year was found to be the best combination. The FFNN, ANFIS and LSTM produced average RMSEs of 19.78, 20.89505 and 15.42295, respectively, while the averages of R^2 were found to be 0.990985, 0.98875 and 0.9945, respectively. The final predicted output was considered as the ensembled (average) value of the outputs resulting from the three AI models. The ensemble was validated with a new measurement of the year 2021, and the calculated percentage error was found to be 6.8675% with an accuracy of 93.1325%, which implies that the model is highly accurate. According to the ensemble forecasting values, the rate of CO_2 emissions is expected to decrease in Saudi Arabia from 9.4976 million tonnes per year in the period 1954:2020 to 6.1707 million tonnes per year in the period 2020–2030. This implies that the increment rate will decrease by 35.03%. Accordingly, this result will definitely encourage and help policymakers to take the right action not only for the near future but also for the far future. The findings of this study suggest that keeping the decrement rate within 35% every 10 years will help in reaching net zero CO_2 emissions in Saudi Arabia by 2060 or even before. In future work, another different ML tool, such as a regression support vector machine (RSVM), can be studied and included in the ensemble to increase the entire model's prediction accuracy. Furthermore, the model's predictability can be enhanced by restraining these models with any new annual measurements that might be added to the database.

Author Contributions: Conceptualization, A.M.N. and H.R.; methodology, A.M.N. and H.R.; software, A.M.N. and H.R.; validation, A.M.N., A.G.O., H.R. and M.A.A.; formal analysis, A.M.N., A.G.O., H.R. and M.A.A.; investigation, A.M.N., A.G.O., H.R. and M.A.A.; resources, A.M.N.; data curation, A.M.N. and H.R.; writing—original draft preparation, A.M.N., A.G.O., H.R. and M.A.A.; writing—review and editing, A.M.N., A.G.O., H.R. and M.A.A.; visualization, A.M.N., A.G.O., H.R. and M.A.A.; supervision, A.M.N.; project administration, A.M.N.; funding acquisition, A.M.N. All authors have read and agreed to the published version of the manuscript.

Funding: The authors extend their appreciation to the Deputyship for Research and Innovation, Ministry of Education in Saudi Arabia for funding this research work through the project number (IF2/PSAU/2022/01/22030).

Conflicts of Interest: The authors declare no conflict of interest.

Nomenclature

Abbreviation	Description
ACO	Ant colony optimization
AI	Artificial intelligent
ANFIS	Adaptive neuro-fuzzy inference system
ANN	Artificial neural networks
ANOVA	Analysis of variance
ARIMA	Autoregressive integrated moving average
BP	Back-propagation
CNN	Convolution neural network
CO_2	Carbon dioxide
DL	Deep learning
FFNN	Feed-forward neural network
GA	Genetic algorithm
GDP	Gross domestic product
GPs	Gaussian processes

LMDI	Logarithmic mean division index
MAPE	Mean absolute percentage error
ML	Machine learning
MLPANN	Multi-layer perceptron artificial neural network
MSE	Mean squared errors
PSO	Particle swarm optimization
R^2	Coefficient of determination
RMSE	Root mean squared errors
RNN	Recurrent neural network
RSVM	Regression support vector machine
SARIMAX	Seasonal autoregressive integrated moving average with exogenous factors
STSM	Structural time-series model
UEDT	Underlying energy demand trend

Appendix A

Table A1. Original and AI models' predictions data for the CO_2 emissions in Saudi Arabia from 1954–2030.

Original Data			Predicted CO_2(t + 1)			
Inputs		Output	AI Models' Predictions			
t(Year)	CO_2(t)	CO_2(t + 1)	FFNN	ANFIS	LSTM	AVERAGE
1954	1.238	2.114	12.50611	−5.08987	−2.01652	1.799911
1955	2.114	1.63	12.679	−2.75545	−1.02799	2.965184
1956	1.63	2.099	12.89792	−1.4773	−9.73667	0.561316
1957	2.099	1.997	13.2161	0.542316	−10.5423	1.072047
1958	1.997	1.854	13.63345	2.119091	−8.34127	2.470423
1959	1.854	2.674	14.19032	3.665127	−5.0274	4.276015
1960	2.674	3.568	14.95435	5.961256	−0.62737	6.762744
1961	3.568	6.25	15.96297	8.316796	4.074516	9.451427
1962	6.25	6.939	17.34324	12.06533	9.549446	12.986
1963	6.939	7.041	19.06351	14.2668	14.49436	15.94156
1964	7.041	4.216	21.25589	16.01483	19.63215	18.96763
1965	4.216	6.407	23.88634	15.48812	24.05496	21.14314
1966	6.407	25.485	27.48476	18.87039	30.33186	25.56233
1967	25.485	29.079	33.15059	35.42178	41.40516	36.65918
1968	29.079	35.271	38.87615	39.9241	46.89222	41.89749
1969	35.271	45.251	45.87024	46.47104	54.6504	48.99723
1970	45.251	59.757	54.38627	56.00684	63.27654	57.88989
1971	59.757	70.282	64.68987	69.13748	73.45562	69.09432
1972	70.282	95.054	75.49803	79.21233	82.22738	78.97924
1973	95.054	98.702	89.86426	100.6304	96.3252	95.60662
1974	98.702	83.261	100.8192	105.461	102.1858	102.822
1975	83.261	101.46	104.6741	95.16006	102.069	100.6344
1976	101.46	118.07	118.4744	111.7087	115.6099	115.2643
1977	118.07	115.025	131.8265	127.1765	127.5863	128.8631

Table A1. *Cont.*

Original Data			Predicted CO_2(t + 1)			
Inputs		Output	AI Models' Predictions			
t(Year)	CO_2(t)	CO_2(t + 1)	FFNN	ANFIS	LSTM	AVERAGE
1978	115.025	138.005	135.3067	126.9572	132.2368	131.5002
1979	138.005	169.24	151.2927	147.8069	147.6745	148.9247
1980	169.24	175.305	172.9232	175.0902	165.833	171.2822
1981	175.305	157.893	180.114	182.1811	172.8314	178.3755
1982	157.893	160.837	168.4436	170.9286	169.5612	169.6445
1983	160.837	155.512	169.7221	175.6413	177.098	174.1538
1984	155.512	172.421	163.3358	173.8981	183.0917	173.4419
1985	172.421	204.604	175.3805	189.1726	200.2302	188.2611
1986	204.604	190.447	206.4798	214.9121	224.8404	215.4108
1987	190.447	202.252	187.7666	206.7261	227.6711	207.3879
1988	202.252	203.424	198.4083	217.1242	241.2667	218.9331
1989	203.424	208.497	196.9009	219.8459	251.2149	222.6539
1990	208.497	288.546	200.3989	225.2428	262.3358	229.3259
1991	288.546	316.175	294.0228	281.6761	305.9605	293.8864
1992	316.175	340.642	303.7727	302.1944	318.6143	308.1938
1993	340.642	328.139	302.1878	320.8028	308.4464	310.479
1994	328.139	262.421	292.5973	314.3748	270.5121	292.4948
1995	262.421	258.275	249.5026	271.1697	215.8058	245.4927
1996	258.275	215.807	246.4183	270.1953	212.6085	243.074
1997	215.807	207.24	227.3038	242.8541	214.6285	228.2622
1998	207.24	225.986	239.6036	238.7858	234.3291	237.5728
1999	225.986	296.364	265.2283	253.5679	265.5935	261.4632
2000	296.364	296.588	293.3954	305.8328	313.5931	304.2738
2001	296.588	325.698	312.3551	308.6746	328.8756	316.6351
2002	325.698	326.517	334.3511	334.7226	341.3091	336.7943
2003	326.517	394.585	360.4692	339.2313	343.7582	347.8196
2004	394.585	395.855	379.8552	405.7337	377.9214	387.8368
2005	395.855	431.292	408.6755	413.6198	399.3748	407.2234
2006	431.292	386.507	429.8014	452.2525	432.9277	438.3272
2007	386.507	432.336	472.5769	419.0285	441.094	444.2331
2008	432.336	465.844	491.8752	468.4847	479.0602	479.8067
2009	465.844	517.716	510.1533	500.8319	526.1445	512.3766
2010	517.716	497.659	514.9955	538.7184	573.7942	542.5027
2011	497.659	563.18	551.6828	535.1195	587.5349	558.1124
2012	563.18	540.805	546.4432	575.9426	600.6204	574.3354
2013	540.805	609.206	580.0512	572.1777	593.1379	581.7889

Table A1. *Cont.*

Original Data			Predicted $CO_2(t+1)$			
Inputs		Output	AI Models' Predictions			
t(Year)	$CO_2(t)$	$CO_2(t+1)$	FFNN	ANFIS	LSTM	AVERAGE
2014	609.206	674.878	571.6647	611.921	591.4464	591.6774
2015	674.878	639.056	561.8672	649.3612	593.2609	601.4964
2016	639.056	639.378	598.6502	639.9262	585.4526	608.0096
2017	639.378	621.953	613.3382	647.1517	578.1209	612.8703
2018	621.953	622.413	627.1504	646.1793	572.3751	615.2349
2019	622.413	625.508	634.0576	653.4732	569.892	619.1409
2020	625.508	672.38	**638.4638**	**674.8708**	**565.2777**	**626.204**
2021	-	-	**642.2474**	**691.7801**	**563.9153**	**632.6476**
2022	-	-	**645.0641**	**706.6142**	**566.0118**	**639.23**
2023	-	-	**647.1015**	**720.4899**	**569.6484**	**645.7466**
2024	-	-	**648.4952**	**733.9226**	**573.9244**	**652.1141**
2025	-	-	**649.3532**	**747.15**	**578.4414**	**658.3149**
2026	-	-	**649.7565**	**760.2819**	**583.0188**	**664.3524**
2027	-	-	**649.762**	**773.3691**	**587.5727**	**670.2346**
2028	-	-	**649.4067**	**786.4351**	**592.0653**	**675.9691**
2029	-	-	**648.712**	**799.491**	**596.4827**	**681.5619**
2030	-	-	**647.6882**	**812.542**	**600.8256**	**687.0186**

References

1. Tollefson, J. Carbon emissions hit new high: Warning from COP27. *Nature* **2022**. [CrossRef] [PubMed]
2. Naddaf, M. 'Actions, not just words': Egypt's climate scientists share COP27 hopes. *Nature* **2022**. [CrossRef] [PubMed]
3. Yao, S.; Yang, L.; Shi, S.; Zhou, Y.; Long, M.; Zhang, W.; Cai, S.; Huang, C.; Liu, T.; Zou, B. A Two-in-One Annealing Enables Dopant Free Block Copolymer Based Organic Solar Cells with over 16% Efficiency. *Chin. J. Chem.* **2023**, *41*, 672–678. [CrossRef]
4. Liu, T.; Zhou, K.; Ma, R.; Zhang, L.; Huang, C.; Luo, Z.; Zhu, H.; Yao, S.; Yang, C.; Zou, B.; et al. Multifunctional all-polymer photovoltaic blend with simultaneously improved efficiency (18.04%), stability and mechanical durability. *Aggregate* **2022**, *e308*. [CrossRef]
5. Manvitha, M.S.; Vani Pujitha, M.; Prasad, N.H.; Yashitha Anju, B. A Predictive Analysis on CO_2 Emissions in Automobiles using Machine Learning Techniques. In Proceedings of the 2023 International Conference on Intelligent Data Communication Technologies and Internet of Things (IDCIoT), Bengaluru, India, 5–7 January 2023; pp. 394–401.
6. Li, Y.; Huang, S.; Miao, L.; Wu, Z. Simulation analysis of carbon peak path in China from a multi-scenario perspective: Evidence from random forest and back propagation neural network models. *Environ. Sci. Pollut. Res. Int.* **2023**, *30*, 46711–46726. [CrossRef]
7. Li, G.; Yang, Z.; Yang, H. A new hybrid short-term carbon emissions prediction model for aviation industry in China. *Alex. Eng. J.* **2023**, *68*, 93–110. [CrossRef]
8. Khajavi, H.; Rastgoo, A. Predicting the carbon dioxide emission caused by road transport using a Random Forest (RF) model combined by Meta-Heuristic Algorithms. *Sustain. Cities Soc.* **2023**, *93*, 104503. [CrossRef]
9. Ma, N.; Shum, W.Y.; Han, T.; Lai, F. Can Machine Learning be Applied to Carbon Emissions Analysis: An Application to the CO_2 Emissions Analysis Using Gaussian Process Regression. *Front. Energy Res.* **2021**, *9*, 756311. [CrossRef]
10. Rahman, S.M.A.; Nassef, A.M.; Al-Dhaifallah, M.; Abdelkareem, M.A.; Rezk, H. The Effect of a New Coating on the Drying Performance of Fruit and Vegetables Products: Experimental Investigation and Artificial Neural Network Modeling. *Foods* **2020**, *9*, 308. [CrossRef]
11. Rezk, H.; Mohammed, R.H.; Rashad, E.; Nassef, A.M. ANFIS-based accurate modeling of silica gel adsorption cooling cycle. *Sustain. Energy Technol. Assess.* **2022**, *50*, 101793. [CrossRef]
12. Deraz, A.A.; Badawy, O.; Elhosseini, M.A.; Mostafa, M.; Ali, H.A.; El-Desouky, A.I. Deep learning based on LSTM model for enhanced visual odometry navigation system. *Ain Shams Eng. J.* **2022**, 102050. [CrossRef]
13. Camastra, F.; Capone, V.; Ciaramella, A.; Riccio, A.; Staiano, A. Prediction of environmental missing data time series by Support Vector Machine Regression and Correlation Dimension estimation. *Environ. Model. Softw.* **2022**, *150*, 105343. [CrossRef]

14. Kumari, S.; Singh, S.K. Machine learning-based time series models for effective CO_2 emission prediction in India. *Environ. Sci. Pollut. Res.* **2022**. [CrossRef]
15. Vlachas, P.R.; Byeon, W.; Wan, Z.Y.; Sapsis, T.P.; Koumoutsakos, P. Data-driven forecasting of high-dimensional chaotic systems with long short-term memory networks. *Proc. R. Soc. A Math. Phys. Eng. Sci.* **2018**, *474*, 20170844. [CrossRef] [PubMed]
16. Amarpuri, L.; Yadav, N.; Kumar, G.; Agrawal, S. Prediction of CO_2 emissions using deep learning hybrid approach: A Case Study in Indian Context. In Proceedings of the 2019 Twelfth International Conference on Contemporary Computing (IC3), Noida, India, 8–10 August 2019; pp. 1–6.
17. Meng, Y.; Noman, H. Predicting CO_2 Emission Footprint Using AI through Machine Learning. *Atmosphere* **2022**, *13*, 1871. [CrossRef]
18. Komeili Birjandi, A.; Fahim Alavi, M.; Salem, M.; Assad, M.E.H.; Prabaharan, N. Modeling carbon dioxide emission of countries in southeast of Asia by applying artificial neural network. *Int. J. Low-Carbon Technol.* **2022**, *17*, 321–326. [CrossRef]
19. Qi, Y.; Peng, W.; Yan, R.; Rao, G.; Hassanien, A.E.I.B. Use of BP Neural Networks to Determine China's Regional CO_2 Emission Quota. *Complexity* **2021**, *2021*, 6659302. [CrossRef]
20. Abdullah, A.M.; Usmani, R.S.A.; Pillai, T.R.; Marjani, M.; Hashem, I.A.T. An Optimized Artificial Neural Network Model using Genetic Algorithm for Prediction of Traffic Emission Concentrations. *Int. J. Adv. Comput. Sci. Appl.* **2021**, *12*, 794–803. [CrossRef]
21. Cansiz, O.F.; Unsalan, K.; Unes, F. Prediction of CO_2 emission in transportation sector by computational intelligence techniques. *Int. J. Glob. Warm.* **2022**, *27*, 271. [CrossRef]
22. Khan, M.Z.; Khan, M.F. Application of ANFIS, ANN and fuzzy time series models to CO_2 emission from the energy sector and global temperature increase. *Int. J. Clim. Chang. Strateg. Manag.* **2019**, *11*, 622–642. [CrossRef]
23. Faruque, M.O.; Rabby, M.A.J.; Hossain, M.A.; Islam, M.R.; Rashid, M.M.U.; Muyeen, S.M. A comparative analysis to forecast carbon dioxide emissions. *Energy Rep.* **2022**, *8*, 8046–8060. [CrossRef]
24. Zhang, X.; Zhang, W. A Hybrid Daily Carbon Emission Prediction Model Combining CEEMD, WD and LSTM. In *Intelligent Computing Methodologies: 18th International Conference, ICIC 2022, Xi'an, China, 7–11 August 2022, Proceedings, Part III*; Springer International Publishing: Cham, Switzerland, 2022; pp. 557–571.
25. Mutascu, M. CO_2 emissions in the USA: New insights based on ANN approach. *Environ. Sci. Pollut. Res. Int.* **2022**, *29*, 68332–68356. [CrossRef]
26. Wang, C.; Li, M.; Yan, J. Forecasting carbon dioxide emissions: Application of a novel two-stage procedure based on machine learning models. *J. Water Clim. Chang.* **2023**, *14*, 477–493. [CrossRef]
27. Hamieh, A.; Rowaihy, F.; Al-Juaied, M.; Abo-Khatwa, A.N.; Afifi, A.M.; Hoteit, H. Quantification and analysis of CO_2 footprint from industrial facilities in Saudi Arabia. *Energy Convers. Manag. X* **2022**, *16*, 100299. [CrossRef]
28. Alam, T.; AlArjani, A. A Comparative Study of CO_2 Emission Forecasting in the Gulf Countries Using Autoregressive Integrated Moving Average, Artificial Neural Network, and Holt-Winters Exponential Smoothing Models. *Adv. Meteorol.* **2021**, *2021*, 8322590. [CrossRef]
29. Alajmi, R.G. Carbon emissions and electricity generation modeling in Saudi Arabia. *Environ. Sci. Pollut. Res.* **2022**, *29*, 23169–23179. [CrossRef]
30. Habadi, M.I.; Tsokos, C.P. Statistical Forecasting Models of Atmospheric Carbon Dioxide and Temperature in the Middle East. *J. Geosci. Environ. Prot.* **2017**, *5*, 11–21. [CrossRef]
31. Althobaiti, Z.F.; Shabri, A. Prediction of CO_2 emissions in Saudi Arabia using Nonlinear Grey Bernoulli Model NGBM (1,1) compared with GM (1,1) model. *J. Phys. Conf. Ser.* **2022**, *2259*, 012011. [CrossRef]
32. Alkhathlan, K.; Javid, M. Carbon emissions and oil consumption in Saudi Arabia. *Renew. Sustain. Energy Rev.* **2015**, *48*, 105–111. [CrossRef]
33. Joseph, V.R. Optimal ratio for data splitting. *Stat. Anal. Data Min. ASA Data Sci. J.* **2022**, *15*, 531–538. [CrossRef]
34. Ritchie, H.; Roser, M.; Rosado, P. CO_2 and Greenhouse Gas Emissions. Available online: https://ourworldindata.org/co2-and-other-greenhouse-gas-emissions (accessed on 24 December 2022).
35. Herculano-Houzel, S. The human brain in numbers: A linearly scaled-up primate brain. *Front. Hum. Neurosci.* **2009**, *3*, 31. [CrossRef] [PubMed]
36. Haykin, S.A. *Neural Networks: A Comprehensive Foundation*; Prentice Hall: Hoboken, NJ, USA, 2008.
37. Haykin, S.S. *Neural Networks and Learning Machines/Simon Haykin*; Prentice Hall: New York, NY, USA, 2009.
38. Zadeh, L.A. Fuzzy sets. *Inf. Control* **1965**, *8*, 338–353. [CrossRef]
39. Nassef, A.M.; Rahman, S.M.A.; Rezk, H.; Said, Z. ANFIS-Based Modelling and Optimal Operating Parameter Determination to Enhance Cocoa Beans Drying-Rate. *IEEE Access* **2020**, *8*, 45964–45973. [CrossRef]
40. Jang, J.S.R. ANFIS: Adaptive-network-based fuzzy inference system. *IEEE Trans. Syst. Man Cybern.* **1993**, 23, 665–685. [CrossRef]
41. Hochreiter, S.; Schmidhuber, J. Long short-term memory. *Neural Comput.* **1997**, *9*, 1735–1780. [CrossRef]
42. Hochreiter, S. The Vanishing Gradient Problem during Learning Recurrent Neural Nets and Problem Solutions. *Int. J. Uncertain. Fuzziness Knowl.-Based Syst.* **1998**, *6*, 107–116. [CrossRef]
43. Luo, J.; Zhang, Z.; Fu, Y.; Rao, F. Time series prediction of COVID-19 transmission in America using LSTM and XGBoost algorithms. *Results Phys.* **2021**, *27*, 104462. [CrossRef]
44. Rahman, S.M.A.; Rezk, H.; Abdelkareem, M.A.; Hoque, M.E.; Mahbub, T.; Shah, S.K.; Nassef, A.M. Predicting Drying Performance of Osmotically Treated Heat Sensitive Products Using Artificial Intelligence. *Comput. Mater. Contin.* **2021**, *67*, 3143–3160. [CrossRef]

45. Gholamy, A.; Kreinovich, V.; Kosheleva, O. *Why 70/30 or 80/20 Relation between Training and Testing Sets: A Pedagogical Explanation*; University of Texas: El Paso, TX, USA, 2018.
46. Lewis, C.D. *Industrial and Business Forecasting Methods: A Practical Guide to Exponential Smoothing and Curve Fitting/Colin D. Lewis*; Butterworth Scientific: London, UK, 1982.

sustainability

Article

A Study of Treatment of Reactive Red 45 Dye by Advanced Oxidation Processes and Toxicity Evaluation Using Bioassays

Muhammad Imran Kanjal [1], Majid Muneer [1,*], Muhammad Asghar Jamal [1], Tanveer Hussain Bokhari [1], Abdul Wahid [2], Shafqat Ullah [3], Abdeltif Amrane [4,*], Amina Hadadi [5], Hichem Tahraoui [6] and Lotfi Mouni [5]

1 Department of Chemistry, Government College University Faisalabad, Faisalabad 38000, Pakistan
2 Department of Environmental Sciences, Bahauddin Zakariya University, Multan 59000, Pakistan
3 Pakistan Meteorological Department, Multan 59000, Pakistan
4 Univ Rennes, Ecole Nationale Supérieure de Chimie de Rennes, CNRS, ISCR—UMR6226, F-35000 Rennes, France
5 Laboratory of Management and Valorization of Natural Resources and Quality Assurance, SNVST Faculty, Akli Mohand Oulhadj University, Bouira 10000, Algeria
6 Laboratory of Biomaterials and Transport Phenomena (LBMTP), University Yahia Fares, Médéa 26000, Algeria
* Correspondence: majid.chemist@yahoo.com (M.M.); abdeltif.amrane@univ-rennes.fr (A.A.)

Abstract: Advanced oxidation processes (AOPs) hold great promise to degrade and detoxify industrial-based effluents. The Reactive Red 45 dye aqueous solutions were treated with AOP using UV and gamma radiation alone and then in the presence of H_2O_2. The dye initial concentration, UV exposure time, and gamma-ray absorbed dose were optimized for maximum degradation. The degradation of dye was 88.85% and 77.7% using UV/H_2O_2 (1 mL/L) at a UV exposure time of 180 min for 50 mg/L and 100 mg/L, respectively. The degradation was noted as 100% and 93.82% as the solutions were subjected to a gamma/H_2O_2 (1 mL/L) absorbed dose of 2 kGy. The chemical oxygen demand was reduced to 77% and 85% by treating the dye samples with UV/H_2O_2 and gamma/H_2O_2, respectively. The removal efficiency (G-value), dose constant (k), $D_{0.50}$, $D_{0.90}$, and $D_{0.99}$ for gamma-irradiated samples were also calculated. The reduction in toxicity for treated samples was monitored by using the *Allium cepa*, Hemolytic, and brine shrimp (*Artemia salina*) tests while the Ames test was performed for mutagenic assessment. The *A. cepa* test showed 39.13%, 36.36%, and 47.82% increases in root length (RL), root count (RC), and mitotic index (MI), respectively, in UV/H_2O_2-treated samples while 48.78%, 48.14%, and 57.14% increases were shown with gamma-ray in conjunction with H_2O_2. The hemolytic test showed 21.25% and 23.21% hemolysis after UV/H_2O_2 and gamma/H_2O_2 treatments, respectively. The brine shrimp (*Artemia salina*) test showed 84.09% and 90.90% decreases in the nauplii death after UV/H_2O_2 and gamma/H_2O_2 treatments, respectively. The mutagenicity of UV/H_2O_2-treated solutions was reduced up to 84.41% and 77.87%, while it was 87.83% and 80.88% using gamma/H_2O_2 using TA98 and TA100 bacterial strains, respectively. The advanced oxidation processes based on UV and gamma radiation in conjunction with H_2O_2 can be applied for the degradation and detoxification of textile waste effluents efficiently.

Keywords: gamma radiation; UV radiation; cytotoxicity; Reactive Red 45 dye; advanced oxidation process

check for updates

Citation: Kanjal, M.I.; Muneer, M.; Jamal, M.A.; Bokhari, T.H.; Wahid, A.; Ullah, S.; Amrane, A.; Hadadi, A.; Tahraoui, H.; Mouni, L. A Study of Treatment of Reactive Red 45 Dye by Advanced Oxidation Processes and Toxicity Evaluation Using Bioassays. *Sustainability* **2023**, *15*, 7256. https://doi.org/10.3390/su15097256

Academic Editors: Silvia Fiore and Kaan Yetilmezsoy

Received: 21 February 2023
Revised: 21 April 2023
Accepted: 25 April 2023
Published: 27 April 2023

1. Introduction

Dyes are chemically stable compounds and are a serious problem when dealing with textile waste effluent. Being highly soluble and environmentally problematic compounds, reactive dyes are the most discussed dyes in the literature [1]. Textile waste effluent reduces water transparency and sunlight penetration which affect aquatic life when mixed with ground and surface waters [2]. The adverse pollution effects of dyes in water systems require treatment of samples before being discharged into any water bodies [3]. Physical methods do not remove the dyes from the water completely and may convert them into a secondary pollutant. The chemical oxidation method using hydrogen peroxide, potassium permanganate, chlorine, ozone, etc., to remove pollutants also has some limitations as it is a time-consuming process [4]. The AOP is an effective method that has gained more attention in decontaminating water-containing dyes [5]. During AOP, a strong oxidizing species ($^{\bullet}OH$) is produced which converts the toxic organic compounds into simple and less toxic compounds via a chain reaction [6]. The AOP is advantageous to other treatment methods as it is reliable, easy to handle, reduces manpower, and has the potential to degrade and detoxify the organic matter without producing a secondary pollutant [7]. Various AOPs based on ozonation, peroxone, UV/O_3, UV/H_2O_2, gamma/H_2O_2, UV/TiO_2, and UV/ZnO have been employed to degrade textile waste effluents [5–8]. UV radiation in the presence of H_2O_2 can decompose the dye molecules effectively according to the following Equations (1) and (2) [9].

$$H_2O_2 + UV \rightarrow 2^{\bullet}OH \quad (1)$$

$$^{\bullet}OH + Dye \rightarrow Further\ Oxidation \quad (2)$$

Gamma radiation along with H_2O_2 is believed to be an efficient technique to decompose organic matter. H_2O_2 is a strong oxidizing agent that promotes the AOP by producing hydroxyl radicals ($^{\bullet}OH$), which not only scavenge the species $^{\bullet}H$ and e_{aq}^{-} produced as a result of radiolysis of water but also enhance the production of $^{\bullet}OH$ (Equations (3)–(5)), which is needed to enhance the degradation rate [10,11].

$$H_2O \xrightarrow{\gamma} {}^{\bullet}OH + {}^{\bullet}H + e_{aq^-} + H_2 + H_2O_2 + H_3O^+ \quad (3)$$

$$H_2O_2 + e_{aq}^{-} \rightarrow {}^{\bullet}OH + OH^- \quad (4)$$

$$H_2O_2 + {}^{\bullet}H \rightarrow {}^{\bullet}OH + H_2O \quad (5)$$

To avoid the generation of toxic compounds, biological safety is an important measure when using radiation to degrade wastewater [12–17]. The bioassays including *A. cepa*, hemolytic, and brine shrimp (*Artemia salina*) tests are used as quick and reliable methods. The mutagenic behavior of the probe compound can effectively be measured by the Ames test [18–20].

This study was designed to treat Reactive Red 45 dye solutions by AOPs using UV and gamma radiation along with H_2O_2. The effect of radiation on pH, COD, and degradation was monitored. The G-value, k, $D_{0.50}$, $D_{0.90}$, and $D_{0.99}$ of gamma-irradiated samples were investigated. The *A. cepa*, hemolytic, and brine shrimp (*Artemia salina*) tests were used for toxicity assessment, whereas the mutagenicity evaluation was carried out using the Ames test.

2. Materials and Methods

2.1. Chemicals

The Reactive Red 45 dye was purchased from the local market and was used without prior purification. The characteristics of the dye are given in Table 1 and its structure is shown in Figure 1.

2.2. Radiation Treatment of Dye Samples

The dye aqueous solutions (50 mg/L and 100 mg/L) were subjected to UV and UV/H_2O_2 for 30 min, 60 min, 90 min, 120 min, 150 min, and 180 min of exposure time at the Department of Chemistry, GCUF Pakistan, and also with gamma alone and in conjunction with H_2O_2 (0.1 mL/L, 0.2 mL/L, 0.3 mL/L, 0.4 mL/L, 0.6 mL/L, 0.8 mL/L, and 1 mL/L) for the gamma absorbed dose of 0.25 kGy, 0.50 kGy, 0.75 kGy, 1 kGy, 1.50 kGy, and 2 kGy using a Cs-137 gamma radiation source at the Nuclear Institute for Agriculture and Biology (NIAB), Faisalabad, Pakistan. The change in absorption at λ_{max} = 512 nm was noted with a UV/Vis Spectrophotometer in the Department of Chemistry, GCUF, Pakistan.

Table 1. Characteristics of Reactive Red 45 dye.

Dye Name	Reactive Red 45
Molecular formula	$Na_3C_{27}H_{19}ClN_7O_{10}S_3$
Molecular weight (g/mol)	802.107
Chemical nature	Anionic red 45
Color index name	Red
λmax (nm)	512
Reactive group	Azo group

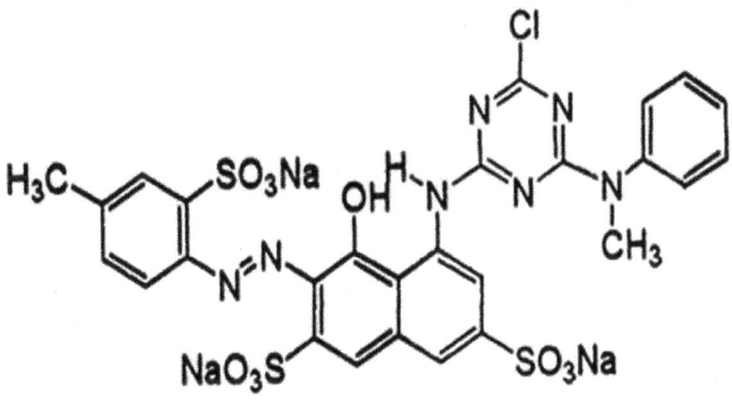

Figure 1. Chemical structure of Reactive Red 45 dye.

2.3. Determination of Chemical Oxygen Demand (COD)

The COD change was investigated before and after the irradiation of the samples [21]. The 1 g $HgSO_4$ and 50 mL samples were mixed in a flask and then 25 mL of 0.042 M $K_2Cr_2O_7$ was added after shaking and cooling. A few drops of ferroin indicator were added and titrated against Mohr's salt after refluxing the mixture for 2 h. The color changed from bluish-green to reddish brown and the endpoint was achieved. To get the readings for the blank, the same process was repeated for 50 mL of water as a the blank. The following expression was used for the COD calculations.

$$\text{COD (mg/L)} = \frac{(\text{A} - \text{B}) * \text{M} * 8000}{\text{mL of sample}} \tag{6}$$

Here,
A = FAS volume (mL) for blank
B = FAS volume (mL) for sample
M = FAS concentration (molarity)

2.4. Determination of G-Value, Dose Constant (k), $D_{0.50}$, $D_{0.90}$, and $D_{0.99}$

The following expression was used to calculate the G-value

$$G = (R)\, N_A / D\, (6.4 \times 10^{20}) \tag{7}$$

Here,

R is the change in concentration, D and N_A are the absorbed dose, and Avogadro's number, respectively. The following expressions were employed to calculate the $D_{0.50}$, $D_{0.90}$, and $D_{0.99}$.

$$D_{0.50} = \frac{\ln 2}{k} \quad (8)$$

$$D_{0.90} = \frac{\ln 10}{k} \quad (9)$$

$$D_{0.99} = \frac{\ln 100}{k} \quad (10)$$

2.5. Toxicity Evaluation

To check the cytotoxicity, 0.80 mg/mL of MnO_2 was added to irradiated solutions to eliminate the effect of residues of hydrogen peroxide before the toxicity evaluation [22]. The solutions after the reaction duration of 1 h were subjected to cytotoxicity evaluation using the reported methods, i.e., *A. cepa* test [23], hemolytic assay [24], brine shrimp (*Artemia salina*) test [25], and Ames test [26].

3. Results and Discussion

3.1. Effect of Hydrogen Peroxide on the Degradation of Reactive Red 45 Dye

In this study, the Reactive Red 45 dye aqueous solutions were treated with H_2O_2, and the effect was observed on degradation, pH, and COD before and after treatment. The degradation was noted as 29.29 to 75.33% for 50 mg/L while it was 23.99 to 63.25% for 100 mg/L using 0.1–1 mL/L of oxidant, respectively (Figure 2A). The pH of the solution before adding H_2O_2 was 8.2 which decreased after treatment to 6.8 for 50 mg/L while it was 7.8 for 100 mg/L (Figure 2B). The COD reduction was from 15 to 63% for 50 mg/L, whereas it was from 8 to 52% for 100 mg/L at 0.1–1 mL of concentration of H_2O_2 (Figure 2C).

The increase in the degradation rate and reduction in pH was noted by enhancing the oxidant concentration. At an optimum concentration of H_2O_2, the benzene ring in the dye was preferentially attacked by the $^{\bullet}OH$. When the H_2O_2 concentration exceeded the optimum concentration, probably due to competition between H_2O_2 and the substrate, the H_2O_2 started to scavenge the $^{\bullet}OH$-produced perhydroxyl radical ($^{\bullet}O_2H$), which has a lower oxidation capability than $^{\bullet}OH$ [13]. Our results are consistent with those of Costa et al. [27] whose study on the degradation of textile dyes by H_2O_2 (30%), solar energy, and UV radiation showed 93% degradation. No significant reduction in COD was observed by adding hydrogen peroxide; this indicates that a higher amount of pollutant was present in the water which cannot be reduced by using an oxidant alone. This study suggested that the COD removal was linearly dependent upon the concentration of the oxidant [28].

3.2. Effect of Radiation on the Degradation

The dye aqueous solutions (50 mg/L and 100 mg/L) were prepared and treated with UV radiation exposure times of 30 min, 60 min, 90 min, 120 min, 150 min, and 180 min while a gamma radiation absorbed dose of 0.25–2 kGy and degradation efficiency were investigated. The degradation was 33.13%, 44.24%, 48.21%, 56.80%, 64.10%, and 76.02% for 50 mg/L while 27.89%, 40.13%, 45.79%, 51.27%, 57.94%, and 55.78% was observed for 100 mg/L of solution (Figure 3A). Degradation amounts of 41.20%, 59.19%, 67.79%, 70.64%, 80.40%, and 85.54% for 50 mg/L and 31.73%, 52.22%, 60.59%, 62.7%1, 69.96%, and 77.70% for 100 mg/L were obtained (Figure 3B) using UV/H_2O_2. It was revealed from the results that degradation increased as the exposure time of UV radiation enhanced, and a maximum degradation (85.54%) was obtained at 180 min. After that, the intermediate species were produced and a decrease in the degradation was observed due to the competition between intermediates.

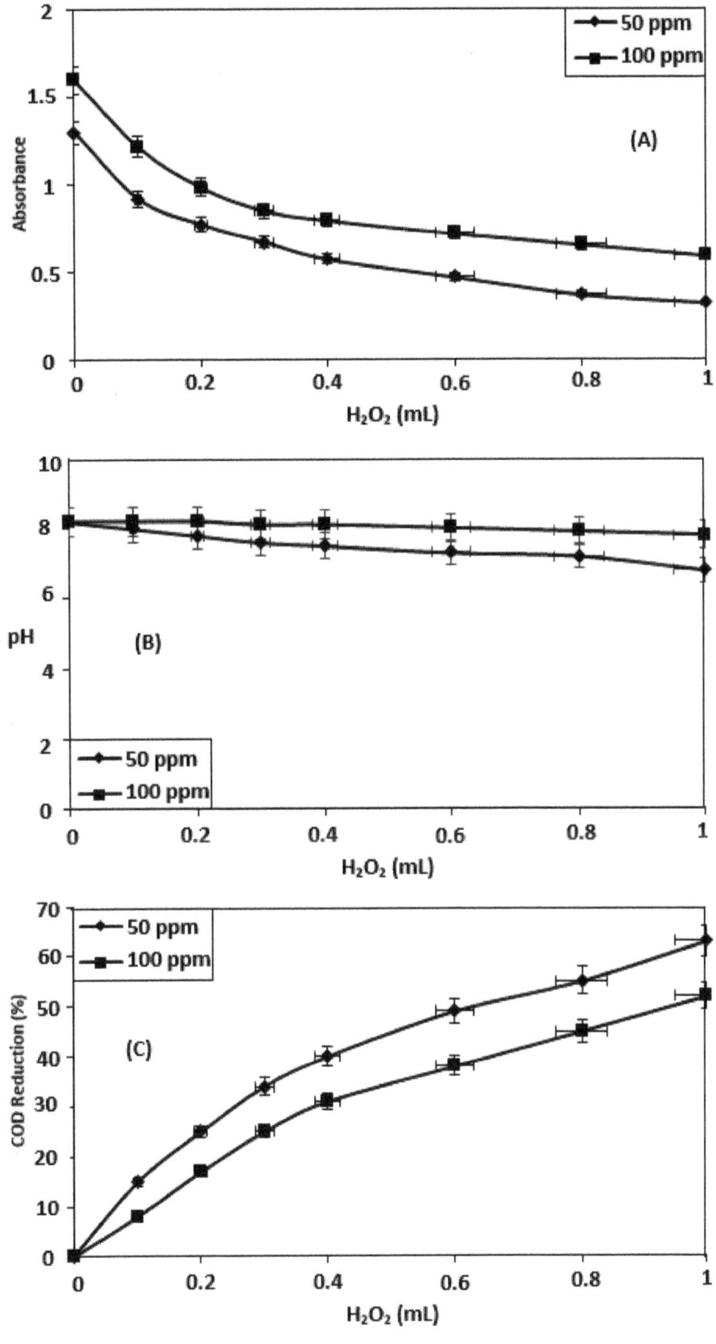

Figure 2. Effect of hydrogen peroxide on Reactive Red 45 dye aqueous solutions: (**A**) Absorbance, (**B**) pH, and (**C**) COD.

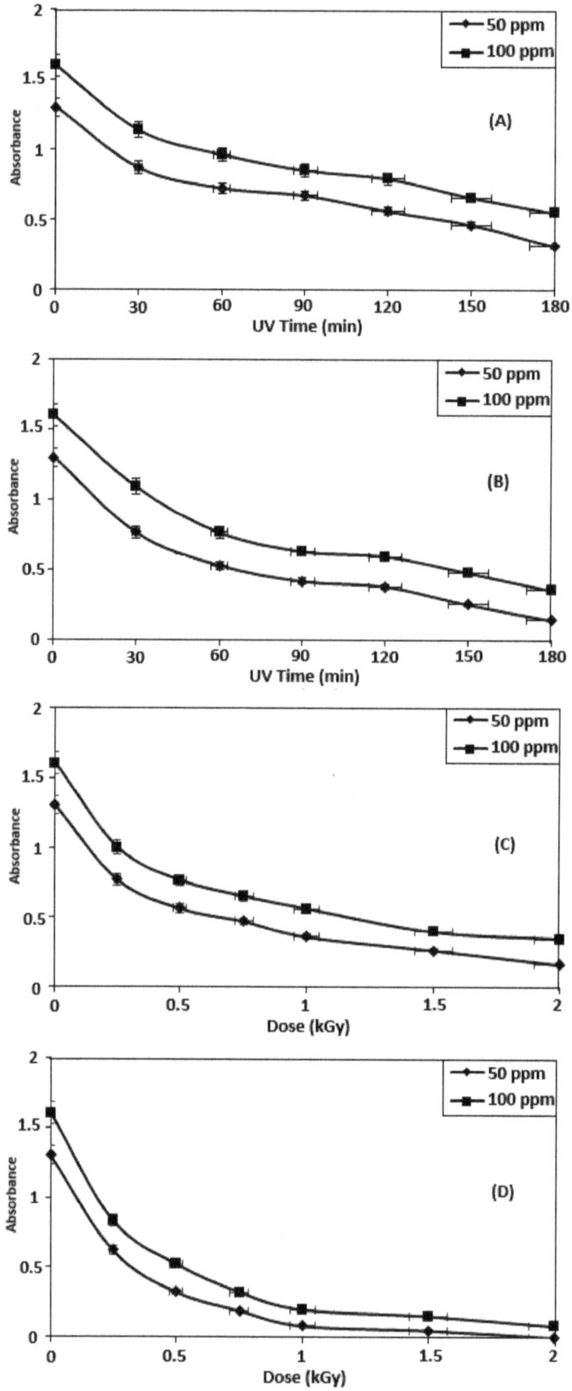

Figure 3. Effect of radiation on the absorbance of Reactive Red 45 dye aqueous solutions: (**A**) UV, (**B**) UV/H_2O_2, (**C**) Gamma, and (**D**) Gamma/H_2O_2.

The degradation was enhanced to 40.81%, 56.57%, 63.64%, 72.18%, 79.48%, and 89.09% for 50 mg/L while it was 37.60%, 52.22%, 59.28%, 64.83%, 75.08%, and 78.25% for 100 mg/L by treating with gamma radiation (Figure 3C). Whereas, 52.27%, 75.02%, 85.32%, 93.62%, 96.08%, and 100% for 50 mg/L and 48.03%, 67.27%, 79.83%, 87.57%, 90.44%, and 94.82% degradation for 100 mg/L (Figure 3D) were achieved by using gamma/H_2O_2 at a gamma absorbed dose of 2 kGy.

The obtained results showed improvement in the degradation when the solutions were subjected to gamma radiation in conjunction with H_2O_2. The optimum dose of H_2O_2 is recommended to obtain better results as the H_2O_2 acts as a promotor and radical scavenger. It was concluded that the ultimate effect of UV, as well as gamma radiation on the dye solution, is the removal of the chromophore group. Similar indications were observed by Ma et al. [29] who treated the wastewater using AOP to degrade the organic matter to purify the water.

3.3. *Effect of Radiation on Chemical Oxygen Demand (COD)*

The dye solutions were treated with UV and gamma radiation and COD evaluation was carried out before and after each radiation treatment. The reduction in COD was 22%, 33%, 45%, 54%, 60%, and 66% for 50 mg/L and 16%, 25%, 36%, 44%, 53%, and 58% for 100 mg/L observed by applying UV radiation for 30–180 min (Figure 4A). The removal in COD was increased to 28%, 42%, 53%, 62%, 70%, and 78% for 50 mg/L and 20%, 32%, 41%, 49%, 58%, and 64% for 100 mg/L by using UV/H_2O_2 (Figure 4B). It was inferred that the COD reduced significantly (78%) after UV/H_2O_2 treatment which is due to the breakdown of larger organic molecules into simple and less toxic compounds. An improvement in COD removal was obtained by treating the samples with gamma radiation. The COD was reduced to 25–70% for 50 mg/L and 17–64% for 100 mg/L at a 0.25–2 kGy gamma absorbed dose (Figure 4C). The COD reduction was enhanced by treating the samples with gamma/H_2O_2 which was 31–85% for 50 mg/L and 25–77% for 100 mg/L (Figure 4D). The rate of COD removal was prominent in the presence of H_2O_2 because it produced more hydroxyl radicals which ultimately increased the rate of the oxidation of dye.

3.4. *Effect of Radiation on pH of Reactive Red 45 Dye Aqueous Solutions*

The dye aqueous solutions were treated with UV and gamma radiation alone and in conjunction with H_2O_2 and the change in pH was determined for untreated and treated samples. The pH of the untreated sample was 8.2 which reduced to 7.8, 7.5, 7.3, 7.1, 6.9, and 6.7 for 50 mg/L and 8, 7.9, 7.7, 7.5, 7.3, and 7.1 for 100 mg/L at 30–180 min of UV exposure time (Figure 5A). The pH was further reduced to 7.6, 7.4, 4.3, 7.1, 6.8, and 6.5 for 50 mg/L and 7.9, 7.6, 7.5, 7.3, 7.1, and 6.8 for 100 mg/L after UV/H_2O_2 treatment (Figure 5B). The pH decreased after gamma radiation and the pH reduced to 7.6, 7.4, 7.1, 6.8, 6.6, and 6.3 for 50 mg/L and 7.8, 7.6, 7.3, 7.1, 6.8, and 6.5 for 100 mg/L (Figure 5C). The effect was more severe for gamma/H_2O_2 and the pH decreased to 7.3, 7.1, 6.8, 6.5, 6.2, and 5.7 for 50 mg/L and 7.5, 7.3, 7.1, 6.7, 6.4, and 6.1 for 100 mg/L after gamma/H_2O_2 treatment (Figure 5D).

There was negligible change in pH by using UV and H_2O_2 alone which indicated that the UV and H_2O_2 alone are not capable of degrading the sample but reduced the pH when using the UV/H_2O_2 process. The reduction in pH was more prominent using gamma/H_2O_2 which might be due to the breakdown of complex molecules into smaller molecules of low molecular weight aliphatic acids such as formic acid, maleic acid, etc., shifting the pH to the lower side. Similar results were reported by a number of authors [3,30,31] showing that the structure of the chromophore group was decomposed into small acidic compounds such as aliphatic carboxylic acid during the degradation of the dye solution to lower the pH.

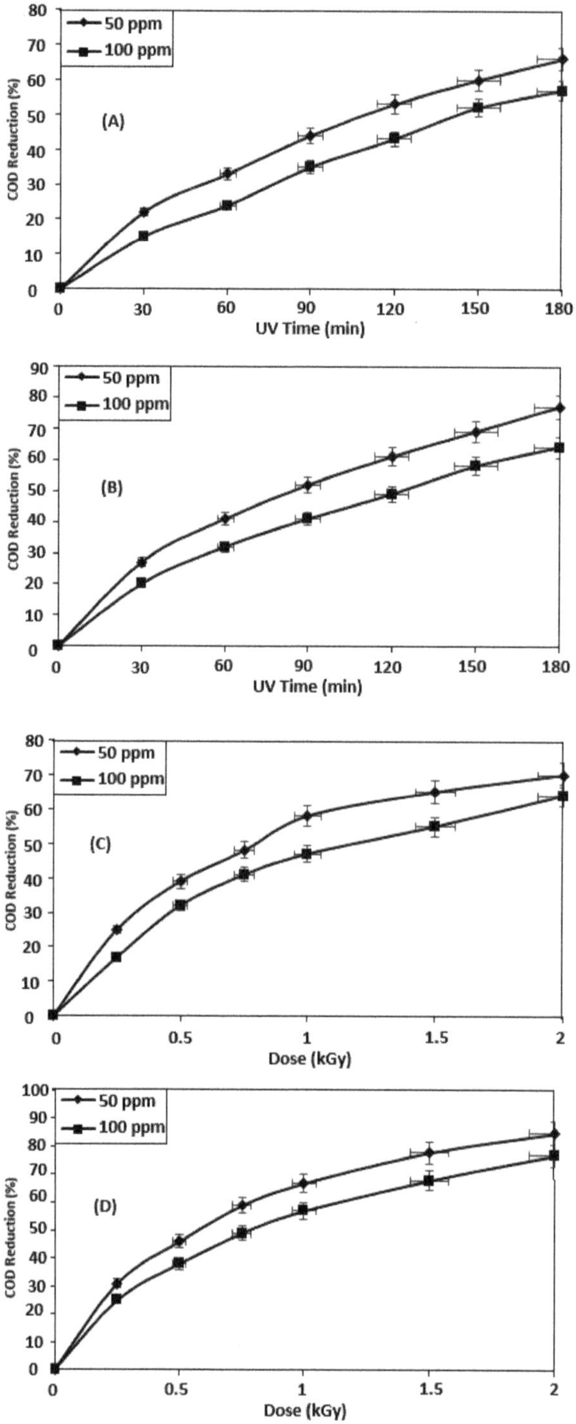

Figure 4. Effect of radiation on the COD of Reactive Red 45 dye aqueous solutions: (**A**) UV, (**B**) UV/H_2O_2, (**C**) Gamma, and (**D**) Gamma/H_2O_2.

3.5. Dose Constant (k), Removal Efficiency (G-Value), $D_{0.50}$, $D_{0.90}$, and $D_{0.99}$

Table 2 shows the observed data of the G-value, k, $D_{0.50}$, $D_{0.90}$, and $D_{0.99}$ of the gamma-ray-treated samples along with H_2O_2. The results indicated a reduction in G-value but significant improvement in $D_{0.50}$, $D_{0.90}$, and $D_{0.99}$ by increasing the dye initial concentration. The results suggest that an elevated absorbed dose of gamma radiation is required to remove 50%, 90%, and 99% dye initial concentration to overcome the competition between intermediates and parent compounds [28]. At a constant dye initial concentration, a significant improvement in G-value was observed for gamma-irradiated samples along with H_2O_2.

Table 2. Removal efficiency, dose constant, $D_{0.50}$, $D_{0.90}$, and $D_{0.99}$ for degradation of Reactive Red 45 dye aqueous solutions.

Treatment Dye	Absorbed Dose		G-Value	Dose Constant	$D_{0.50}$	$D_{0.90}$	$D_{0.99}$
	(ppm)	(kGy)	(μmol/J)	(μM/kGy)	(kGy)	(kGy)	(kGy)
Gamma	50	0.25–2	0.9814–0.0228	0.944	0.734	2.439	4.878
	100	0.25–2	1.8098–0.0181	0.714	0.969	3.221	6.443
Gamma/H_2O_2	50	0.25–2	1.2577–0.0117	2.028	0.034	1.135	2.270
	100	0.02–2	2.3119–0.0263	1.415	0.489	1.626	3.252

3.6. Cytotoxicity of Reactive Red 45 Dye

During the treatment of dye aqueous solutions, more toxic compounds may be generated which might be dangerous for aquatic life. Therefore, careful treatment of the effluent is a major concern [32–35]. The bioassays such as hemolytic, brine shrimp (*Artemia salina*), and *A. cepa* tests were carried out for the cytotoxicity evaluation before and after UV and gamma radiation, and the results obtained in the case of the *A. cepa* tests are mentioned in Tables 3 and 4.

Table 3. *Allium cepa* test exhibiting cytotoxicity reduction after UV-radiation treatment.

Sample	Exposure Time (min)	*A. cepa* Test					
		Root Length (cm)		Root Count		Mitotic Index	
Untreated		4.2 ± 0.52		14 ± 0.63		12 ± 0.28	
Positive control		9.3 ± 0.77		29 ± 0.45		32 ± 0.74	
Negative control		3.5 ± 0.63		11 ± 0.23		14 ± 0.66	
		UV alone	UV + H_2O_2	UV alone	UV + H_2O_2	UV alone	UV + H_2O_2
Sample 1	30	4.9 ± 0.68	5.4 ± 0.44	15 ± 0.44	17 ± 0.68	15 ± 0.47	18 ± 0.69
Sample 2	90	5.5 ± 0.31	6.2 ± 0.65	18 ± 0.31	20 ± 0.74	19 ± 0.31	21 ± 0.74
Sample 3	180	6.3 ± 0.67	6.9 ± 0.57	20 ± 0.44	22 ± 0.76	22 ± 0.44	23 ± 0.13

Table 4. *Allium cepa* test exhibiting cytotoxicity reduction after gamma radiation treatment.

Sample	Absorbed Dose (kGy)	*A. cepa* Test					
		Root Length (cm)		Root Count		Mitotic Index	
Untreated		4.2 ± 0.52		14 ± 0.63		12 ± 0.28	
Positive control		9.3 ± 0.77		29 ± 0.45		32 ± 0.74	
Negative control		3.5 ± 0.63		11 ± 0.23		14 ± 0.66	
		Gamma alone	Gamma +H_2O_2	Gamma alone	Gamma +H_2O_2	Gamma alone	Gamma +H_2O_2
Sample 1	0.25	5.9 ± 0.62	6.5 ± 0.44	19 ± 0.42	21 ± 0.61	20 ± 0.41	21 ± 0.62
Sample 2	0.75	6.7 ± 0.31	7.4 ± 0.65	21 ± 0.51	25 ± 0.24	23 ± 0.61	24 ± 0.44
Sample 3	2	7.8 ± 0.65	8.2 ± 0.55	24 ± 0.43	27 ± 0.73	25 ± 0.42	28 ± 0.12

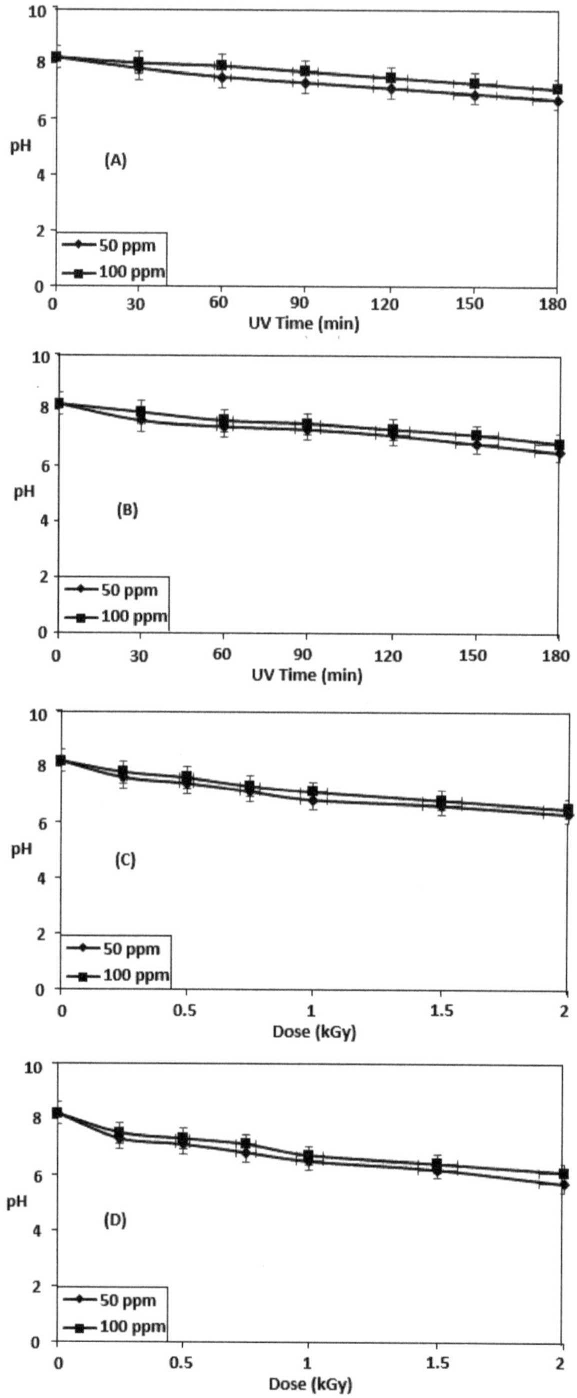

Figure 5. Effect of radiation on the pH of Reactive Red 45 dye aqueous solutions: (**A**) UV, (**B**) UV/H_2O_2, (**C**) Gamma, and (**D**) Gamma/H_2O_2.

Before treatment, RC, RL, and MI values were measured to be 14, 4.2 cm, and 12, respectively, while the observed increase was up to 36%, 39.11%, and 47.82% in the *A. cepa* roots using UV/H_2O_2, and a clear increment of up to 42%, 46.36%, and 56% in RC, RL, and MI was noted in the case of gamma/H_2O_2. Our results correlate with the findings of Jadhav et al. [35] who determined that the mitotic index before the radiation treatment was 11.68% which was increased up to 14.25% after the radiation treatment. Recently, Iqbal and Nisar [31] treated the textile effluent photo-catalytically and found a 41.17% and 41.12% increase in root count and root length, respectively, in the *A. cepa* test [31]. The cytotoxicity of Reactive Red 45 dye can also be checked by using the brine shrimp (*Artemia salina*) test [31]. This test is based on the killing ability of nauplii cultured in the laboratory at a given time. Before the radiation treatment, the death rate of cultured brine shrimp nauplii was 88% when exposed to dye solution which was decreased to 21% and 16% when using UV/H_2O_2 and gamma/H_2O_2 processes, respectively (Tables 5 and 6).

Table 5. Brine shrimp (*Artemia salina*) test exhibiting the reduction in cytotoxicity after UV radiation treatment.

Sample	Exposure Time (min)	Brine Shrimp Test	
		% Age Death (after 24 h)	
Untreated		88 ± 0.27	
Positive control		100 ± 0.0	
Negative control		0	
		UV alone	UV + H_2O_2
Sample 1	30	26 ± 0.67	20 ± 0.85
Sample 2	90	22 ± 0.53	16 ± 0.35
Sample 3	180	19 ± 0.81	14 ± 1.04

Guelli Souza et al. [36] determined that the sample treated with peroxidase enzyme showed a significant reduction (four times) in toxicity than untreated textile effluent using bioassays. The hemolytic activity of RBCs was carried out for the untreated and treated samples by UV/H_2O_2 and gamma/H_2O_2 for toxicity assessment (Tables 7 and 8).

Table 6. Brine shrimp (*Artemia salina*) test exhibiting the reduction in cytotoxicity after gamma radiation treatment.

Sample	Absorbed Dose (kGy)	Brine Shrimp (*Artemia salina*) Test	
		% Age Death (after 24 h)	
Untreated		88 ± 0.27	
Positive control		100 ± 0.0	
Negative control		0	
		Gamma alone	Gamma + H_2O_2
Sample 1	0.25	23 ± 0.68	17 ± 0.85
Sample 2	0.75	28 ± 0.63	12 ± 0.35
Sample 3	2	25 ± 0.82	8 ± 1.01

Table 7. Hemolytic test exhibiting cytotoxicity reduction after UV radiation treatment.

Sample	Exposure Time (min)	Hemolytic Test (% Age Hemolysis)	
Positive control		96.23 ± 0.82	96.23 ± 0.82
Negative control		2.11 ± 0.32	2.11 ± 0.32
		UV alone	UV + H_2O_2
Sample 1	30	16.32 ± 0.55	18.52 ± 0.28
Sample 2	120	17.56 ± 0.75	19.36 ± 0.76
Sample 3	180	19.23 ± 0.56	21.25 ± 0.53

Table 8. Hemolytic test exhibiting cytotoxicity reduction after gamma radiation treatment.

Sample	Absorbed Dose (kGy)	Hemolytic Test (% Age Hemolysis)	
Positive control		96.23 ± 0.82	96.23 ± 0.82
Negative control		2.11 ± 0.32	2.11 ± 0.32
		Gamma alone	Gamma + H_2O_2
Sample 1	0.25	17.57 ± 0.28	19.56 ± 0.58
Sample 2	0.75	18.62 ± 0.66	21.34 ± 0.37
Sample 3	2	20.25 ± 0.51	23.21 ± 0.69

The obtained data showed a clear reduction in toxicity after UV/H_2O_2 and gamma/H_2O_2 treatments. However, a greater reduction was observed in the case of gamma/H_2O_2. Up to 21.25% and 23.21% hemolysis were observed after UV/H_2O_2 and gamma/H_2O_2 treatments, respectively. The dye aqueous solutions were found to be toxic to *A. cepa* root tip cells, RBCs, and brine shrimp (*Artemia salina*) nauplii reduced significantly after UV/H_2O_2 and gamma/H_2O_2 treatments. However, the samples treated with gamma/H_2O_2 showed a considerable reduction in toxicity compared with UV/H_2O_2. The obtained results reveal that gamma/H_2O_2 is more suitable than UV/H_2O_2 for the degradation and detoxification of dye effluents. Recently, Iqbal and Nisar [31] treated the textile effluent by using gamma radiation along with H_2O_2 and found 42.12% hemolysis in the hemolytic test.

3.7. Mutagenicity Evaluation of Reactive Red 45 Dye

The Ames test is a validated and reference test that can be used for mutagenic evaluation [37]. Tables 9 and 10 show the mutagenic evaluation for untreated and treated dye solutions using TA98 and TA100 bacterial strains.

Table 9. The Ames test exhibiting mutagenicity reduction after UV radiation treatment.

Sample	Exposure Time (min)	AMES Test (% Age Mutagenicity)			
-		TA98 Count		TA100 Count	
Standard		77.08 ± 0.46		70.83 ± 0.91	
Background		33 ± 0.71		29.17 ± 0.85	
-		UV alone	UV + H_2O_2	UV alone	UV + H_2O_2
Blank	-	39.58 ± 0.33	31.08 ± 0.65	35.42 ± 0.55	32.24 ± 0.41
Sample 1	30	31.25 ± 0.19	28.13 ± 0.48	33.33 ± 0.42	29.12 ± 0.25
Sample 2	120	27.08 ± 0.43	16.67 ± 0.47	28.13 ± 0.18	19.79 ± 0.81
Sample 3	180	20.83 ± 0.45	12.5 ± 0.58	21.88 ± 0.66	15.67 ± 0.76

Table 10. The Ames test exhibited mutagenicity reduction after gamma radiation treatment.

Sample	Radiation Dose (kGy)	AMES Test (% Age Mutagenicity)			
-		TA98 Count		TA100 Count	
Standard		77.08 ± 0.46		70.83 ± 0.91	
Background		33 ± 0.71		29.17 ± 0.85	
-		Gamma alone	Gamma + H_2O_2	Gamma alone	Gamma + H_2O_2
Blank	-	39.58 ± 0.35	31.08 ± 0.82	35.42 ± 0.29	32.24 ± 0.41
Sample 1	0.25	29.17 ± 0.25	19.79 ± 0.51	30.24 ± 0.25	23.96 ± 0.35
Sample 2	0.75	23.96 ± 0.91	13.54 ± 0.26	25.32 ± 0.79	18.75 ± 0.48
Sample 3	2	17.71 ± 0.34	9.38 ± 0.85	20.83 ± 0.34	13.54 ± 0.37

After UV and gamma radiation treatment, the mutagenicity reduced significantly. The non-mutagenic behavior against TA98 and TA100 strains was observed in the solutions which were treated with UV and UV/H_2O_2 (31.25%, 27.08%, and 20.83% for TA98, and 33.33%, 28.13%, and 21.88% for TA100 with UV radiation alone while 28.13%, 16.67% and 12.5% for TA98 and 29.12%, 19.79%, and 15.67% along with H_2O_2 at 30, 120 and 180-min) respectively. The mutagenicity was decreased for the gamma-irradiated samples along with oxidant towards TA98 as well as the TA100 strain (29.17%, 23.96%, and 17.71% for TA98, while 30.24%, 25.32%, and 20.83% for TA100 with gamma radiation alone, and 19.79%, 13.54%, and 9.38% for TA98 while 23.96%, 18.75%, and 13.54% with H_2O_2 at 0.25, 0.75 and 2 kGy absorbed dose, respectively). The obtained data revealed that the treated solutions with UV and gamma radiation along with H_2O_2 showed more non-mutagenic behavior towards TA98 and TA100 strains and can be released into the environment without causing harm [31]. This is due to the metabolized reaction in the living cells to convert them into chemical products to react with cellular macromolecules for participation in redox reactions [38,39].

4. Conclusions

It was concluded from the results of this study that the toxic compounds present in the textile effluent can efficiently be degraded using gamma radiation along with H_2O_2 compared with UV radiation. The toxicity of the treated samples was reduced significantly. A high dose of gamma radiation or electron beam accelerator can treat the wastewater in a very low time frame without storage of wastewater. The *A. cepa*, brine shrimp (*Artemia, salina*), and hemolytic tests proved to be better tools for the measurement of toxicity, and mutagenicity was efficiently evaluated using the Ames test. Furthermore, it was observed that dye aqueous solutions were cytotoxic and mutagenic; however, these can efficiently be detoxified by the degradation using advanced oxidation processes (Gamma/H_2O_2 and UV/H_2O_2), and these treatment methods might be an interesting alternative, especially for the degradation of effluents contaminated with cytotoxic and mutagenic agents.

Author Contributions: Conceptualization, M.I.K., M.M., M.A.J. and T.H.B.; methodology, M.I.K., A.W. and S.U.; validation, L.M. and A.A.; formal analysis, M.I.K., M.M., M.A.J., T.H.B., H.T. and A.W.; investigation, M.I.K.; resources, L.M., A.A. and S.U.; data curation, M.I.K., M.M., A.H. and M.A.J.; writing—original draft preparation, M.I.K.; writing—review and editing, T.H.B., A.W., S.U., A.H., L.M. and A.A.; visualization, H.T., L.M. and A.A.; supervision, L.M. and A.A.; project administration, L.M. and A.A. All authors have read and agreed to the published version of the manuscript.

Funding: This research received no external funding.

Institutional Review Board Statement: Not applicable.

Informed Consent Statement: Not applicable.

Data Availability Statement: Not applicable.

Acknowledgments: The authors are highly grateful to the Nuclear Institute of Agriculture and Biology (NIAB) in Faisalabad, Pakistan, for providing a gamma radiation source facility. We are also thankful to the University of Agriculture, Faisalabad, Pakistan, for providing analytical services.

Conflicts of Interest: The authors declare no conflict of interest.

References

1. Khan, N.-U.; Bhatti, H.N.; Iqbal, M.; Nazir, A.; Ain, H. Kinetic Study of Degradation of Basic Turquise Blue X-GB and Basic Blue X-GRRL using Advanced Oxidation Process. *Z. Phys. Chem.* **2019**, *234*, 1803–1817. [CrossRef]
2. Ezzatzadeh, E. Green synthesis of α-aminophosphonates using ZnO nanoparticles as an efficient catalyst. *Z. Naturforsch. B* **2018**, *73*, 179–184. [CrossRef]
3. Kanjal, M.I.; Muneer, M.; Bhatti, I.A.; Saeed, M.; Atta-Ul-Haq, A.-U.; Den, N.Z.U.; Haq, E.U.; Nisar, J.; Iqbal, M. Gamma and UV radiation induced degradation of methotrexate (anti-rheumatic drug) in aqueous solution and conditions optimization. *Desalinat. Water Treat.* **2020**, *191*, 332–341. [CrossRef]
4. Ali, S.; Muneer, M.; Khosa, M.K.K.; Alfryyan, N.; Iqbal, M. Ionizing radiation based advanced oxidation process for reactive orange 122 dye degradation and kinetics studies. *Z. Phys. Chem.* **2022**, *236*, 1321–1338. [CrossRef]
5. Irfan, A.; Abbas, G. Exploring the Photovoltaic Properties of Metal Bipyridine Complexes (Metal = Fe, Zn, Cr, and Ru) by Density Functional Theory. *Z. Naturforsch. A* **2018**, *73*, 337–344. [CrossRef]

6. Rauf, M.; Marzouki, N.; Körbahti, B.K. Photolytic decolorization of Rose Bengal by UV/H2O2 and data optimization using response surface method. *J. Hazard. Mater.* **2008**, *159*, 602–609. [CrossRef]
7. Zamouche, M.; Tahraoui, H.; Laggoun, Z.; Mechati, S.; Chemchmi, R.; Kanjal, M.I.; Amrane, A.; Hadadi, A.; Mouni, L. Optimization and Prediction of Stability of Emulsified Liquid Membrane (ELM): Artificial Neural Network. *Processes* **2023**, *11*, 364. [CrossRef]
8. Kanjal, M.I.; Muneer, M.; Abdelhaleem, A.; Chu, W. Degradation of methotrexate by UV/peroxymonosulfate: Kinetics, effect of operational parameters and mechanism. *Chin. J. Chem. Eng.* **2020**, *28*, 2658–2667. [CrossRef]
9. Qashqoosh, M.T.A.; Alahdal, F.A.M.; Kadaf Manea, Y.; Zakariya, S.M.; Naqvi, S. Synthesis, characterization, spectroscopic and docking studies of antiulcer and nanoantiulcer drugs (LPZ and NLPZ) with bovine serum albumin. *Chem. Phys.* **2019**, *527*, 110462. [CrossRef]
10. Kubra, K.T.; Salman, S.; Znad, H.; Hasan, N. Efficient encapsulation of toxic dye from wastewater using biodegradable polymeric adsorbent. *J. Mol. Liq.* **2021**, *329*, 115541. [CrossRef]
11. Noreen, S.; Ismail, S.; Ibrahim, S.M.; Kusuma, H.S.; Nazir, A.; Yaseen, M.; Iqbal, M. ZnO, CuO and Fe_2O_3 green synthesis for the adsorptive removal of direct golden yellow dye adsorption: Kinetics, equilibrium and thermodynamics studies. *Z. Phys. Chem.* **2021**, *235*, 1055–1075. [CrossRef]
12. Muneer, M.; Kanjal, M.I.; Saeed, M.; Jamal, M.A.; Haq, A.U.; Iqbal, M.; Haq, E.U.; Ali, S. Degradation of moxifloxacin by ionizing radiation and toxicity assessment. *Z. Phys. Chem.* **2021**, *235*, 1629–1643. [CrossRef]
13. Shaheen, M.; Bhatti, I.A.; Ashar, A.; Mohsin, M.; Nisar, J.; Almoneef, M.M.; Iqbal, M. Synthesis of Cu-doped MgO and its enhanced photocatalytic activity for the solar-driven degradation of disperse red F3BS with condition optimization. *Z. Phys. Chem.* **2021**, *235*, 1395–1412. [CrossRef]
14. Verma, P.; Samanta, S.K. Microwave-enhanced advanced oxidation processes for the degradation of dyes in water. *Environ. Chem. Lett.* **2018**, *16*, 969–1007. [CrossRef]
15. Shafiquea, A.; Bhattia, I.A.; Ashara, A.; Mohsina, M.; Ahmadc, S.A.; Nisard, J.; Iqbal, M. nanoparticles synthesis, characterization and photocatalytic activity evaluation for the degradation of 2-chlorophenol. *Desalinat. Water Treat.* **2020**, *187*, 399–409. [CrossRef]
16. Mohsin, M.; Bhatti, I.A.; Ashar, A.; Mahmood, A.; ul Hassan, Q.; Iqbal, M. Fe/ZnO@ ceramic fabrication for the enhanced photocatalytic performance under solar light irradiation for dye degradation. *J. Mater. Res. Technol.* **2020**, *9*, 4218–4229. [CrossRef]
17. Bokhari, T.H.; Ahmad, N.; Jilani, M.I.; Saeed, M.; Usman, M.; Haq, A.U.; Javed, T. UV/H_2O_2, UV/H_2O_2/SnO_2 and Fe/H_2O_2 based advanced oxidation processes for the degradation of disperse violet 63 in aqueous medium. *Mater. Res. Express* **2020**, mboxemph7, 015531. [CrossRef]
18. Zeng, W.; Yin, Z.; Gao, M.; Wang, X.; Feng, J.; Ren, Y.; Wei, T.; Fan, Z. In-situ growth of magnesium peroxide on the edge of magnesium oxide nanosheets: Ultrahigh photocatalytic efficiency based on synergistic catalysis. *J. Colloid Interface Sci.* **2020**, *561*, 257–264. [CrossRef] [PubMed]
19. Stojadinović, S.; Radić, N.; Vasilić, R. ZnO Particles Modified MgAl Coatings with Improved Photocatalytic Activity Formed by Plasma Electrolytic Oxidation of AZ31 Magnesium Alloy in Aluminate Electrolyte. *Catalysts* **2022**, *12*, 1503. [CrossRef]
20. Hasan, N.; Salman, S.; Hasan, M.; Kubra, K.T.; Sheikh, C.; Rehan, A.I.; Rasee, A.I.; Awual, E.; Waliullah, R.; Hossain, M.S.; et al. Assessing sustainable Lutetium(III) ions adsorption and recovery using novel composite hybrid nanomaterials. *J. Mol. Struct.* **2023**, *1276*, 134795. [CrossRef]
21. Muneer, M.; Bhatti, I.A.; Bhatti, H.N.; Khalil-Ur-Rehman. Treatment of Dyes Industrial Effluents by Ionizing Radiation. *Asian J. Chem.* **2011**, *23*, 2392–2394.
22. USEPA. Lettuce seed germination (*Lactuca sativa*). In *Protocol for Short-Term Toxicity Screening of Hazardous Waste Sites*; EPA/600/3-88/029; Environmental Research Laboratory: Corvallis, OR, USA, 1989.
23. Leme, D.M.; Marin-Morales, M.A. Allium cepa test in environmental monitoring: A review on its application. *Mutat. Res. Mol. Mech. Mutagen.* **2009**, *682*, 71–81. [CrossRef]
24. Hasan, M.; Kubra, K.T.; Hasan, N.; Awual, E.; Salman, S.; Sheikh, C.; Rehan, A.I.; Rasee, A.I.; Waliullah, R.; Islam, S.; et al. Sustainable ligand-modified based composite material for the selective and effective cadmium(II) capturing from wastewater. *J. Mol. Liq.* **2023**, *371*, 121125. [CrossRef]
25. Ata, S.; Shaheen, I.; Ayne, Q.U.; Ghafoor, S.; Sultan, M.; Majid, F.; Bibi, I.; Iqbal, M. Graphene and silver decorated ZnO composite synthesis, characterization and photocatalytic activity evaluation. *Diam. Relat. Mater.* **2018**, *90*, 26–31. [CrossRef]
26. Maron, D.M.; Ames, B.N. Revised methods for the Salmonella mutagenicity test. *Mutat. Res. Environ. Mutagen. Relat. Subj.* **1983**, *113*, 173–215. [CrossRef]
27. Costa, F.A.; dos Reis, E.M.; Azevedo, J.C.; Nozaki, J. Bleaching and photodegradation of textile dyes by H_2O_2 and solar or ultraviolet radiation. *Sol. Energy* **2004**, *77*, 29–35. [CrossRef]
28. Muneer, M.; Kanjal, M.I.; Saeed, M.; Javed, T.; Haq, A.U.; Ud Den, N.Z.; Jamal, M.A.; Ali, S.; Iqbal, M. High energy radiation induced degradation of reactive yellow 145 dye: A mechanistic study. *Radiat. Phys. Chem.* **2020**, *177*, 109115. [CrossRef]
29. Ma, D.; Yi, H.; Lai, C.; Liu, X.; Huo, X.; An, Z.; Yang, L. The use of advanced oxidation processes in organic wastewater treatment. *Chemosphere* **2021**, *275*, 130104. [CrossRef] [PubMed]
30. Kanjal, M.I.; Muneer, M.; Saeed, M.; Chu, W.; Alwadai, N.; Iqbal, M.; Abdelhaleem, A. Oxone activated TiO_2 in presence of UV-LED light for the degradation of moxifloxacin: A mechanistic study. *Arab. J. Chem.* **2022**, *15*, 104061. [CrossRef]

31. Iqbal, M.; Nisar, J. Cytotoxicity and mutagenicity evaluation of gamma radiation and hydrogen peroxide treated textile effluents using bioassays. *J. Environ. Chem. Eng.* **2015**, *3*, 1912–1917. [CrossRef]
32. Iqbal, M. *Vicia faba* bioassay for environmental toxicity monitoring: A review. *Chemosphere* **2016**, *144*, 785–802. [CrossRef] [PubMed]
33. Janssens, R.; Cristovao, M.; Bronze, M.; Crespo, J.; Pereira, V.; Luis, P. Coupling of nanofiltration and UV, UV/TiO_2 and UV/H_2O_2 processes for the removal of anti-cancer drugs from real secondary wastewater effluent. *J. Environ. Chem. Eng.* **2019**, *7*, 103351. [CrossRef]
34. Huang, Y.; Luo, M.; Xu, Z.; Zhang, D.; Li, L. Catalytic ozonation of organic contaminants in petrochemical wastewater with iron-nickel foam as catalyst. *Sep. Purif. Technol.* **2019**, *211*, 269–278. [CrossRef]
35. Jadhav, J.; Kalyani, D.; Telke, A.; Phugare, S.; Govindwar, S. Evaluation of the efficacy of a bacterial consortium for the removal of color, reduction of heavy metals, and toxicity from textile dye effluent. *Bioresour. Technol.* **2010**, *101*, 165–173. [CrossRef]
36. Bilal, M.; Asgher, M.; Iqbal, M.; Hu, H.; Zhang, X. Chitosan beads immobilized manganese peroxidase catalytic potential for detoxification and decolorization of textile effluent. *Int. J. Biol. Macromol.* **2016**, *89*, 181–189. [CrossRef] [PubMed]
37. Iqbal, M.; Abbas, M.; Arshad, M.; Hussain, T.; Ullah Khan, A.; Masood, N.; Ahmad Khera, R. Short Communication Gamma Radiation Treatment for Reducing Cytotoxicity and Mutagenicity in Industrial Wastewater. *Pol. J. Environ. Stud.* **2015**, *24*, 2745–2750. [CrossRef] [PubMed]
38. Muneer, M.; Kanjal, M.I.; Iqbal, M.; Saeed, M.; Khosa, M.K.; Den, N.Z.U.; Ali, S.; Nazir, A. Gamma and UV radiations induced treatment of anti-cancer methotrexate drug in aqueous medium: Effect of process variables on radiation efficiency evaluated using bioassays. *Appl. Radiat. Isot.* **2020**, *166*, 109371. [CrossRef]
39. Iqbal, M.; Muneer, M.; Iqbal, M.; Kanjal, M.I. Tannery Wastewater Treatment Using Gamma Radiation: Kasur Tanneries Waste Management Agency, Pakistan. *Phys. Chem.* **2015**, *17*, 41–46.

sustainability

MDPI

Article

A Hybrid Fuzzy AHP-TOPSIS Approach for Implementation of Smart Sustainable Waste Management Strategies

Bihter Gizem Demircan and Kaan Yetilmezsoy *

Department of Environmental Engineering, Faculty of Civil Engineering, Yildiz Technical University, Davutpasa, Esenler, Istanbul 34220, Turkey
* Correspondence: yetilmez@yildiz.edu.tr

Abstract: The integration of smart city technologies into waste management is a challenging field for decision makers due to its multivariate, multi-limiting, and multi-stakeholder structure, despite its contribution to the ecological and economic sustainability understanding of cities. The success of smart sustainable waste management strategies depends on many environmental, technical, economic, and social variables, and many stakeholders are involved in these processes. Using fuzzy multi-criteria decision-making (MCDM) methods helps decision makers determine effective, affordable, and acceptable smart waste management strategies. Although MCDM methods are widely used in various environmental engineering applications, the determination of smart sustainable waste management strategies using these methods has not yet received enough attention in the literature. This study aims to contribute to this gap in the literature by evaluating four different smart waste management strategies using a hybrid fuzzy MCDM method. The performance of the proposed strategy alternatives according to fifteen sub-criteria (under four main criteria selected from the literature) was evaluated using a combined application of fuzzy analytic hierarchy process (fuzzy AHP) and fuzzy technique for order preference by similarity to obtain the ideal solution (fuzzy TOPSIS). For this evaluation, the subjective opinions of ten different experts working in academia, in the private sector, or in the public sector were obtained using prepared questionnaires. As a result, the sub-criteria of fewer atmospheric emissions (0.42), operational feasibility (0.64), initial investment costs (0.56), and increased awareness of sustainable cities (0.53) had the highest weight values in their main criteria groups. The performance ranking of the alternatives according to the closeness coefficient (CC_i) values was obtained as A2 (0.458) > A3 (0.453) > A4 (0.452) > A1 (0.440), with A3 being slightly ahead of A4 due only to a 0.001 higher CC_i value. To test the reliability and stability of the obtained performance ranking results, a sensitivity analysis was also performed using eighteen different scenarios, in which the weights of the different sub-criteria were increased by 25% or decreased by 50%, or they were assumed to be 1 and 0, or all sub-criteria in the same group had equal weight values. Since the performance ranking of the alternatives did not change, the ranking obtained at the beginning was found to be robust against the sub-criterion weight changes.

Keywords: smart sustainable waste management; multi-criteria decision making; fuzzy AHP; fuzzy TOPSIS; smart city technologies

Citation: Demircan, B.G.; Yetilmezsoy, K. A Hybrid Fuzzy AHP-TOPSIS Approach for Implementation of Smart Sustainable Waste Management Strategies. *Sustainability* **2023**, *15*, 6526. https://doi.org/10.3390/su15086526

Academic Editor: Agostina Chiavola

Received: 25 January 2023
Revised: 5 April 2023
Accepted: 11 April 2023
Published: 12 April 2023

1. Introduction

The high trend of urbanization makes cities key players in both the global climate crisis and the global water crisis. It has been reported that two-thirds of the world's energy is consumed in cities, and cities are responsible for 70% of global carbon emissions [1]. In addition to emissions from the use of large quantities of fossil fuels, cement-related emissions from the creation and use of urban infrastructure in cities are also serious (9.2 $GtCO_2e$ and 9.6 $GtCO_2e$, respectively) [2]. Increasing per capita water consumption, in parallel with rising income levels in cities, poses a drought threat. On the other hand, in cities where people's wealth levels are low, there are problems, such as unhygienic water

consumption and inability to access improved water resources. According to a recent report [3], 700 million urban residents lack improved sanitation facilities, and 156 million urban residents lack improved water resources.

Another consequence of urban overcrowding is that issues, such as convenient travel without long waits in traffic, effortless parking of vehicles, fast and hygienic disposal of garbage and sewage, and lower electricity and water bills, require more attention from city authorities. They are expected to use existing natural and economic resources as efficiently as possible and to cause as little harm to the environment as possible while meeting the basic needs of people living in cities, such as water, housing, transportation, education, and health care. For this reason, it is more urgent than ever to use innovative and technological applications to deliver urban services faster and more efficiently. Since the 1990s, this urgency has brought a new concept known as "smart city" into urban life [4,5].

In smart cities, where it is essential to provide city services in a greener, more economical, and faster manner and to meet the expectations of citizens, the integration of cutting-edge technologies into sustainable waste management approaches is becoming increasingly important [6,7]. However, this integration is directly related to the climatic, geophysical, economic, and socio-cultural characteristics of cities. Therefore, the implementation and maintenance of smart sustainable waste management strategies depends on many technical, economic, political, and social variables. These variables affect and change each other, further complicating management processes. On the other hand, the number of stakeholders involved in the relevant processes is usually quite high. It is often difficult for public authorities, private sector managers, researchers, and residents to act in a coördinated and coherent manner in a seamless and interconnected structure.

A fuzzy multi-criteria decision-making approach can be used as a solution to all the above-mentioned problems, while identifying smart sustainable waste management strategies that are technically, scientifically, and economically feasible and responsive to the needs of residents. In this approach, it is possible to rank, compare, or prioritize different smart sustainable waste management strategy alternatives through various mathematical operations after subjective expert evaluation based on a predetermined set of criteria. Containing fuzzy sets that allow for partial membership degrees, fuzzy MCDM makes it possible to work with incomplete, unmeasurable, or imprecise information. In this respect, it would be more appropriate to use this approach instead of traditional MCDM methods, which are still being developed in the field of smart waste management strategies and have not yet reached a satisfactory level of knowledge.

Although different types of multi-criteria decision-making methods are widely used in various environmental engineering applications, the number of studies using these methods for the determination of smart sustainable waste management strategies is very limited. It is seen that studies applying multi-criteria decision-making techniques in this field focused on either smart waste collection or smart waste disposal strategies, or they examined various smart waste management approaches for a particular type of waste. The strategy alternatives proposed in this study cover all processes from waste reduction to waste collection and disposal for all waste types, with a holistic view of smart sustainable waste management. On the other hand, although there are studies using different AHP and TOPSIS methods in this area, a hybrid study integrating the results obtained by using these two methods simultaneously has not yet been conducted. This study aims to fulfill this gap in the literature by evaluating four different smart waste management strategies using a hybrid fuzzy AHP-TOPSIS approach that can support local authorities in determining sustainable waste management strategies in smart cities.

To construct the hierarchical structure of the problem, as a first step, four different smart sustainable waste management strategies, which combine different sustainable waste management strategies and smart waste management technologies in the literature and practice, are selected. Then, in order to evaluate the performance of the strategies, fifteen sub-criteria were determined under four main criteria from the relevant literature. After constructing the hierarchical structure of the problem, a decision-making expert group of ten experts from academia, the private sector, and the public sector was formed for linguistic assessments. A weight criterion was made by using the verbal and subjective evaluations of the decision makers through the questionnaires prepared in the fuzzy AHP procedure. Decision makers also evaluated the performance of each strategy alternative in the questionnaires verbally and subjectively according to each sub-criterion. Finally, the performance ranking of the alternatives was obtained with the fuzzy TOPSIS method, applied by using these evaluation results and the sub-criteria weights obtained by fuzzy AHP.

The study also performed a sensitivity analysis using eighteen different scenarios to explore whether different weightings of the sub-criteria would alter the ranking of the alternatives. In these scenarios, the existing sub-criteria weight values were increased or decreased by different amounts, assumed as 1 or 0, and assumed as equal in a main criteria group.

2. Literature Review

2.1. Sustainable Waste Management

The term sustainability has been used in relation to the environment since the 1970s, with an understanding of proactive solutions to environmental problems. It was first mentioned in the literature [8] as "sustainable development", which oversees the protection of ecological life and natural resources. Sustainable development has been shown to have three dimensions, namely social, economic, and environmental [9], and these dimensions have been examined separately or together in various studies.

Considering the environment from these three pillars, it is seen that environmental sustainability is defined as "a set of rules to be followed on the use of recyclable and non-recyclable resources and waste and pollution removal" by Goodland [10], who introduced this concept into the literature. Another perspective [11] on environmental sustainability defines the term as meeting the needs of resources and services for present and future generations without compromising the health of the ecosystems that provide them.

Public administrations and businesses, which play a major role in ensuring environmental sustainability, have, over time, moved away from traditional waste management approaches to plans and strategies that follow the principle of viewing the processes of waste generation, collection, and disposal as parts of a whole that influence each other. A strong sustainable waste management system focuses on processes rather than products, and its products, functions, and organizational structures must also be adaptable and versatile [12]. As a contribution to this view, it could be added that the most effective management of waste has to relate to local environmental, economic, and social priorities and must go beyond traditional consultative approaches that require the use of experts [13]. To put it more simply, for a waste management system to be sustainable, it must be environmentally effective, economically affordable, and socially acceptable [14].

2.2. Sustainable Waste Management in Smart Cities

In the literature, there are many studies that propose different approaches for sustainable waste management. It is possible to come across multidisciplinary studies in different fields, such as resource reduction, waste minimization, and energy recovery from waste. Although there have been many studies on the sustainable waste management topic, the concept of internet of things (IoT)-enabled waste management is quite new, and the number of publications in this field is still growing. It has been accepted by researchers through various studies that the integration of smart city technologies with sustainable waste management approaches contributes to the understanding of environmental sustainability in terms of the efficient use of resources, reducing pollution loads, and saving energy and time. Since environmental sustainability is one of the key features of smart cities, it is quite understandable that the use of smart city technologies for sustainable waste management is a valid trend that is becoming more widespread every day.

Looking at the literature, it is seen that IoT-based smart waste management studies mainly focus on the management of solid waste. However, there are studies using information communication technology tools for the separation, collection, transportation, and disposal of different types of waste. Depending on the technologies used and the scope of application, a classification can be found in [15] for smart sustainable solid waste management.

Dynamic route optimization applications for improvements in household waste collection are one of the oldest and most common solid waste management approaches in smart cities. They are based on the simultaneous transmission of data, produced by different types of sensors placed in containers, to a database via different communication protocols and processing using decision support tools and mathematical models.

Another common IoT-enabled waste management strategy is using smart containers to make waste collection and separation for source activities more effective. In a 2019 study [16], smart management of e-waste on campus was enabled by level sensors in smart collection boxes installed at Monash University in Malaysia and a mobile application connected to a cloud database and Wi-Fi module. The mobile application guides users to the nearest e-waste collection box on campus based on their current location using GPS.

IoT tools were used in a 2020 study [17] that aimed to contribute to the decision-making process by examining the consumption patterns of city residents according to the waste they throw away. Data, such as the amount of daily collected waste and incorrect waste sorting behaviors, are stored in the cloud system through QR codes owned by the residents of a neighborhood. To test the effectiveness of the approach, a case study was conducted in Shanghai. Significant results were obtained in terms of the frequency of waste disposal, hours of waste disposal per day, and the change in waste disposal behavior depending on weather conditions.

As can be seen from the examples, the integration of IoT technologies into waste management is mostly in the field of solid waste management. However, as mentioned before, it is possible to come across studies involving other types of waste. Two examples of this can be seen in 2020. The first one [18] aimed to manage landfill leachate using IoT technology and developed a mechanical waste segregator that segregates incoming waste according to whether it is dry or wet. The device includes an Arduino microcontroller, an infrared light proximity sensor, and a soil moisture sensor. In the study, a leachate reactor—upflow anaerobic sludge blanket reactor system—was proposed to treat the leachate-containing microchips for imaging and a Wi-Fi module for data transmission to a cloud server. A fuzzy control system was developed based on real-time data on turbidity, suspended solids, dissolved oxygen, and chemical oxygen demand collected by sensors. In a second study [19], an air quality monitoring system using fog computing-based IoT was proposed. The sensor module, the fog computing device, and the IoT cloud platform were integrated to evaluate the effectiveness. Experiments were conducted in different environments for 15 days. As an air quality metric, the air quality index was calculated by measuring six major pollutants.

Recently, it has become possible to come across deep-learning- and machine-learning-based waste management studies. An artificial intelligence tool was used to label the images based on a specific trained set. In a study conducted in Vietnam [20], the objective was to estimate the amount of waste generated in waste bins and to optimize the distance between bins and make the emptying of bins as short as possible. For this purpose, two different algorithms were used in machine learning for the waste data obtained from garbage cans with IoT tools. The performance of the proposed system was evaluated using operational tests at Ton Duc Thang University. In another study [21], deep learning enabled the smart bin to classify the waste to be disposed of and, thanks to servo motors, open the appropriate compartment for each type of waste (e.g., plastic, metal, paper, or general waste). Data received via the Long-Range communication protocol, object detection, and waste classification were utilized in the TensorFlow framework using a pre-trained object detection model. This object detection model was trained with waste images to generate a frozen inference graph used for object detection, which is performed using a camera connected to the Raspberry Pi 3 Model BCas, the main processing unit. A scrap-metal-producing company and a waste management company collaborated in a field study [22] as part of a European project aimed at making operations more efficient by developing an integrated information management system conducted in the same year. Thanks to the simultaneous monitoring and notification provided by the sensors installed in the containers, the elevator company's scrap metal waste management was more effective, while the private company's waste management activities were effectively planned thanks to the developed online analytics platform. With the help of the deep learning method, it was possible to make price estimates for possible future waste types based on past company data.

Currently, there are various initiatives aimed at further developing IoT-based applications for waste management through integration with blockchain technology. In a 2021 study [23], researchers created a framework based on blockchain technology combined with the IoT developed using smart contracts to improve the management of hospital waste and wastewater. Data from waste containers and a wastewater treatment plant were collected by smart IoT tools. They were transmitted to the blockchain through existing communication technologies. Peripheral nodes then grouped the collected information into data blocks, which were temporarily stored. After verification, they were added to the blockchain, allowing stakeholders to process these data and enabling real-time decisions. In another study [24] conducted in the same year, researchers proposed a smart waste management system and utilized smart contracts on the Ethereum blockchain while designing it. They aimed to prove the superiority of their proposed system over traditional waste management by developing different algorithms for each step, from the time the waste is thrown into the smart waste container to its collection and disposal, and rewarding the waste producer.

2.3. *Multi-Criteria Decision Making in Sustainable Waste Management*

Problems that arise in academic research and in the public or private sector often complicate decision-making processes due to their multidimensional and complex nature. It is not always easy for decision makers to choose between different alternatives or scenarios, to rank them according to their expectations, or to compare them with each other. In such cases, regardless of the subject and area of the problem to be solved, the use of MCDM techniques helps decision makers reach the optimal result. In its most general definition, this method is the ranking, selection, or comparison of different scenarios, alternatives, or factors of a problem using various mathematical models according to the subjective evaluation of relevant stakeholders based on a predetermined set of criteria.

Although a variety of MCDM methods are used in the literature and in practice, and it is possible to classify them differently according to their distinct characteristics, a basic classification for MCDM according to the alternatives and criteria used would be multi-attribute decision making and multi-objective decision making. In multi-objective decision-making-based problems, the alternatives are infinite, and the selection among the available criteria is represented by continuous functions [25]. This method is generally used in operational applications such as route optimization studies that require taking action in a relatively short time. On the other hand, the multi-attribute decision-making approach requires choosing among decision alternatives defined according to their characteristics. There are a limited number of decision alternatives given in multi-attribute decision-making-related problems [26].

Multi-attribute decision making can be basically divided into four main categories as follows:

1. Value-based approaches:
    - Multi-attribute utility theory
    - Analytic hierarchy process
    - Weighted sum model
2. Outranking approaches:

  - Preference ranking organization and method for enrichment evaluation
  - Elimination and choice expressing reality
3. Goal-based approaches:
  - Technique for order preference by similarity
  - Data envelopment analysis
4. Evaluation-based approaches:
  - Structural equation modeling
  - Interpretive structural modeling
  - Decision-making trial and evaluation laboratory

In the field of sustainability, which is inherently multivariate, multirestrictive, and multi-stakeholder, researchers benefit from MCDM methods that allow them to address problems with a systematic approach. Transportation and logistics, energy, construction and infrastructure, and supply chain management are the most common areas where MCDM methods are used to find solutions to sustainability problems [27]. In addition to the supply chain in the manufacturing sector, MCDM methods are also used in the areas of sustainable product design, sustainable material, sustainable technology, or sustainable project selection.

A recent study [28] in the field of sustainable transport examined the barriers to sustainable transport systems in India using the gray-decision-making trial and evaluation laboratory approach. In another study [29], the preference ranking organization and method for enrichment evaluation method was used to try to identify composite indices that can be used to compare and monitor sustainability in seaport regions in the Mediterranean.

In one of the latest MCDM studies in the field of sustainable energy, Dhiman and Deb [30] used fuzzy TOPSIS and fuzzy COPRAS methods to evaluate the performances of four different alternatives for a hybrid wind farm in terms of penalty costs. In India, researchers [31] evaluated seven different energy sources in terms of sustainability concepts by applying fuzzy AHP and fuzzy TOPSIS together. Spherical fuzzy AHP and TOPSIS were used simultaneously in another study [32] proposing a fuzzy MCDM model for sustainable energy source selection for an industrial complex in Vietnam.

Considering the studies in the field of sustainable construction and infrastructure; the preference ranking organization and method for enrichment evaluation and AHP methods were used together [33] to select the best-performing bridge in terms of sustainability among three different bridge alternatives planned for the construction of the intersection of a two-lane highway. Researchers used the decision-making trial and evaluation laboratory method [34] to identify and rank sustainability indicators for green building manufacturing assessment considering the green building index.

In one of the studies using MCDM techniques in the field of sustainable production, Chu and Lin [35] used fuzzy TOPSIS to select the most appropriate robot from three different alternatives. In another study [36] addressing the problem of selecting green suppliers with the goal of sustainability, researchers used the intuitionistic fuzzy TOPSIS model for selecting the most appropriate green supplier. For sustainable supplier selection in the food processing industry in Vietnam, researchers [37] proposed a hybrid model, and fuzzy AHP was used as the MCDM method. A hybrid MCDM approach, including data envelopment analysis, spherical fuzzy AHP, and spherical fuzzy weighted aggregated sum product assessment, was used, all together [38], for sustainable supplier selection in the steel industry.

When combining MCDM with sustainable waste management applications, researchers generally aim to [39]: (1) investigate optimal decision making among various alternative waste management strategies (e.g., landfill disposal, incineration, recycling, reuse, etc.); (2) explore possible technologies (e.g., mass incineration, pyrolysis, gasification, plasma incineration in cement kilns, etc.); and (3) determine the optimal location of a waste management facility (e.g., landfill, waste treatment facility, recycling facility, etc.).

A 2015 study [40] using fuzzy AHP and fuzzy TOPSIS together evaluated three different options for landfill sites in Istanbul, Turkey. Four criteria were used, including soil conditions, topography, and climatologic and hydrologic conditions. In a technology selection study [41] in the same year, ten different solid waste disposal technologies were evaluated with fuzzy TOPSIS according to eighteen criteria set by experts in order to determine the most suitable alternative for Istanbul, Turkey.

A hybrid fuzzy multi-criteria decision analysis method was applied in a 2017 study [42] evaluating six different wastewater treatment technologies, including membrane bioreactor, moving-bed bio-film reactor, and activated sludge process. Fuzzy AHP was used to weight the criteria, and fuzzy TOPSIS was used to rank the alternatives according to these criteria. In a 2021 study [43] aiming to determine the best disposal method for healthcare waste during and after the recent COVID-19 pandemic, the results obtained using fuzzy TOPSIS were compared with the results obtained with another fuzzy MCDM method. Four main criteria with ten subcategories were used to determine the most optimal one out of nine different disposal alternatives for healthcare waste.

In one of the most recent studies in this field [44], a hybrid fuzzy MCDM application was applied to evaluate eight different solid waste management scenarios for India. Fuzzy AHP was used for criterion weighting, and fuzzy TOPSIS was used to rank alternatives. Looking at recent MCDM studies in the field of sustainable waste management, it is seen that in addition to evaluating traditional waste management approaches, more specific waste management strategies adapted to socioeconomic conditions of the countries were also evaluated using this methodology. For example, in a study [45] conducted in Pakistan in 2022, fuzzy TOPSIS and fuzzy AHP were used together to identify the barriers that may be encountered in the transition to a circular economy (CE) in the management of food waste in that country.

In recent years, it has been observed that waste management strategies have been developed in accordance with the understanding of using waste for the benefit of a sustainable environment and economy, rather than seeing it as a problem that needs to be eliminated. It is seen that MCDM techniques are also used in the evaluation of these strategies. Two recent studies used fuzzy MCDM methods to determine the most suitable option for the location of a waste-to-energy plant. In the study of [46], which was used on a renewable energy project in Vietnam, fuzzy AHP was used to determine the solid waste-to-energy plant location in the country. In the second study [47], four possible waste-to-energy plant locations in Kırıkkale, Turkey, were evaluated using fuzzy AHP and fuzzy TOPSIS according to four criteria determined by expert opinion and literature research.

As can be seen from the above examples, fuzzy MCDM methods are applied to problems both in sustainability and waste management fields seperately. However, since smart sustainable waste management approaches are still a growing research area, the evaluation of these approaches with these methods is not very common in the literature. In a study [48] in which four different alternatives for smart solid waste collection using information and communication technologies were proposed for the Tepebasi region in Turkey, alternatives were evaluated using the type 2 fuzzy TOPSIS method, taking into account seven criteria set by experts. In another recent study [49], seven different smart strategies selected for medical waste disposal were prioritized using the decision-making trail and evaluation laboratory method. A hybrid MCDM method (modified entropy and combinative distance-based assessment under q-level interval-valued fuzzy sets) was used in a study [50] investigating different smart waste collection methods for a municipality in Istanbul, Turkey. In another study [51] covering smart waste management approaches in a more general way, researchers used the spherical AHP method to evaluate three different alternatives for smart waste management.

It should be noted that the examples given above are related to either the collection or disposal of waste. However, the present study covers broader strategies for smart waste management from production to disposal. It also differs from the above-mentioned studies using the fuzzy hybrid AHP-TOPSIS approach. A smart sustainable waste management strategy evaluation using this approach has not yet been reported.

It would be inaccurate to say that there is only one ideal method for solving a particular problem using fuzzy MCDM. By applying different methods for the same problem, different rankings of alternatives can occur. To overcome this shortcoming, a hybrid approach that integrates the results for final decision making was used in the study. The effectiveness and reliability of a hybrid fuzzy AHP-TOPSIS approach have been previously proven and are still used in current research, such as selecting cloud services [52], evaluating the service quality of hospitals [53], selecting suppliers for a construction company [54], evaluating hotel websites [55], evaluating the financial performance of banks [56], selecting computer-integrated manufacturing technologies [57], selecting aquaculture species [58], and studying the long-term growth in the online market for food delivery services [59].

3. Methodology

3.1. Constructing the Hierarchical Structure of the Problem

When working with a multi-criteria decision-making method, the problem to be solved must first be clearly defined. Then, the hierarchical structure of the problem in question is created. The hierarchical structure includes the alternatives, scenarios, or constraints of the problem to be evaluated or prioritized, as well as the criteria to be considered in the evaluation or prioritization. In determining the criteria, databases, expert opinions, or studies from the literature on the area in question may be consulted.

In this study, the problem was set to evaluate smart sustainable waste management strategies. Four smart waste management strategies were selected as alternatives for the present problem. Fifteen sub-criteria belonging to five main criteria were selected from the literature. The smart waste management strategy alternatives for the problem are presented in Table 1.

Table 1. Smart waste management strategy alternatives evaluated in the analysis.

Alternative Number	Alternative Name
A1	Integrating informal recyclable waste collection into a formal smart system
A2	A pay as you throw application leveraging blockchain technology
A3	IoT-Based community composting
A4	Preventing illegal sewage discharge by utilizing IoT

The main criteria and sub-criteria used in the evaluation of the strategy alternatives of the problem are given in Table 2.

Table 2. Main criteria and sub-criteria considered in the analysis.

Main Criteria	Sub-Criteria No.	Sub-Criteria Name	Reference
Environmental criteria (C1)	C1.1	Less atmospheric emissions	[60]
	C1.2	Less soil pollution	[60]
	C1.3	Less surface water pollution	[60]
	C1.4	Energy recovery	[61]
	C1.5	Natural resources recovery	[62]
Technical criteria (C2)	C2.1	Operational feasibility	[48]
	C2.2	Innovativeness	[48]
	C2.3	Need for qualified personnel	[63]
Economic criteria (C3)	C3.1	Maintenance costs	[64]
	C3.2	Transportation costs	[63]
	C3.3	Operational costs	[63]
	C3.4	Initial invesment costs	[63]
Social criteria (C4)	C4.1	Increased awareness on sustainable cities	[64]
	C4.2	Increased quality of life in the city	[60]
	C4.3	New employment opportunities	[65]

3.2. Fuzzy Analytic Hierarchy Process (Fuzzy AHP)

Although there are various fuzzy AHP approaches in the literature, and van Laarhoven and Pedrycz [66] were the first to propose this approach, Chang's Extent Analysis [67], which is widely used due to its ease of implementation, was also used in this study. The method is based on pairwise comparisons of criteria and alternatives. For these comparisons, linguistic terms selected for this analysis and the corresponding triple fuzzy numbers are presented in Table 3. The steps of the extent analysis method [68–70] are summarized in Appendix A.

Table 3. Linguistic terms and corresponding triple fuzzy numbers.

Linguistic Term	Triangular Fuzzy Number Equivalent
Equally important	(1, 1, 1)
Slightly more important	(2/3, 1, 3/2)
Strongly more important	(3/2, 2, 5/2)
Very strongly more important	(5/2, 3, 7/2)
Definitely more important	(7/2, 4, 9/2)

3.3. Fuzzy Technique for Order Preference by Similarity to Ideal Solution (Fuzzy TOPSIS)

This method, firstly introduced by Yoon and Hwang [71], is an example of multi-attribute decision making that is widely used in all fields. It is based on the assumption that the chosen alternative is closest to the positive ideal solution and farthest from the negative ideal solution.

In the first step of this method, a decision-maker group is formed of people who have expertise on the problem, and the cluster, consisting of K decision makers, is shown as $E = \{DM_1, DM_2, \ldots, DM_K\}$. The alternatives to the problem $A = \{A_1, A_2, \ldots, A_m\}$ and the criteria to be used in the evaluation of the alternatives $K = \{K_1, K_2, \ldots, K_n\}$ are determined. Then, the linguistic expressions that decision makers will use when weighting the criteria and evaluating the alternatives according to the relevant criteria are selected [61]. The steps in the fuzzy TOPSIS method [72] used in this study are summarized in Appendix B. The linguistic terms and corresponding triangular fuzzy number equivalents used in this method can be found in [73].

3.4. Obtaining Decision-Maker Opinions

In a fuzzy MCDM problem, subjective expert opinions are needed to determine the weighting of the selected criteria and to evaluate the proposed strategy alternatives according to these criteria. As long as there is at least one decision maker, there is no limit to the number of experts whose opinions are obtained [74]. In this study, a group of decision makers consisting of ten experts was formed. Eight of the experts work in the field of environmental engineering and two in the field of smart cities.

Three of the experts in the field of environmental engineering were working in academia (two professors and one research assistant). Among these, one professor and the research assistant were conducting studies that address waste management problems using various MCDM methods. The fourth specialist in the field was an environmental engineer working for the Ministry of Environment, Urbanization, and Climate in Turkey. The fifth expert was working as an environmental engineer at ÇEVKO (Environmental Protection and Packaging Waste Utilization Foundation), a non-profit foundation promoting recycling of packaging waste in Turkey. The sixth expert was an electrical and electronics engineer working at the Bursa Metropolitan Municipality, Turkey, and holds a master's degree in environmental engineering. The seventh professional worked as an environmental engineer in a private sustainability company in Turkey, and the last one was also an environmental engineer, working for a company that operates an airport in Istanbul, Turkey.

The first of the decision makers with smart city expertise was a geomatics engineer, working as a project manager in an international mapping and technology company in

Ghent, Belgium. The other was an electronics and communications engineer who was involved in the implementation of thirty different smart city projects in Turkey and is currently a project manager at a defense electronics company in the country.

Figure 1 shows the work areas of the experts who make up the decision-making group and the type of institution they work for.

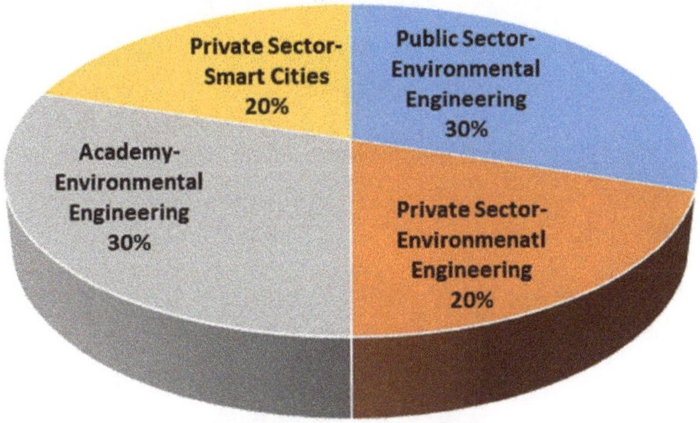

Figure 1. Distribution of expertise and work area of the decision-maker group.

In the first questionnaire prepared for the criteria weighting, the decision makers were asked to evaluate the 15 selected sub-criteria with linguistic expressions in the AHP method chosen in the study. In the second questionnaire, the decision makers were asked to rate the performance of the four proposed strategy alternatives for each criterion using verbal expressions in the TOPSIS method chosen in the study.

The experts were reached via e-mail and LinkedIn, and the questionnaire prepared with Google Forms was sent to them via the same means. The results of the linguistic expert evaluations were drawn from the survey in Google Forms. Microsoft® Excel® 2016 (Microsoft Inc., Redmond, WA, USA), running under Windows 10 system on a HP Pavilion (Intel® Core™ i5-7200U CPU, 2.50 GHz, 8 GB of RAM, 64-bit) PC platform, was used for the calculations of the fuzzy AHP and fuzzy TOPSIS.

3.5. Dealing with the Experts' Subjectivity

In traditional multi-attribute decision-making approaches, the criteria present and their weights are given with precise and clear values. Thus, it is assumed that the selection or ranking of alternatives to be made with them is absolutely correct. In real problems, however, the nature of the problem makes it impossible to know exactly what the goals or constraints of the problem are. This uncertainty can be caused by reasons [75], such as unmeasurable information, incomplete information, inaccessible information, and partial ignorance. On the other hand, it would be unrealistic to claim that the decisions made by decision makers are absolutely correct or complete. Human subjectivity should also be taken into account in decision-making processes. One solution to these problems is to use fuzzy MCDM methods that allow for partial membership and work with uncertain and imprecise inputs or information. With the integration of fuzzy sets into the field of MCDM by Belman and Zadeh [76] and Zimmerman [77], fuzzy decision approaches now form a branch of fuzzy set theory. Another method that makes it possible to work with uncertain data is the grey numbers in the grey theory introduced by Deng [78]. As with fuzzy numbers, there are ambiguities in the data, but existing knowledge and experience allow the data in question to fall within a certain range of values. In other words, while subjective uncertainty is in question in fuzzy theory, objective uncertainty can be spoken of in grey theory [79].

The integration of smart city technologies into sustainable waste management complicates the administrative steps that need to be taken, especially in terms of increasing expertise and budgetary demands and difficulties in acceptance by city residents. Moreover, because this integration is a relatively new area of research, practical experience does not yet provide the desired level of data in the decision-making processes. Considering these difficulties, to avoid erroneous results, the data in this study were used in fuzzy form.

4. Results

4.1. Determination of Criterion Weights

The weights of the sub-criteria within each main criteria group were determined using pairwise comparison matrices based on verbal and subjective expert evaluations. These four comparison matrices were created through the experts' interpretation of the superiority of the sub-criteria over the others using the linguistic expressions. Since there were ten experts in the group of decision makers, the geometric mean of the triple fuzzy number equivalents of the linguistic expressions was taken to reduce the result to one. Then, the criteria weights were determined by following the calculations given below. Table 4 presents the pairwise comparison matrix obtained for environmental criteria.

Table 4. Pairwise comparison matrix obtained for environmental criteria.

Environmental Criteria	Less Atmospheric Emissions	Less Soil Pollution	Less Surface Water Pollution	Energy Recovery	Natural Resources Recovery
Less atmospheric emissions	(1.00, 1.00, 1.00)	(1.36, 1.69, 2.09)	(1.20, 1.58, 2.06)	(1.56, 1.97, 2.43)	(1.46, 1.74, 2.05)
Less soil pollution	(0.48, 0.59, 0.74)	(1.00, 1.00, 1.00)	(1.09, 1.32, 1.59)	(1.37, 1.76, 2.24)	(1.28, 1.52, 1.76)
Less surface water pollution	(0.49, 0.63, 0.84)	(0.63, 0.76, 0.92)	(1.00, 1.00, 1.00)	(1.50, 1.97, 2.54)	(1.35, 1.69, 2.08)
Energy recovery	(0.41, 0.51, 0.64)	(0.45, 0.57, 0.73)	(0.39, 0.51, 0.67)	(1.00, 1.00, 1.00)	(1.18, 1.52, 1.91)
Natural resources recovery	(0.49, 0.57, 0.69)	(0.57, 0.66, 0.78)	(0.48, 0.59, 0.74)	(0.52, 0.66, 0.84)	(1.00, 1.00, 1.00)

In order to obtain the fuzzy synthetic extent value of a criterion, the vector obtained by the sum of the vectors in the row belonging to the relevant criterion and the inverse of the vector, which was the sum of all row sums, were multiplied as follows:

$S_{Less\ atmospheric\ emissions} = (6.57, 7.98, 9.63) \otimes (0.03, 0.04, 0.04) = (0.20, 0.29, 0.41)$

$S_{Less\ soil\ pollution} = (5.22, 6.19, 7.32) \otimes (0.03, 0.04, 0.04) = (0.16, 0.22, 0.31)$

$S_{Less\ surface\ water\ pollution} = (4.97, 6.05, 7.36) \otimes (0.03, 0.04, 0.04) = (0.15, 0.22, 0.32)$

$S_{Energy\ recovery} = (3.44, 4.10, 4.95) \otimes (0.03, 0.04, 0.04) = (0.10, 0.15, 0.21)$

$S_{Natural\ resources\ recovery} = (3.06, 3.48, 4.05) \otimes (0.03, 0.04, 0.04) = (0.09, 0.13, 0.17)$

Using these vectors, the values of $V(M_2 \geq M_1)$ and $V(M_1 \geq M_2)$ were calculated according to Equation (A6) (see Appendix A). The results are summarized in Appendix C. Thus, the weight vector for the environmental criteria was calculated as follows: $W_{Environmental\ criteria} = (0.42, 0.27, 0.27, 0.04, 0.00)$.

The results showed that the most important criterion among environmental criteria was that of fewer atmospheric emissions. Less soil pollution and less surface water pollution criteria were equally important. The energy recovery criterion was of much lower importance than all these criteria. The weight value for the criterion of natural resources was given as 0, implying that the mentioned criterion was not important at all compared to the other environmental criteria. Table 5 presents the pairwise comparison matrix obtained for technical criteria.

Table 5. Pairwise comparison matrix obtained for technical criteria.

Technical Criteria	Operational Feasibility	Innovativeness	Need for Qualified Personnel
Operational feasibility	(1.00, 1.00, 1.00)	(1.59, 2.08, 2.63)	(1.34, 1.62, 1.93)
Innovativeness	(0.38, 0.48, 0.63)	(1.00, 1.00, 1.00)	(1.52, 2.00, 2.56)
Need for qualified personnel	(0.52, 0.62, 0.75)	(0.39, 0.50, 0.66)	(1.00, 1.00, 1.00)

The fuzzy synthetic extent values were obtained accordingly:

$S_{Operational\ feasibility}$ = (3.93, 4.71, 5.55) ⊗ (0.08, 0.10, 0.11) = (0.32, 0.46, 0.64)
$S_{Innovativeness}$ = (2.90, 3.48, 4.19) ⊗ (0.08, 0.10, 0.11) = (0.24, 0.34, 0.48)
$S_{Need\ for\ qualified\ personnel}$ = (1.91, 2.12, 2.41) ⊗ (0.08, 0.10, 0.11) = (0.16, 0.21, 0.28)

The values of $V(M_2 \geq M_1)$ and $V(M_1 \geq M_2)$ are summarized in Appendix C. Thus, the weight vector for the technical criteria was calculated as follows: $W_{Technical\ criteria}$ = (0.64, 0.36, 0.00).

The results showed that operational feasibility was the most important criterion among the technical criteria. The criterion of innovativeness had lower importance compared to the criterion of operational feasibility. However, the weight value of the criterion for the need for qualified personnel was given as 0, indicating that the mentioned criterion in question was not important at all compared to the other technical criteria. Table 6 tabulates the pairwise comparison matrix obtained for economic criteria.

Table 6. Pairwise comparison matrix obtained for economic criteria.

Economic Criteria	Initial Invesment Costs	Operational Costs	Maintenance Costs	Transportation Costs
Initial invesment costs	(1.00, 1.00, 1.00)	(1.37, 1.64, 1.96)	(1.25, 1.58, 1.97)	(1.53, 1.94, 2.41)
Operational costs	(0.51, 0.61, 0.73)	(1.00, 1.00, 1.00)	(1.35, 1.58, 1.82)	(1.25, 1.47, 1.73)
Maintenance costs	(0.51, 0.63, 0.80)	(0.55, 0.63, 0.74)	(1.00, 1.00, 1.00)	(1.25, 1.58, 1.97)
Transportation costs	(0.41, 0.51, 0.65)	(0.58, 0.68, 0.80)	(0.51, 0.63, 0.80)	(1.00, 1.00, 1.00)

The fuzzy synthetic extent values were obtained accordingly:

$S_{Initial\ investment\ costs}$ = (5.15, 6.17, 7.35) ⊗ (0.05, 0.06, 0.07) = (0.25, 0.35, 0.49)
$S_{Operational\ costs}$ = (4.11, 4.66, 5.28) ⊗ (0.05, 0.06, 0.07) = (0.20, 0.27, 0.35)
$S_{Maintenance\ costs}$ = (3.30, 3.85, 4.52) ⊗ (0.05, 0.06, 0.07) = (0.16, 0.22, 0.30)
$S_{Transportation\ costs}$ = (2.50, 2.83, 3.26) ⊗ (0.05, 0.06, 0.07) = (0.12, 0.16, 0.22)

The values of $V(M_2 \geq M_1)$ and $V(M_1 \geq M_2)$ are summarized in Appendix C. Thus, the weight vector for the economic criteria was calculated as follows: $W_{Economic\ criteria}$ = (0.56, 0.30, 0.14, 0.00).

The results revealed that the initial investment cost criterion is very important compared to other economic criteria. This criterion is followed by the operational cost criterion. The criterion of maintenance costs is insignificant compared to these two criteria. The criterion of transportation costs did not make any sense compared to other economic criteria. Table 7 shows the pairwise comparison matrix obtained for social criteria.

Table 7. Pairwise comparison matrix obtained for social criteria.

Social Criteria	Increased Awareness on Sustainable Cities	Increased Quality of Life in the City	New Employment Opportunities
Increased awareness on sustainable cities	(1.00, 1.00, 1.00)	(1.48, 1.89, 2.35)	(1.10, 1.47, 1.95)
Increased quality of life in the city	(0.42, 1.00, 0.67)	(1.00, 1.00, 1.00)	(1.54, 2.02, 2.58)
New employment opportunities	(0.51, 0.68, 0.91)	(0.39, 0.49, 0.65)	(1.00, 1.00, 1.00)

The fuzzy synthetic extent values were obtained accordingly:

$S_{Increased\ awareness\ on\ sustainable\ cities}$ = (3.59, 4.36, 5.31) ⊗ (0.08, 0.09, 0.12) = (0.30, 0.41, 0.63)

$S_{Increased\ quality\ of\ life\ in\ the\ city}$ = (2.97, 4.02, 4.26) ⊗ (0.08, 0.09, 0.12) = (0.24, 0.38, 0.50)

$S_{New\ employment\ opportunities}$ = (1.90, 2.17, 2.55) ⊗ (0.08, 0.09, 0.12) = (0.16, 0.21, 0.30)

The values of $V(M_2 \geq M_1)$ and $V(M_1 \geq M_2)$ are summarized in Appendix C. Thus, the weight vector for the social criteria was calculated as follows: $W_{Social\ criteria}$ = (0.53, 0.46, 0.01).

Although the two criteria were very close to each other, it was seen that the criterion of increasing awareness about sustainable cities is slightly more important than the criterion of increasing the quality of life in the city. The new employment opportunities criterion was rather insignificant compared to the other two criteria.

4.2. *Ranking of the Strategy Alternatives according to the Weighted Criteria*

In this part of the study, the performance ranking of the strategy alternatives was performed using the fuzzy TOPSIS method. For this purpose, weighted criteria with the fuzzy AHP method from the previous section were used. Expert opinions were taken to evaluate the performance of each proposed strategy alternative according to each sub-criterion using linguistic terms [80]. Since there were ten experts in the decision-making group, the arithmetic mean of the triple fuzzy number equivalents of linguistic terms was taken, and the result was reduced to one. The fuzzy decision matrix (not shown here due to space constraints) was obtained accordingly.

Equations (A13) and (A14) were used to obtain a normalized fuzzy decision matrix (see Appendix B). The normalization was performed differently depending on whether the criterion is a benefit criterion or a cost criterion. For example, since C1.3 is a benefit criterion, normalization of the corresponding vector to the second alternative (A2) ((5.70, 7.20, 8.30)) is performed by dividing the vector elements by 8.90, which is the third-largest element in the vectors in the relevant column, and the new vector is obtained as (5.70/8.90, 7.20/8.90, 8.30/8.90) = (0.64, 0.81, 0.93).

Let us take the C3.1 criterion. Since the criterion is a cost criterion, the first vector element with the smallest value in the column is determined. This value is 5.80. Accordingly, the normalization of the vector (6.60, 8.30, 9.40) corresponding to the second alternative (A2) of this criterion is calculated as (5.80/9.40, 5.80/8.30, 5.80/6.60) = (0.62, 0.70, 0.88). Accordingly, the normalized fuzzy decision matrix (not shown here due to the lack of space) was created within the scope of the present analysis.

To obtain the weighted normalized fuzzy decision matrix, the matrix elements were multiplied by the sub-criteria weights obtained by the fuzzy AHP method in the previous section. The weighted normalized fuzzy decision matrix (not shown here due to space limitations) was obtained by multiplying the vectors in the column of each criterion by the weight of that criterion.

When determining the fuzzy positive ideal solution values (A^*), the vector elements were assigned a value of 1 if the criterion in question was a benefit criterion and 0 if it was a cost criterion. The opposite was considered as true when determining the fuzzy negative ideal solution values (A^-). The fuzzy positive/negative ideal solution values were determined using the weighted fuzzy decision matrix as follows:

A^* = [(1, 1, 1), (1, 1, 1), (1, 1, 1), (1, 1, 1), (1, 1, 1), (1, 1, 1), (1, 1, 1), (0, 0, 0), (0, 0, 0), (0, 0, 0), (0, 0, 0), (0, 0, 0), (1, 1, 1), 1, 1, 1), 1, 1, 1)]

$A^- = [(0, 0, 0), (0, 0, 0), (0, 0, 0), (0, 0, 0), (0, 0, 0), (0, 0, 0), (0, 0, 0), (1, 1, 1), (1, 1, 1), (1, 1, 1), (1, 1, 1), (1, 1, 1), (0, 0, 0), (0, 0, 0), (0, 0, 0))]$

The distances (d^* and d^-) (not presented here due to limited place) of the alternatives from A^* and A^- were calculated using Equations (A20) and (A21), respectively. Based on the above values, the closeness coefficient (CC_i) and the ranking of the alternatives were determined using Equation (A22) (see Appendix B). The ranking of the proposed smart waste management strategy alternatives according to the above closeness coefficients (values are given in the respective parentheses) was A2 (0.458) > A3 (0.453) > A4 (0.452) > A1 (0.440).

The results indicated that considering the selected criteria, the application pay as you throw (PAYT), which uses blockchain technology, had the best performance among the proposed smart sustainable waste management strategies. Among the two strategies with very close CC_i values, IoT-based community composting was slightly ahead of IoT-based illegal sewage discharge prevention by only 0.001. The strategy of integrating illegal waste collection workers into a formal smart system came in last with the lowest CC_i value compared to others.

4.3. Sensitivity Analysis

In multi-criteria decision making, sensitivity analysis was used to test the consistency of the ranking obtained and to be sure of the result obtained. It is noted that in studies where hybrid MCDM is used, researchers usually perform sensitivity analysis using scenarios in which the original criteria weights are changed to varying degrees [81].

In this study, a sensitivity analysis was performed with eighteen different scenarios, in which the weighting of the criteria was changed [82]. In the first ten scenarios, the weight of each of the ten benefit criteria was increased by 25%. In Scenarios 11–15, the weight of each of the five cost criteria was increased by 25%. In Scenario 16, the weight of the criterion with the highest weight in each main criteria group was reduced by 50%. In Scenario 17, the weight of the criterion with the highest weight in each main criterion group was assumed to be 1, and all other criterion weights were assumed to be 0. Finally, in Scenario 18, the weights of the criteria were chosen to be the same in each main criteria group. Accordingly, the new closeness coefficient values (CC_i) of the alternatives after the sensitivity analysis were obtained. The results of the sensitivity analysis showed that the ranking of the alternatives obtained with the fuzzy AHP-fuzzy TOPSIS hybrid approach remained unchanged when the criteria weights were varied. The obtained CC_i values of the alternatives are summarized in Appendix D, Table A1.

In all eighteen scenarios, the application PAYT, which uses blockchain technology, (A1) took first place in the ranking of alternatives, with the highest closeness coefficient value. It was followed by IoT-based community composting (A2). Although they had very close CC_i values, as in the present results, preventing illegal sewage discharge by utilizing IoT (A3) performed slightly better than integrating informal recyclable waste collection into an Iot-based formal system (A4). Although no significant change in the closeness coefficients occurred in the first sixteen scenarios in which the criterion weights were increased or decreased at specific rates, it was found that the weights changed in the 16th scenario, in which the criterion weights were assigned values of 1 and 0, and in the 17th scenario, in which the criterion weights were assigned values of 1 and 0. However, this situation did not affect the ranking of the alternatives. This confirmed the reliability and the stability of the results obtained from the hybrid fuzzy AHP-TOPSIS approach.

5. Discussion

5.1. Managerial Implication

The results of this research will assist decision makers in determining the sustainable waste management strategies to be implemented. It also motivates and guides managers and researchers in conducting pilot studies that will place smart sustainable waste management concepts on a more realistic footing.

The operational feasibility (C2.1) and initial investment costs (C3.4) criteria had the highest weighting values after the criterion weighting. This result confirmed the necessity of conducting detailed and realistic feasibility and budget studies before implementing smart sustainable waste management strategies.

Among the four proposed strategy alternatives, a PATY application leveraging blockchain technology took first place, with the highest closeness coefficient (CC_i). The long-term success and acceptance of this strategy are directly related to the conditions and needs of the region in which the strategy is implemented. The main disadvantage of such a system may be the illegal dumping of waste. To prevent this, legal regulations and strict controls are needed. On the other hand, detailed cost analysis is required depending on the blockchain infrastructure to be used (e.g., public or private blockchain).

The second-ranked strategy alternative (IoT-based community composting (A3)) has a closeness coefficient value that is only 0.001 higher than the third-ranked alternative (preventing illegal sewage discharge by utilizing IoT (A4)). In such cases where the two alternatives have very close closeness coefficient values, a multi-tiered approach in decision making can be carried out by the decision makers. Just as Multi-Group Confirmatory Factor Analysis used in the social sciences helps researchers [83] confirm the accuracy of results obtained with MCDM, the use of methods, such as Life Cycle Assesment (LCA) and/or SWOT (strengths, weaknesses, opportunities, threats), in combination with a hybrid fuzzy MCDM method can do the same for smart sustainable waste management problems.

5.2. *Limitations of the Study*

Although the smart sustainable waste management strategies proposed in the study are comprehensive and multi-level, no specific city, residential area, or pilot region was chosen as the application site for the strategies. For this reason, the proposed strategies were determined to encompass more general approaches without sharp boundaries. However, the needs and expectations of city residents, as well as local government plans and goals for a sustainable environment and economy, require that these strategies be implemented at different scales and with different adaptations.

Since the proposed strategies in the study were handled in a more general way, the evaluations of the decision makers were also made in this way. However, take, for example, the inclusion of illegal waste collection workers in a formal smart system. Since the human factor is involved here, the data obtained through preliminary interviews with potential workers to be included in the system will help make the expert evaluations in the decision-making processes healthier. Alternatively, consider the PAYT application based on blockchain. Before implementing this strategy, knowing the main goals of the authorities (e.g., reducing the amount of household waste generated annually, increasing the amount of waste recycled annually, etc.) could put the decision makers' assessments on a more realistic footing in terms of costs, operational feasibility, and acceptance.

Local governments are primarily responsible for ensuring environmental and economic sustainability in cities. In addition, most waste management services are provided by municipalities. This makes them among the most important stakeholder groups in the integration of smart city technologies into sustainable waste management. In forming the decision-making group in the study, experts working in the smart city departments of various municipalities were reached, but although some of the experts provided positive feedback, they did not participate in the mailed questionnaires. For this reason, in the part of the study where the criteria and alternatives are evaluated verbally, there is no opinion of a smart city expert for working in the public sector.

6. Conclusions

The long-term success of smart sustainable waste management strategies depends on the extent to which they meet the needs and expectations of the city in which they are implemented and how compatible they are with the sustainability goals of local governments. This makes these strategies multivariate, multi-restrictive, and multi-stakeholder, which complicates the determination processes. In this study, it was shown that decision makers can benefit from a hybrid fuzzy AHP-TOPSIS approach when determining sustainable waste management strategies in smart cities. The present study also revealed that authorities should give due consideration to operational feasibility and initial investment costs before implementing such strategies. In future studies, piloting the proposed smart waste management strategies will allow decision makers to make more concrete assessments by identifying the strengths and weaknesses of the strategies and testing their applicability. Furthermore, decision-making processes can be more strongly supported in this way by applying an LCA or SWOT analysis in parallel with a hybrid fuzzy AHP-TOPSIS approach.

Author Contributions: Conceptualization, B.G.D. and K.Y.; methodology, B.G.D. and K.Y.; software, B.G.D.; validation, B.G.D. and K.Y.; formal analysis, B.G.D.; investigation, B.G.D.; resources, B.G.D. and K.Y.; data curation, B.G.D.; writing—original draft preparation, B.G.D. and K.Y.; writing—review and editing, K.Y.; visualization, B.G.D. and K.Y.; supervision, K.Y. All authors have read and agreed to the published version of the manuscript.

Funding: This research received no external funding.

Institutional Review Board Statement: Not applicable.

Informed Consent Statement: Informed consent was obtained from all subjects involved in the study.

Data Availability Statement: Not applicable.

Acknowledgments: This work was carried out as a part of an M.Sc. thesis by Bihter Gizem Demircan, completed under the supervision of Kaan Yetilmezsoy. The authors also acknowledge the valuable insights of the reviewers and academic editor who contributed extensively to improving the article's quality.

Conflicts of Interest: The authors declare that there are no conflicts of interest, including any financial, personal, or other relationships, with other people or organizations.

Appendix A. Methodological Steps Used in Fuzzy AHP

The steps in the extent analysis method are summarized below:

Step 1: The value of fuzzy synthetic extent with respect to the i-th object is defined as in Equation (A1):

$$S_i = \sum_{j=1}^{m} M_{g_i}^{j} \otimes \left[\sum_{i=1}^{n} \sum_{j=1}^{m} M_{g_i}^{j} \right]^{-1} \tag{A1}$$

To obtain $\sum_{j=1}^{m} M_{g_i}^{j}$, the fuzzy addition operation of m extent analysis values for a particular matrix is performed as in Equation (A2):

$$\sum_{j=1}^{m} M_{g_i}^{j} = \left(\sum_{j=1}^{m} l_j, \sum_{j=1}^{m} m_j, \sum_{j=1}^{m} u_j \right) \tag{A2}$$

and to obtain $\left[\sum_{i=1}^{n} \sum_{j=1}^{m} M_{g_i}^{j} \right]^{-1}$, the fuzzy edition operation of m extent analysis values for a particular matrix is performed as in Equation (A3):

$$\left[\sum_{i=1}^{n} \sum_{j=1}^{m} M_{g_i}^{j} \right]^{-1} = \left(\sum_{i=1}^{n} l_i, \sum_{i=1}^{n} m_i, \sum_{i=1}^{n} u_i \right) \tag{A3}$$

and then the inverse of the vector in Equation (A3) is computed as in Equation (A4):

$$\left[\sum_{i=1}^{n} \sum_{j=1}^{m} M_{g_i}^{j} \right]^{-1} = \left(\frac{1}{\sum_{i=1}^{n} u_i} \right), \left(\frac{1}{\sum_{i=1}^{n} m_i} \right), \left(\frac{1}{\sum_{i=1}^{n} l_i} \right) \tag{A4}$$

Step 2: The degree of possibility of $M_2 = (l_2, m_2, u_2) \geq M_1 = (l_1, m_1, u_1)$ is defined as in Equation (A5):

$$V(M_2 \geq M_1) = \sup_{y \geq x}[\min(\mu_{M1}(x), \mu_{M2}(y))] \tag{A5}$$

and is equivalently shown as in Equation (A6):

$$V(M_2 \geq M_1) = hgt(M_1 \cap M_2) = \mu_{M1}(d) \begin{cases} 1, & if\ m_2 \geq m_1 \\ 0, & if\ \ l_1 \geq u_2 \\ \frac{l_1 - u_2}{(m_2 - u_2) - (m_1 - l_1)}, & other \end{cases} \tag{A6}$$

where hgt and d represent the highest intersection point and x coordinate of the intersection point D, respectively, between μ_{M1} and μ_{M2} (Figure A1). To compare two fuzzy numbers, M_1 and M_2, both the values of $V(M_1 \geq M_2)$ and $V(M_2 \geq M_1)$ are needed.

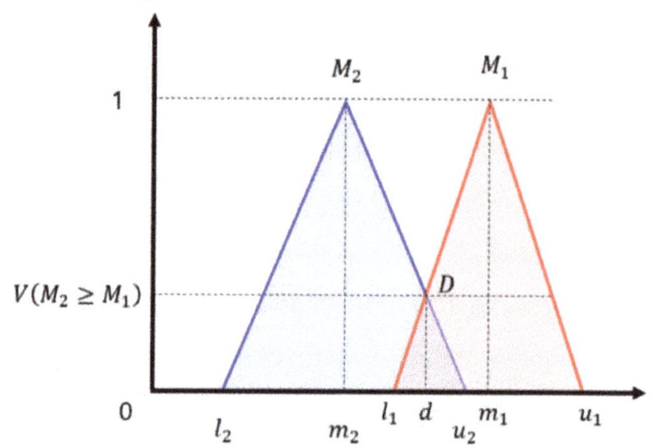

Figure A1. The intersection between M_1 and M_2.

Step 3: The degree possibility for a convex fuzzy number to be greater than convex fuzzy numbers M_i ($i = 1, 2, \ldots, k$) can be defined as in Equation (A7):

$$\begin{aligned} V(M \gg M_1, M_2, \ldots, M_k) &= V[(M \gg M_1)\ and\ (M \gg M_2)\ and\ \ldots\ and\ (M \gg M_k)] \\ &= min V(M \gg Mi), \quad i = 1, 2, \ldots, k \end{aligned} \tag{A7}$$

Assume that $d\prime(A_i) = min\ V(S_i \gg S_k)$ for $k = 1, 2, \ldots, n;\ k \neq 1$. Then, the weight factor for k_i is shown by Equation (A8):

$$W' = (d'(A_1), d'(A_2), \ldots, d'(A_n))^T, \tag{A8}$$

where $A_i (i = 1, 2, \ldots, n)$ are n elements.

Step 4: Via normalization, the normalized weight vectors would be as in shown Equation (A9):

$$W = (d(A_1), d(A_2), \ldots, d(A_n))^T \tag{A9}$$

where W is a non-fuzzy number.

Appendix B. Methodological Steps Used in Fuzzy TOPSIS

In the group consisting of K decision makers, where $\tilde{x}_{ij}$ represents the criterion value of the alternative, which was evaluated by the decision makers, and where $\tilde{w}_j$ represents the criterion weight, the criterion values and criteria weights, respectively, are calculated as in Equations (A10) and (A11):

$$\tilde{x}_{ij} = \frac{1}{K}\left[\ \tilde{x}_{ij}{}^{1} + \tilde{x}_{ij}{}^{2} + \ldots +\ \tilde{x}_{ij}{}^{K}\right] \tag{A10}$$

$$\tilde{w}_j = \frac{1}{K}\left[\tilde{w}_j{}^{1} + \tilde{w}_j{}^{2} + \ldots + \tilde{w}_j{}^{K}\right] \tag{A11}$$

A fuzzy multi-criteria group decision-making problem that can be succinctly expressed in matrix form as follows:

$$\tilde{\boldsymbol{D}} = \begin{bmatrix} \tilde{x}_{11} & \tilde{x}_{12} & \cdots & \tilde{x}_{1n} \\ \tilde{x}_{21} & \tilde{x}_{22} & \cdots & \tilde{x}_{2n} \\ \vdots & \vdots & \cdots & \vdots \\ \tilde{x}_{m1} & \tilde{x}_{m2} & \cdots & \tilde{x}_{mn} \end{bmatrix} \tag{A12}$$

$$\tilde{\boldsymbol{W}} = [\tilde{w}_1,\ \tilde{w}_2, \ldots, \tilde{w}_n] \tag{A13}$$

where $\tilde{x}_{ij}\ \forall\ i, j$ and $\tilde{w}_j$, $j = 1,\ 2,\ \ldots, n$ are linguistic variables. They could be described by triangular fuzzy numbers, $\tilde{x}_{ij} = (a_{ij}, b_{ij}, c_{ij})$ and $\tilde{w}_j = (w_{j1}, w_{j2}, w_{j3})$.

The normalized fuzzy decision matrix ($\tilde{\boldsymbol{R}}$) is shown as in Equation (A14):

$$\tilde{\boldsymbol{R}} = [\tilde{r}_{ij}]_{m\times n} \tag{A14}$$

With B and C being the benefit and cost criteria, respectively, the normalization is performed as in Equations (A15) and (A16):

$$\tilde{r}_{ij} = \left(\frac{a_{ij}}{c_j^*}, \frac{b_{ij}}{c_j^*}, \frac{c_{ij}}{c_j^*}\right), j \in B; \qquad c_j^* = \max_i c_{ij} \text{ if } j \in B \tag{A15}$$

$$\tilde{r}_{ij} = \left(\frac{a_j^-}{c_{ij}}, \frac{a_j^-}{b_{ij}}, \frac{a_j^-}{a_{ij}}\right), j \in C; \qquad a_j^- = \min_i a_{ij} \text{ if } j \in C \tag{A16}$$

Considering the different importance of each criterion, the weighted normalized fuzzy decision matrix is constructed as in Equation (A17):

$$\tilde{V} = [\tilde{v}_{ij}]_{mxn} \qquad i = 1, 2, \ldots, m \quad j = 1,\ 2, \ldots, n \tag{A17}$$

where $\tilde{v}_{ij} = \tilde{r}_{ij}\ (\cdot)\ \tilde{w}_j$. Fuzzy positive ideal solution (A^*) and fuzzy negative ideal solution (A^-) are defined as in Equations (A18) and (A19):

$$A^* = (\tilde{v}_1{}^*, \tilde{v}_2{}^*, \ldots, \tilde{v}_n{}^*) \tag{A18}$$

$$A^- = (\tilde{v}_1{}^-, \tilde{v}_2{}^-, \ldots, \tilde{v}_n{}^-) \tag{A19}$$

where $\tilde{v}_j^* = (1,\ 1,\ 1)$ and $\tilde{v}_j^- = (0,\ 0,\ 0)$, $j = 1,\ 2, \ldots, n$. The distance of the alternatives from A^* and A^- is calculated as in Equations (A20) and (A21):

$$d_i^* = \sum_{j=1}^{n} d(\tilde{v}_{ij},\ \tilde{v}_j^*),\ i = 1,\ 2, \ldots, m, \tag{A20}$$

$$d_i^- = \sum_{j=1}^{n} d(\tilde{v}_{ij},\ \tilde{v}_j^-),\ i = 1,\ 2, \ldots, m, \tag{A21}$$

where $d(.\ ,.)$ is the measurement of distance between two fuzzy numbers. After the d_i^* and d_i^- of alternatives have been calculated, the closeness coefficient of each alternative is calculated as in Equation (A22). The ranking of the alternatives is determined with this value.

$$CC_i = \frac{d_i^-}{d_i^* + d_i^-} \quad \text{(A22)}$$

Appendix C. Summary of the Values of $V(M_2 \geq M_1)$ and $V(M_1 \geq M_2)$

For the "environmental criteria", the values of $V(M_2 \geq M_1)$ and $V(M_1 \geq M_2)$ are presented as follows:

$V(S_{Less\ atmospheric\ emissions} \geq S_{Less\ soil\ pollution}) = 1.00$
$V(S_{Less\ atmospheric\ emissions} \geq S_{Less\ surface\ water\ pollution}) = 1.00$
$V(S_{Less\ atmospheric\ emissions} \geq S_{Energy\ recovery}) = 1.00$
$V(S_{Less\ atmospheric\ emissions} \geq S_{Natural\ resources\ recovery}) = 1.00$
$V(S_{Less\ soil\ pollution} \geq S_{Less\ atmospheric\ emissions}) = 0.65$
$V(S_{Less\ soil\ pollution} \geq S_{Less\ surface\ water\ pollution}) = 1.00$
$V(S_{Less\ soil\ pollution} \geq S_{Energy\ recovery}) = 1.00$
$V(S_{Less\ soil\ pollution} \geq S_{Natural\ resources\ recovery}) = 1.00$
$V(S_{Less\ surface\ water\ pollution} \geq S_{Less\ atmospheric\ emissions}) = 0.63$
$V(S_{Less\ surface\ water\ pollution} \geq S_{Less\ soil\ pollution}) = 1.00$
$V(S_{Less\ surface\ water\ pollution} \geq S_{Energy\ recovery}) = 1.00$
$V(S_{Less\ surface\ water\ pollution} \geq S_{Natural\ resources\ recovery}) = 1.00$
$V(S_{Energy\ recovery} \geq S_{Less\ atmospheric\ emissions}) = 0.10$
$V(S_{Energy\ recovery} \geq S_{Less\ soil\ pollution}) = 0.43$
$V(S_{Energy\ recovery} \geq S_{Natural\ resources\ recovery}) = 0.48$
$V(S_{Energy\ recovery} \geq S_{Natural\ resources\ recovery}) = 1.00$
$V(S_{Natural\ resources\ recovery} \geq S_{Less\ atmospheric\ emissions}) = 0.00$
$V(S_{Natural\ resources\ recovery} \geq S_{Less\ soil\ pollution}) = 0.15$
$V(S_{Natural\ resources\ recovery} \geq S_{Less\ surface\ water\ pollution}) = 0.21$
$V(S_{Natural\ resources\ recovery} \geq S_{Energy\ recovery}) = 0.76$

For the "technical criteria", the values of $V(M_2 \geq M_1)$ and $V(M_1 \geq M_2)$ are presented as follows:

$V(S_{Operational\ feasibility} \geq S_{Innovativeness}) = 1.00$
$V(S_{Operational\ feasibility} \geq S_{Need\ for\ qualified\ personnel}) = 1.00$
$V(S_{Innovativeness} \geq S_{Operational\ feasibility}) = 0.57$
$V(S_{Innovativeness} \geq S_{Need\ for\ qualified\ personnel}) = 1.00$
$V(S_{Need\ for\ qualified\ personnel} \geq S_{Operational\ feasibility}) = 0.00$
$V(S_{Need\ for\ qualified\ personnel} \geq S_{Innovativeness}) = 0.22$

For the "economic criteria", the values of $V(M_2 \geq M_1)$ and $V(M_1 \geq M_2)$ are presented as follows:

$V(S_{Initial\ investment\ costs} \geq S_{Operational\ costs}) = 1.00$
$V(S_{Initial\ investment\ costs} \geq S_{Maintenance\ costs}) = 1.00$
$V(S_{Initial\ investment\ costs} \geq S_{Transportation\ costs}) = 1.00$
$V(S_{Operational\ costs} \geq S_{Initial\ investment\ costs}) = 0.53$
$V(S_{Operational\ costs} \geq S_{Maintenance\ costs}) = 1.00$
$V(S_{Operational\ costs} \geq S_{Transportation\ costs}) = 1.00$
$V(S_{Maintenance\ costs} \geq S_{Initial\ investment\ costs}) = 0.26$
$V(S_{Maintenance\ costs} \geq S_{Operational\ costs}) = 0.68$
$V(S_{Maintenance\ costs} \geq S_{Transportation\ costs}) = 1.00$
$V(S_{Transportation\ costs} \geq S_{Initial\ investment\ costs}) = 0.00$
$V(S_{Transportation\ costs} \geq S_{Operational\ costs}) = 0.12$

$V(S_{Transportation\ costs} \geq S_{Maintenance\ costs}) = 0.48$

For the "social criteria", the values of $V(M_2 \geq M_1)$ and $V(M_1 \geq M_2)$ are presented as follows:

$V(S_{Increased\ awareness\ on\ sustainable\ cities} \geq S_{Increased\ quality\ of\ life\ in\ the\ city}) = 1.00$
$V(S_{Increased\ awareness\ on\ sustainable\ cities} \geq S_{New\ employment\ opportunities}) = 1.00$
$V(S_{Increased\ quality\ of\ life\ in\ the\ city} \geq S_{Increased\ awareness\ on\ sustainable\ cities}) = 0.87$
$V(S_{Increased\ quality\ of\ life\ in\ the\ city} \geq S_{New\ employment\ opportunities}) = 1.00$
$V(S_{New\ employment\ opportunities} \geq S_{Increased\ awareness\ on\ sustainable\ cities}) = 0.03$
$V(S_{New\ employment\ opportunities} \geq S_{Increased\ quality\ of\ life\ in\ the\ city}) = 0.25$

Appendix D. Summary of New Closeness Coefficient Values

Table A1. New closeness coefficient values (CC_i) of the alternatives after the sensitivity analysis.

Scenario	CC_i (A1)	CC_i (A2)	CC_i (A3)	CC_i (A4)
1	0.440	0.458	0.454	0.451
2	0.441	0.458	0.453	0.452
3	0.440	0.458	0.453	0.452
4	0.441	0.459	0.455	0.453
5	0.440	0.458	0.453	0.452
6	0.440	0.458	0.453	0.451
7	0.440	0.458	0.454	0.453
8	0.439	0.458	0.453	0.451
9	0.440	0.457	0.453	0.452
10	0.440	0.458	0.453	0.452
11	0.440	0.458	0.453	0.452
12	0.440	0.458	0.453	0.452
13	0.440	0.458	0.454	0.452
14	0.440	0.458	0.453	0.452
15	0.440	0.458	0.453	0.452
16	0.410	0.423	0.419	0.417
17	0.437	0.458	0.450	0.443
18	0.408	0.422	0.417	0.413

References

1. United Nations Human Settlements Programme (UN-Habitat). World Cities Report 2020: The Value of Sustainable Urbanization. 2020. Available online: https://unhabitat.org/sites/default/files/2020/10/wcr_2020_report.pdf (accessed on 23 January 2023).
2. United Nations Environment Programme. Emissions Gap Report 2019. Available online: https://www.unep.org/resources/emissions-gap-report-2019 (accessed on 23 January 2023).
3. United Nations. Water and Urbanization. Available online: https://www.unwater.org/app/uploads/2018/10/WaterFacts_water_and_urbanization_sep2018.pdf (accessed on 23 January 2023).
4. Camero, A.; Alba, E. Smart city and information technology: A review. *Cities* **2019**, *93*, 89–94. [CrossRef]
5. Eremia, M.; Toma, L.; Sanduleac, M. The smart city concept in the 21st century. *Procedia Eng.* **2017**, *181*, 12–19. [CrossRef]
6. Syed, A.S.; Sierra-Sosa, D.; Kumar, A.; Elmaghraby, A. IoT in smart cities: A survey of technologies, practices and challenges. *Smart Cities* **2021**, *4*, 429–475. [CrossRef]
7. Chatterjee, S.; Kar, A.K.; Gupta, M.P. Success of IoT in smart cities of India: An empirical analysis. *Gov. Inf. Q.* **2018**, *35*, 349–361. [CrossRef]
8. Giovannoni, E.; Fabietti, G. What Is Sustainability? A Review of the Concept and Its Applications. In *Integrated Reporting*; Busco, C., Frigo, M., Riccaboni, A., Quattrone, P., Eds.; Springer: Cham, Switzerland, 2013; Volume 26. [CrossRef]
9. Goodland, R. The Concept of Environmental Sustainability. *Annu. Rev. Ecol. Syst.* **1995**, *26*, 1–24. [CrossRef]
10. Morelli, R. Environmental Sustainability: A Definition for Environmental Professionals. *J. Environ. Sust.* **2011**, *1*, 2. [CrossRef]
11. Seadon, J.K. Sustainable waste management systems. *J. Clean. Prod.* **2010**, *18*, 1639–1651. [CrossRef]
12. Petts, J. Municipal waste management: Inequities and the role of deliberation. *Risk Anal.* **2000**, *20*, 821–832. [CrossRef]
13. Nilsson-Djerf, J.; McDougall, F. Social factors in sustainable waste management. *Warmer Bull.* **2000**, *73*, 18–20.
14. Esmaeilian, B.; Wang, B.; Lewis, K.; Duarte, F.; Ratti, C.; Behdad, S. The future of waste management in smart and sustainable cities: A review and concept paper. *Waste Manag.* **2018**, *81*, 177–195. [CrossRef]
15. Hannan, M.A.; Al Mamuna, M.D.A.; Hussain, A.; Basri, H.; Begum, R.A. A review on technologies and their usage in solid waste monitoring and manage-ment systems: Issues and challenges. *Waste Manag.* **2015**, *43*, 509–523. [CrossRef] [PubMed]

16. Kang, K.D.; Kang, H.; Ilankoon, I.M.S.K.; Chong, C.Y. Electronic waste collection systems using internet of things (iot): Household electronic waste management in Malaysia. *J. Clean. Prod.* **2020**, *252*, 119801. [CrossRef]
17. Jiang, P.; Van Fan, Y.; Zhou, J.; Zheng, M.; Liu, X.; Klemeš, J.J. Data-driven analytical framework for waste-dumping behaviour analysis to facilitate policy regulations. *Waste Manag.* **2020**, *103*, 285–295. [CrossRef] [PubMed]
18. Gopikumar, S.; Raja, S.; Robinson, Y.H.; Shanmuganathan, V.; Rho, S. A method of landfill leachate management using internet of things for sustainable smart city development. *Sustain. Cities Soc.* **2020**, *66*, 102521. [CrossRef]
19. Senthilkumar, R.; Venkatakrishnan, P.; Balaji, N. Intelligent based novel embedded system based iot enabled air pollution monitoring system. *Microprocess. Microsyst.* **2020**, *77*, 103172. [CrossRef]
20. Khoa, T.A.; Phuc, C.H.; Lam, P.D.; Nhu, L.M.B.; Trong, N.M.; Phuong, N.T.H.; Dung, N.V.; Tan-Y, N.; Nguyen, H.N.; Duc, D.N.M. Waste management system using iot-based machine learning in university. *Wirel. Commun. Mob. Comput.* **2020**, *2020*, 6138637. [CrossRef]
21. Sheng, T.J.; Islam, M.S.; Misran, N.; Baharuddin, M.H.; Arshad, H.; Islam, M.D.R.; Chowdhury, M.E.H.; Rmili, H.; Islam, M.T. An internet of things based smart waste management system using lora and tensorflow deep learning model. *IEEE Access* **2020**, *8*, 148793–148811. [CrossRef]
22. Mastos, T.D.; Nizamis, A.; Vafeiadis, T.; Alexopoulos, N.; Ntinas, C.; Gkortzis, D.; Papadopoulos, A.; Ioannidis, D.; Tzovaras, D. Industry 4.0 sustainable supply chains: An application of an IoT enabled scrap metal management solution. *J. Clean. Prod.* **2020**, *269*, 122377. [CrossRef]
23. Kassou, M.; Bourekkadi, S.; Khoulji, S.; Slimani, K.; Chikri, H.; Kerkeb, M.L. Blockchain-based medical and water waste management conception. *E3S Web Conf.* **2021**, *234*, 00070. [CrossRef]
24. Gupta, Y.S.; Mukherjee, S.; Dutta, R.; Bhattacharya, S. A blockchain-based approach using smart contracts to develop a smart waste management system. *Int. J. Environ. Sci. Technol.* **2022**, *19*, 7833–7856. [CrossRef]
25. Kahraman, C. Multi-Criteria Decision Making Methods and Fuzzy Sets. In *Fuzzy Multi-Criteria Decision Making*; Springer Optimization and Its Applications; Kahraman, C., Ed.; Springer: Boston, MA, USA, 2008; Volume 16. [CrossRef]
26. Aouam, T.; Chang, S.I.; Lee, E.S. Fuzzy MADM: An outranking method. *Eur. J. Oper. Res.* **2003**, *145*, 317–328. [CrossRef]
27. Stojčić, M.; Zavadskas, E.K.; Pamučar, D.; Stević, Ž.; Mardani, A. Application of MCDM methods in sustainability engineering: A literature review 2008–2018. *Symmetry* **2019**, *11*, 350. [CrossRef]
28. Rajak, S.; Parthiban, P.; Dhanalakshmi, R. Analysing barriers of sustainable transportation systems in India using Grey-DEMATEL approach: A supply chain perspective. *Int. J. Sustain. Eng.* **2021**, *14*, 419–432. [CrossRef]
29. Stanković, J.J.; Marjanović, I.; Papathanasiou, J.; Drezgić, S. Social, Economic and Environmental Sustainability of Port Regions: MCDM Approach in Composite Index Creation. *J. Mar. Sci. Eng.* **2021**, *9*, 74. [CrossRef]
30. Dhiman, H.S.; Deb, D. Fuzzy TOPSIS and fuzzy COPRAS based multi-criteria decision making for hybrid wind farms. *Energy* **2020**, *202*, 117755. [CrossRef]
31. Saraswat, S.K.; Digalwar, A.K. Evaluation of energy sources based on sustainability factors using integrated fuzzy MCDM approach. *Int. J. Environ. Sec. Manag.* **2021**, *15*, 246–266. [CrossRef]
32. Van Thanh, N. Sustainable Energy Source Selection for Industrial Complex in Vietnam: A Fuzzy MCDM Approach. *IEEE Access* **2022**, *10*, 50692–50701. [CrossRef]
33. Gervásio, H.; Simões da Silva, L. A probabilistic decision-making approach for the sustainable assessment of infrastructures. *Exp. Syst. Appl.* **2012**, *39*, 7121–7131. [CrossRef]
34. Yadegaridehkordi, E.; Hourmand, M.; Nilashi, M.; Alsolami, E.; Samad, S.; Mahmoud, M.; Alarood, A.A.; Zainol, A.; Hamsa, D.M.; Shuib, L. Assessment of sustainability indicators for green building manufacturing using fuzzy multi-criteria decision making approach. *J. Clean. Prod.* **2020**, *277*, 122905. [CrossRef]
35. Chu, T.C.; Lin, Y.C. A fuzzy TOPSIS method for robot selection. *J. Adv. Manuf. Technol.* **2003**, *21*, 284–290. [CrossRef]
36. Rouyendegh, B.D.; Yildizbasi, A.; Üstünyer, P. Intuitionistic fuzzy TOPSIS method for green supplier selection problem. *Soft Comput.* **2020**, *24*, 2215–2228. [CrossRef]
37. Thanh, N.V.; Lan, N.T.K. A New Hybrid Triple Bottom Line Metrics and Fuzzy MCDM Model: Sustainable Supplier Selection in the Food-Processing Industry. *Axioms* **2022**, *11*, 57. [CrossRef]
38. Nguyen, T.-L.; Nguyen, P.-H.; Pham, H.-A.; Nguyen, T.-G.; Nguyen, D.-T.; Tran, T.-H.; Le, H.-C.; Phung, H.-T. A Novel Integrating Data Envelopment Analysis and Spherical Fuzzy MCDM Approach for Sustainable Supplier Selection in Steel Industry. *Mathematics* **2022**, *10*, 1897. [CrossRef]
39. Achillas, C.; Moussiopoulos, N.; Karagiannidis, A.; Banias, G.; Perkoulidis, G. The use of multi-criteria decision analysis to tackle waste management problems: A literature review. *Waste Manag. Res.* **2013**, *31*, 115–129. [CrossRef] [PubMed]
40. Sobral, M.M.; Hipel, K.W.; Farquhar, G.J. A multi-criteria model for solid waste management. *J. Environ. Manag.* **1981**, *12*, 97–110.
41. Erkut, E.; Moran, S.R. Locating obnoxious facilities in the public sector: An application of the analytic hierarchy process to municipal landfill siting decisions. *Socio-Econ. Plan. Sci.* **1991**, *25*, 89–102. [CrossRef]
42. Banai, R. Fuzziness in geographical Information Systems: Contributions from the analytic hierarchy process. *Int. J. Geogr. Inf. Syst.* **1993**, *7*, 315–329. [CrossRef]
43. Charnpratheep, K.; Zhou, Q.; Garner, B. Preliminary landfill site screening using fuzzy geographical information systems. *Waste Manag. Res.* **1997**, *15*, 197–215. [CrossRef]

44. Su, J.P.; Chiueh, P.T.; Hung, M.L.; Ma, H.W. Analyzing policy impact potential for municipal solid waste management decision-making: A case study of Taiwan. *Resour. Conserv. Recycl.* **2007**, *51*, 418–434. [CrossRef]
45. Önüt, S.; Soner, S. Transshipment site selection using the AHP and TOPSIS approaches under fuzzy environment. *Waste Manag.* **2008**, *28*, 1552–1559. [CrossRef]
46. Van Thanh, N. Optimal waste-to-energy strategy assisted by fuzzy MCDM model for sustainable solid waste management. *Sustainability* **2022**, *14*, 6565. [CrossRef]
47. Albayrak, K. A hybrid fuzzy decision making approach for sitting a solid waste energy production plant. *Soft Comput.* **2022**, *26*, 575–587. [CrossRef]
48. Topaloglu, M.; Yarkin, F.; Kaya, T. Solid waste collection system selection for smart cities based on a type-2 fuzzy multi-criteria decision technique. *Soft Comput.* **2018**, *22*, 4879–4890. [CrossRef]
49. Chauhan, A.; Jakhar, S.K.; Chauhan, C. The interplay of circular economy with industry 4.0 enabled smart city drivers of healthcare waste disposal. *J. Clean. Prod.* **2021**, *279*, 123854. [CrossRef]
50. Seker, S. IoT based sustainable smart waste management system evaluation using MCDM model under interval-valued q-rung orthopair fuzzy environment. *Technol. Soc.* **2022**, *71*, 102100. [CrossRef]
51. Yörükoğlu, M.; Aydın, S. Assessment of smart waste management systems with spherical ahp method. In *Artificial Intelligence for Knowledge Management, Energy, and Sustainability*; AI4KMES 2021. IFIP Advances in Information and Communication Technology; Mercier-Laurent, E., Kayakutlu, G., Eds.; Springer: Cham, Switzerland, 2022; Volume 637. [CrossRef]
52. Kumar, R.R.; Mishra, S.; Kumar, C. Prioritizing the solution of cloud service selection using integrated MCDM methods under Fuzzy environment. *J. Supercomput.* **2017**, *73*, 4652–4682. [CrossRef]
53. Yucesan, M.; Gul, M. Hospital service quality evaluation: An integrated model based on Pythagorean fuzzy AHP and fuzzy TOPSIS. *Soft Comput.* **2020**, *24*, 3237–3255. [CrossRef]
54. Karabayir, A.N.; Botsali, A.R.; Kose, Y.; Cevikcan, E. Supplier Selection in a Construction Company Using Fuzzy AHP and Fuzzy TOPSIS. In *Intelligent and Fuzzy Techniques in Big Data Analytics and Decision Making*; INFUS 2019. Advances in Intelligent Systems and Computing, 1029; Kahraman, C., Cebi, S., Cevik Onar, S., Oztaysi, B., Tolga, A., Sari, I., Eds.; Springer: Cham, Switzerland, 2019. [CrossRef]
55. Baki, R. Evaluating hotel websites through the use of fuzzy AHP and fuzzy TOPSIS. *Int. J. Contemp. Hospit. Manag.* **2020**, *32*, 3747–3765. [CrossRef]
56. Gülsün, B.; Erdoğmuş, K.N. Bankacılık Sektöründe Bulanık Analitik Hiyerarşi Prosesi ve Bulanık TOPSIS Yöntemleri ile Finansal Performans Değerlendirmesi. *Süleyman Demirel Üniversitesi Fen Bilim. Enstitüsü Derg.* **2021**, *25*, 1–15. [CrossRef]
57. Banadkouki, M.Z.; Lotfi, M.M. Selection of computer-integrated manufacturing technologies using a combined fuzzy analytic hierarchy process and fuzzy TOPSIS. *Int. J. Indust. Eng. Product. Res.* **2021**, *32*, 105–120. [CrossRef]
58. Padma, T.; Shantharajah, S.P.; Ramadoss, P. Hybrid Fuzzy AHP and Fuzzy TOPSIS Decision Model for Aquaculture Species Selection. *Int. J. Inform. Technol. Decis. Mak.* **2022**, *21*, 999–1030. [CrossRef]
59. Ajjipura Shankar, H.U.; Kodipalya Nanjappa, U.K.; Alsulami, M.D.; Prasannakumara, B.C. A Fuzzy AHP-Fuzzy TOPSIS Urged Baseline Aid for Execution Amendment of an Online Food Delivery Affability. *Mathematics* **2022**, *10*, 2930. [CrossRef]
60. Torkayesh, A.E.; Rajaeifar, M.A.; Rostom, M.; Malmir, B.; Yazdani, M.; Suh, S.; Heidrich, O. Integrating life cycle assessment and multi criteria decision making for sustainable waste management: Key issues and recommendations for future studies. *Renew. Sustain. Energy Rev.* **2022**, *168*, 112819. [CrossRef]
61. Zorpas, A.A.; Saranti, A. Multi-criteria analysis of sustainable environmental clean technologies for the treatment of winery's wastewater. *Int. J. Glob. Environ. Issues* **2016**, *15*, 151–168. [CrossRef]
62. Kumar, A.; Dixit, G. A novel hybrid MCDM framework for WEEE recycling partner evaluation on the basis of green competencies. *J. Clean. Prod.* **2019**, *241*, 18017. [CrossRef]
63. Coban, A.; Firtina Ertis, I.; Ayvaz Cavdaroglu, N. Municipal solid waste management via multi-criteria decision making methods: A case study in Istanbul, Turkey. *J. Clean. Prod.* **2018**, *180*, 159–167. [CrossRef]
64. Ilangkumaran, M.; Sasirekha, V.; Anojkumar, L.; Sakthivel, G.; Boopathi Raja, M.; Ruban Sundara Raj, T.; Siddhartha, C.; Nizamuddin, P.; Praveen Kumar, S. Optimization of wastewater treatment technology selection using hybrid MCDM. *Manag. Environ. Qual.* **2013**, *24*, 619–641. [CrossRef]
65. Khan, I.; Kabir, Z. Waste-to-energy generation technologies and the developing economies: A multi-criteria analysis for sustainability assessment. *Renew. Energy* **2020**, *150*, 320–333. [CrossRef]
66. Van Laarhoven, P.J.M.; Pedrycz, W. A fuzzy extension of Saaty's priority theory. *Fuzzy Sets Syst.* **1983**, *11*, 229–241. [CrossRef]
67. Da-Yong, C. Applications of the extent analysis method on fuzzy AHP. *Eur. J. Oper. Res.* **1996**, *95*, 649–655. [CrossRef]
68. Dursun, E. Supplier Selection by Using Fuzzy AHP Method and An Application in Textile Sector. Master's Thesis, The Graduate School of Science Engineering and Technology, Institute of Science and Technology, Istanbul Technical University, Istanbul, Turkey, 2009. (In Turkish).
69. Sahu, N.K.; Sahu, A.K.; Sahu, A.K. Appraisement and benchmarking of third-party logistic service provider by exploration of risk-based approach. *Cogent Bus. Manag.* **2015**, *2*, 1121637. [CrossRef]
70. Kim, C.; Won, J.-S. A fuzzy analytic hierarchy process and cooperative game theory combined multiple mobile robot navigation algorithm. *Sensors* **2020**, *20*, 2827. [CrossRef]

71. Hwang, C.-L.; Yoon, K.P. *Multiple Attribute Decision Making: Methods and Applications*; Springer: Berlin/Heidelberg, Germany; New York, NY, USA, 1981.
72. Chen, C.-T.; Lin, C.-T.; Huang, S.-F. A fuzzy approach for supplier evaluation and selection in supply chain management. *Int. J. Prod. Econ.* **2006**, *102*, 289–301. [CrossRef]
73. Alptekin, N.; Eroglu Hall, E.; Sevim, N. Evaluation of websites quality using fuzzy TOPSIS method. *Int. J. Acad. Res. Bus. Soc. Sci.* **2015**, *5*, 221–242. [CrossRef]
74. 1000minds. Decision-Making/Multi-Criteria Decision Analysis (MCDA/MCDM). Available online: https://www.1000minds.com/decision-making/what-is-mcdm-mcda (accessed on 23 January 2023).
75. Taskin Gumus, A. Evaluation of hazardous waste transportation firms by using a two step fuzzy-AHP and TOPSIS methodology. *Exp. Syst. Appl.* **2009**, *36*, 4067–4074. [CrossRef]
76. Bellman, R.E.; Zadeh, L.A. Decision making in a fuzzy environment. *Manag. Sci.* **1970**, *17*, 141–164. [CrossRef]
77. Zimmermann, H.J. Fuzzy programming and linear programming with several objective functions. *Fuzzy Sets Syst.* **1978**, *1*, 45–55. [CrossRef]
78. Deng, J.L. Introduction to gray system theory. *J. Grey Syst.* **1989**, *1*, 1–24.
79. Khuman, A.S. The similarities and divergences between grey and fuzzy theory. *Exp. Syst. Appl.* **2021**, *186*, 115812. [CrossRef]
80. Chen, C.-T. Extensions of the TOPSIS for group decision-making under fuzzy environment. *Fuzzy Sets Syst.* **2000**, *114*, 1–9. [CrossRef]
81. Goyal, S.; Garg, D.; Luthra, S. Sustainable production and consumption: Analysing barriers and solutions for maintaining green tomorrow by using fuzzy-AHP–fuzzy-TOPSIS hybrid framework. *Environ. Dev. Sustain.* **2021**, *23*, 16934–16980. [CrossRef]
82. Demircan, B.G. Determination of Sustainable Waste Management Strategies in Smart Cities Using Fuzzy Multi-Criteria Decision Approach. Master's Thesis, Institute of Science, Department of Environmental Engineering, Yildiz Technical University, Istanbul, Turkey, 2023.
83. Martín, J.C.; Indelicato, A. Comparing a Fuzzy Hybrid Approach with Invariant MGCFA to Study National Identity. *Appl. Sci.* **2023**, *13*, 1657. [CrossRef]

sustainability

Article

Additive Manufacture of Recycled Poly(Ethylene Terephthalate) Using Pyromellitic Dianhydride Targeted for FDM 3D-Printing Applications

Mohammed Alzahrani [1], Hesham Alhumade [1], Leonardo Simon [1], Kaan Yetilmezsoy [2], Chandra Mouli R. Madhuranthakam [3,*] and Ali Elkamel [1,4,*]

1 Department of Chemical Engineering, University of Waterloo, Waterloo, ON N2L 3G1, Canada; mohadeng@hotmail.com (M.A.); halhumade@uwaterloo.ca (H.A.); lsimon@uwaterloo.ca (L.S.)
2 Department of Environmental Engineering, Faculty of Civil Engineering, Yildiz Technical University, Davutpasa, Esenler, Istanbul 34220, Turkey; yetilmez@yildiz.edu.tr
3 Department of Chemical Engineering, Abu Dhabi University, Abu Dhabi P.O. Box. 59911, United Arab Emirates
4 Department of Chemical Engineering, Khalifa University, Abu Dhabi P.O. Box. 127788, United Arab Emirates
* Correspondence: chandra.mouli@adu.ac.ae (C.M.R.M.); aelkamel@uwaterloo.ca (A.E.)

Abstract: The suitability of recycled poly(ethylene terephthalate) (R-PET) for 3D-printing applications was evaluated by studying the melt flow characteristics of the polymer. R-PET is known to experience significant deterioration in its mechanical properties when recycled due to molecular weight loss that results from reprocessing. Lower molecular weight hinders R-PET from being 3D-printable due to low viscosity and melt strength. The hypothesis was that R-PET can be modified with reasonable effort and resources to a 3D-printable thermoplastic if the low viscosity problem is tackled. Higher viscosity will enhance both the melt strength and the melt flow characteristic of the polymer, making it more suitable for processing and 3D printing. Reactive extrusion was selected as the method for modifying the polymer to achieve the objective via a coupling reaction with chain extender PMDA (pyromellitic dianhydride). A decrease in the melt flow index (MFI) from 90 to 1.2 (g/10 min) was recorded when PMDA was added at 0.75 wt% which lowered the MFI of modified R-PET to a comparable value to commercial 3D-printing filaments. Furthermore, FT-IR analysis was performed to investigate the chemical composition of the product. Finally, a 3D-printing filament was made from the modified R-PET by mimicking the main processing stations that exist in the filament-making process, which are the extrusion stage, water bath cooling stage and spooling stage. With 0.75 wt% PMDA, the melt strength was satisfactory for pulling the filament and, therefore, a filament with on-spec dimension was produced. Finally, a small object was successfully 3D-printed using the filament product at a minimum recommended temperature of 275 °C.

Keywords: 3D printing; poly(ethylene terephthalate); reactive extrusion; recycled thermoplastic

Citation: Alzahrani, M.; Alhumade, H.; Simon, L.; Yetilmezsoy, K.; Madhuranthakam, C.M.R.; Elkamel, A. Additive Manufacture of Recycled Poly(Ethylene Terephthalate) Using Pyromellitic Dianhydride Targeted for FDM 3D-Printing Applications. *Sustainability* **2023**, *15*, 5004. https://doi.org/10.3390/su15065004

Academic Editor: Marc A. Rosen

Received: 5 January 2023
Revised: 26 February 2023
Accepted: 2 March 2023
Published: 11 March 2023

1. Introduction

The additive manufacturing, or 3D-printing, market has been growing fast in recent years as many companies and entrepreneurs see great potential in the market. In 2019, world plastic manufacturing was around 368 million tons as per Plastics Atlas [1], of which a majority of the thermo plastics can potentially be processed for additive manufacturing [2]. Today, 3-D-printing technologies have a wide range of applications from the construction industry [3–5], automobile applications [6–8], the food industry [9], the aerospace industry [10] and in jewelry collections [11], to more sophisticated medical applications and the manufacturing of artificial human organs [12–14]. Among several technologies, fused deposition modeling (FDM) is recognized as the most popular due to its simplicity and low capital cost [15–17]. In the last 15 years, 3D printing has become increasingly popular as people realize the diversity of applications that 3D printing can contribute to. Therefore, more researchers and entrepreneurs dedicated time and effort to developing new technologies as well as enhancing existing ones. As a result, the additive manufacturing industry thrived unprecedentedly with the invention of various robust and commercially viable machineries. The classification of technologies of additive manufacturing processes can be achieved in many different ways based on the specific aspect the consumer is looking for. For example, it is common to classify 3D printers based on the technology baseline of the device, whether it uses a laser source to construct the object or an extrusion-based technology, as is the case in Fused Deposition Modelling (FDM) processes. However, this classification does not consider the type of the raw material (i.e., liquid photo-polymer, powder, solid polymer, etc.), which is an essential piece of information to end-users. Therefore, a more inclusive classification was developed that specifies two important dimensions in any additive manufacturing process: the method of constructing layers and the type of raw material. Firstly, the method of constructing layers was divided into two main subcategories: a one-dimensional source of construction, such as a nozzle, and a two-dimensional source which is essentially an array of one-dimensional sources of construction methods, such as those used in PolyJet technologies. The second dimension recognizes four types of raw material used in additive manufacturing: liquid polymers, discrete particles, molten materials and solid sheets. Among the different technologies, Fusion Deposition Modeling (FDM) technology is by far the most popular 3D-printing technology due to its relatively low cost and simplicity. It was first developed by the American company Stratasys, but competition from many other entities led to lowering the cost of this technology, making it feasible for home uses. In this type of 3D printing, raw materials are supplied as rounded filaments of a thermoplastic, or sometimes, a metal wire. In this technology, the 3D-printing device is equipped with a small nozzle, or in some cases two nozzles, that is connected to a heating element which allows the setup of the temperature of the nozzle at the melting point of the thermoplastic, or metal, raw material. The inlet I.D. of the nozzle comes in two different standard sizes: 3.0 mm and 1.75 mm, whereas the printer's nozzle size is often 0.4 mm (although it can be modified). The principle of this technology is that the raw material is fed to the heated extrusion nozzle where it melts, while maintaining specific melt flow characteristics. Then, it is ejected through the nozzle outlet on the printer's bed. This bed is movable by a motor and is often connected to a heating element to allow the melt to adhere to the heated bed. With the bed being movable in the y-axis and the nozzle movable on the x-axis, the printer uses the object's digital code and the 3D model to construct the first layer on the bed by continuously dispositioning the melt on specific x and y coordinates. Once the first layer is constructed, the nozzle moves in the z-axis to start constructing the second layer on top of the first one (which has already solidified). This move in the z-axis is as small as the thickness of the first layer (often 0.1 mm), and the smaller the thickness of each layer, the higher the object's resolution will be. The nozzle moves to the third layer after completing the second, and continues, until the whole object is constructed.

FDM technology mainly serves home users and enthusiasts as well as serving educational purposes, and it mainly utilizes thermoplastics, most commonly ABS (acrylonitrile butadiene styrene), PLA (poly(lactic acid)), and nylon, as the building materials for printed objects. With FDM 3D-printing technology being the most affordable and widespread, thermoplastics are the most suitable material for this extrusion-based technology. In FDM technology, raw materials should experience a phase change, from solid to viscous and paste-like, when heated in the extrusion chamber and should retain the mechanical properties when cooled down. Furthermore, melting and solidification must be well studied with respect to temperature and time to evaluate the suitability of a thermoplastic for additive manufacturing. It is suggested that amorphous thermoplastics are generally more suitable for FDM 3D printing than crystalline polymers. This is because amorphous polymers do not have a specific melting point; therefore, keeping them in the semisolid (paste-like) phase can be easily achieved by finding the right temperature. Having said that, some semi-crystalline thermoplastics, such as PLA, have been proven very suitable and are popular polymers for FDM printing.

ABS is one of the two most popular thermoplastics used in FDM machines (along with PLA). It is known for being a uniquely tough and ductile amorphous polymer that can withstand heavy uses. Moreover, ABS has excellent temperature resistance as it starts softening at relatively high temperatures (around 230 °C). ABS is also cheaper than other thermoplastics used for 3D printing. On the other hand, ABS has some disadvantages from an environmental and health point of view. Since it is a petroleum-based polymer, it is a non-biodegradable plastic. Furthermore, while 3D printing with ABS, it is anticipated that mild fumes will be released that can be easily smelled. Therefore, it is often recommended to have some ventilation in the room where ABS is being 3D-printed. ABS requires a heated bed in the printing machine to avoid warping. Polylactic acid (PLA) is a biodegradable polymer and is relatively easy to use. PLA has become one of the most favorable materials for 3D printing. Compared to ABS, PLA is considered a healthier and safer material to work with since it is not toxic and produces much safer fumes than ABS when 3D-printed. From the mechanical performance point of view, however, it is not as tough as ABS and is more brittle. Moreover, PLA starts to soften at relatively low temperatures (around 50 °C), which makes it less heat-resistant than ABS. The process of 3D-printing with PLA is carried out at roughly 210 °C, and it does not require a heated bed (although having a heated bed at around 60 °C can enhance the product's quality, depending on the object's size and shape). Since it is derived from renewable resources, such as corn starch, PLA can be used in printing parts with biomedical uses. High-Impact Polystyrene (HIPS) is known to have similar properties to ABS, but it is slightly more expensive. Because HIPS is soluble in Limonene, it is often used as a structural support for 3D printing other objects. Then, it is quickly dissolved and removed, resulting in cleaner finishing of the supported object. This saves users the effort of sanding their built object, which is a comparatively much more tedious job. HIPS is usually printed at 250 °C on a heated bed and is likely to experience less warpage than ABS. Nylon, which is also known as polyamide, is an increasingly used material in FDM printing. It has several desirable properties, such as strength, flexibility and durability. On the other hand, it is one of the most hygroscopic polymers among all kinds of thermoplastics. It absorbs moisture significantly and, as a result, users might need to dry it in an oven before using it to obtain good printing quality. Nylon is printed at around 250 °C and requires a heated bed at around 65 °C.

Poly(ethylene terephthalate), or PET, is the most recycled thermoplastic in Canada and arguably worldwide [18]. Additionally, it has unique desirable properties over other thermoplastics, including thermal and chemical stability and stiffness [19,20]. Therefore, utilizing PET waste in producing a 3D-printing filament can bring both economic opportunity and environmental benefits. On the other hand, like other thermoplastics, R-PET suffers drastic change in its mechanical properties when it is recycled, mainly due to degradation that causes loss in molecular weight [21–23]. Hence, it was expected that making filaments from R-PET and 3D printing with them would be challenging due to the deterioration in the polymer's melt flow characteristics mainly because of loss in molecular weight. However, it is believed that the polymer can be 3D-printable when modified by reactive extrusion.

Reactive extrusion is a popular method for modifying polymers since it offers several advantages over conventional polymerization processes, such as being a continuous process, massively reducing processing times and providing precise thermal control throughout the process. Several reaction types have been conducted by reactive extrusion (REX) including coupling reactions [24–27]. In coupling reactions, a functional coupling agent, or chain extender, is used to link separated polymer chains by reacting with end-groups in polymer chains. This linkage produces longer polymer chains with higher molecular weights (MWs) and, therefore, increases the polymer's viscosity. Qin et al. [28] used triphenyl phosphite as a chain extender in the reactive extrusion of PET. Yang et al. [29] used reactive extrusion with a multi-functional epoxide to produce PET with high melt elasticity. Among the several additives, pyromellitic dianhydride (PMDA) is an important additive that has the promising attribute of increasing the intrinsic viscosity due to an increase in the molecular weight. It has been used to make foamable PET with supercritical carbon dioxide as a blowing agent and had an effect on intrinsic viscosity [30]. In this work, the chain extender PMDA was used as an additive for the reactive extrusion of PET due to its effectiveness and economic feasibility [31]. In addition, we show that actual printing is feasible and successfully carried out using recycled PET.

2. Materials and Methods

2.1. Sample Mixing and Drying

It was important to ensure proper mixing of R-PET flakes and PMDA powder since they were in different forms. Because the volumetric concentration of the additive was negligible in the mixture, a high-intensity mixer was selected for mixing the two components since this was the most suitable type in this case [32]. A single-screw-type extruder was used for the four samples of different wt% of PMDA (0%, 0.25%, 0.50% and 0.75%). Moreover, drying was achieved by allowing the sample mixture to dry in a vacuum oven overnight at 100 °C. For this, each sample was placed in a beaker covered with perforated aluminum foil.

2.2. Extrusion

The extrusion of compounded samples was achieved using two types of extruders: single-screw and twin-screw, for the purpose of comparison. The single-screw mini-extruder was manufactured by Filabot and designed for private home-use where 3D-printing enthusiasts can use the extruder to recycle their 3D-printed objects and produce a filament. This extruder has a fixed screw rotational speed of 35 rpm, whereas the temperature can be varied up to 400 °C. One unique feature of this extruder is the ability to replace the extruder's die with various die sizes that can be purchased from Filabot. Samples are fed into this extruder by simply pouring the sample into the extruder hopper, and the screw draws the material into the barrel until the hopper is empty. Design sheets for the extruder's barrel and screw were obtained for reference. The other extruder is a co-current self-wiping twin-screw mini-extruder manufactured by Thermo Fisher Scientific Company (Mississauga, Ontario, Canada). This extruder is a more advanced piece of equipment than the first and, therefore, provides many additional features such as the ability to manipulate the rotation speed and display the value of the die pressure and the torque and the pressure difference across the barrel. These features are useful as they can indicate the melt behavior inside the barrel. For example, a higher torque reading can be attributed to higher melt viscosity which can also be observed by an increase in the die pressure. The Filabot single-screw extruder was used for the filament manufacturing for two reasons: firstly, being a desk-top extruder, it has the advantage of being moveable and conveniently aligned with the other equipment; secondly, the sample from the hopper to the barrel can be drawn without any interference as opposed to when the twin-screw extruder is used. This latter feature is considered important because when interference is required by pushing the material to the extruder's barrel, it is expected that this might result in inconsistent discharge of the melt which results in inconsistency in the product's dimensions. In other words, if excessive pushing was applied at some point throughout the experiment, then the melt would have been discharged at a faster rate which would create variation in product specifications. Samples were compounded in the extruder, and the product was tested for the melt flow index (MFI) and then re-extruded for filament shaping.

2.3. Lab-Scale Filament Production Process

The filament production process included three main pieces of equipment: an extruder, cooling water bath and a spooler (or puller), which is the equivalent of the tractor system in conventional manufacturing processes. Filabot's spooler and a single-screw extruder were used. Further, a 3D-printed cooling water bath was used for the quenching stage.

The first step in the experiment was to establish a steady flow of warm water into and out of the cooling water bath. The temperature of the cooling water in the bath was in the range of 40~47 °C throughout the experiment.

Once the flow to the cooling water bath was stabilized, extrusion was started at around 280 °C and 35 rpm. As a standard, we waited for about a minute to let the extruder's operation stabilize in order to have a consistent residence time for the polymer inside the extruder. Then, the melt was pulled manually, then immersed in the water bath and fed to the spooler. The spooling speed was adjusted to have the filament at the desirable diameter and also to maintain tension on the melt being pulled from the extruder's die. The filament diameter was measured manually every 30 s using a digital dial caliper tool, and the spooler's rotational speed was adjusted accordingly. The measuring point was chosen to be downstream of the spooler's rollers in order to avoid disturbance of the pulling process. Figures 1 and 2 show the laboratory filament extrusion setup that was used.

Figure 1. Laboratory filament-making setup.

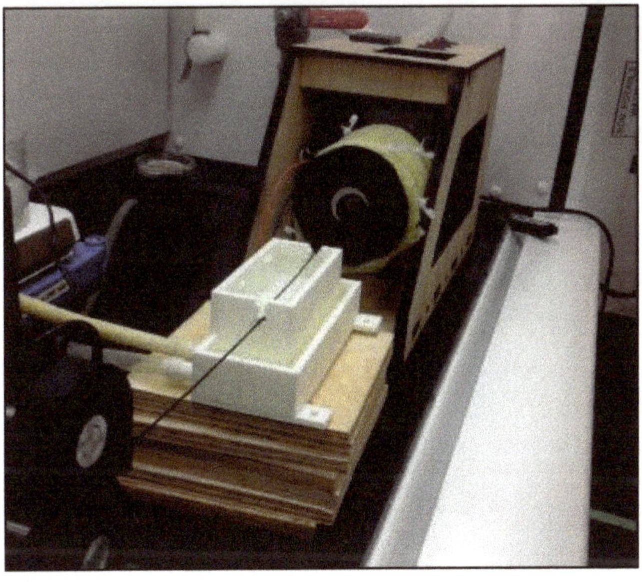

Figure 2. Modified R-PET filament being pulled from the extruder.

We were able to produce few short-segments of modified R-PET that have on-spec filament dimensions when PMDA concentrations was approximately 0.75 wt%. Measurements of the filament diameters were taken at 4 different angles.

FDM filaments are produced commercially with a filament diameter of 1.75 mm and 3.00 mm with a tolerance of ± 0.05 mm. Therefore, the commercial standard (1.75 mm ± 0.05 mm) was considered as the baseline for determining an on-spec product.

3. Results and Discussion

3.1. Benchmarking MFI Values of Commercial Filaments and Unmodified R-PET

The MFI values of several commercial filaments were measured and compared against reprocessed R-PET. The temperature at which the MFI test was conducted for each filament was the printing temperature of that filament as recommended by the manufacturer since this allowed us to investigate the melt flow characteristics that occurred while printing with the filament. Further, the load that was used in the MFI test was kept constant for all materials (2.15 kg). The results are shown in Table 1. MFI values of all commercial filaments fell within a relatively narrow range (from 5 to 30 g/10 min), whereas R-PET (reprocessed in a single-screw extruder) had a significantly higher value.

Table 1. Summary of MFI results of various filaments.

Sample	Material (Manufacturer)	MFI T (°C)	Test 1 (g/10 min)	Test 2 (g/10 min)	Mean	Standard Deviation
1	PLA (Dremel)	215.0	24.64	21.11	22.87	2.50
2	PLA (Wanhao)	215.0	25.49	31.71	28.60	4.40
3	R-PET (reprocessed)	260.0	100.51	80.6	90.56	15.04
4	ABS (HATCHBOX)	245.0	5.18	5.60	5.39	0.29
5	Nylon (Taulman)	240.0	4.19	4.68	4.44	0.34
6	TPE elastomer (Filaments.ca)	240.0	17.31	19.10	18.21	1.26
7	PETG (MG Chemicals)	260.0	5.37	5.26	5.32	0.08

3.2. Effect of Chain Extender Concentration on Decreasing MFI of R-PET

The effect of PMDA concentration on the MFI value was studied at three different concentrations. Furthermore, MFI values were measured for all samples at three different temperatures, 260, 275, and 290 °C, in order to obtain a broader understanding of the melt behavior at various temperatures. These temperatures were carefully selected since they are within the suitable region for 3D printing with R-PET. Table 2 summarizes the MFI test results for modified R-PET.

Table 2. MFI results of unmodified R-PET (sample 1) and modified R-PET with 0.25%, 0.5%, 0.75% PMDA (samples 2, 3 and 4, respectively).

Sample	MFI at 260 °C (g/10 min)				MFI at 275 °C (g/10 min)				MFI at 290 °C (g/10 min)			
	Test 1	Test 2	μ_{260}	SD_{260}	Test 1	Test 2	μ_{275}	SD_{275}	Test 1	Test 2	μ_{290}	σ_{290}
1	100.51	80.6	90.56	14.08	>200	>200			>200	>200		
2	4.33	4.12	4.22	0.15	5.46	5.10	5.28	0.26	15.55	16.27	15.91	0.51
3	1.43	1.51	1.47	0.06	3.13	3.38	3.25	0.18	9.79	9.14	9.47	0.46
4	1.26	1.17	1.21	0.06	2.80	2.83	2.81	0.02	6.33	6.49	6.41	0.11

The results confirmed that the viscosity of the polymer was successfully increased via reactive extrusion with the PMDA chain extender. The figure shows a significant decrease in the MFI value that correlated with PMDA presence. Figure 2 demonstrates the MFI values as a function of temperature at various PMDA concentrations. Finally, the modified product had an MFI value that is comparable to those of commercial 3D-printing filaments.

3.3. Fourier Transform Infrared (FT-IR) Spectroscopy

In this analysis, all polymer samples were tested in a thin-film form that was made using a hot-press machine. The air signal was used as the background signal when testing polymeric materials. For powdered PMDA, the tested sample contained a mixture of PMDA and KBr that was made into a flake form using a cold-press, and pure KBr was the background signal. Figure 3 shows FT-IR spectra of modified R-PET at different PMDA concentrations as well as the IR spectrum of the PMDA additive. In this figure, a mild difference can be seen in transmittance between the samples in the region around 3250 cm^{-1} to 3280 cm^{-1}. This region is important for characterizing PET in general since it is the carboxyl end-group absorption region [33,34]. Figures 3 and 4 show that when PMDA concentration was at the highest level (0.75 wt%), lower signal intensity was captured at around 3270 cm^{-1}, which indicates lower -COOH end-group presence. The difference between the other two samples (0.25% and 0.50%), however, was not clear. In the PMDA spectrum, the strong signals around 1770 cm^{-1} and 1859 cm^{-1} were assigned to the C=O stretching in the O=C-O-C=O that exists in the PMDA molecule [35].

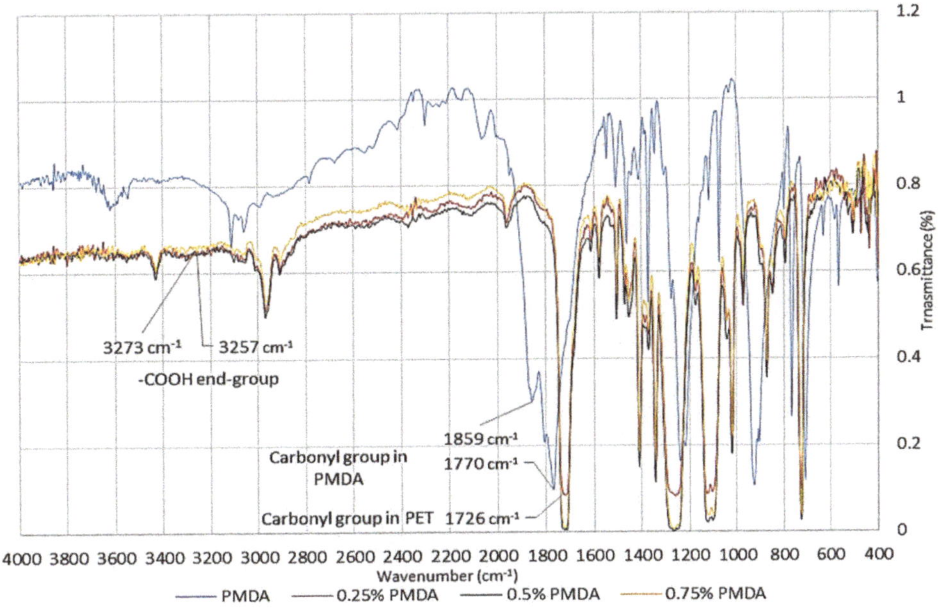

Figure 3. FTIR results of modified R-PET (single-screw) and PMDA.

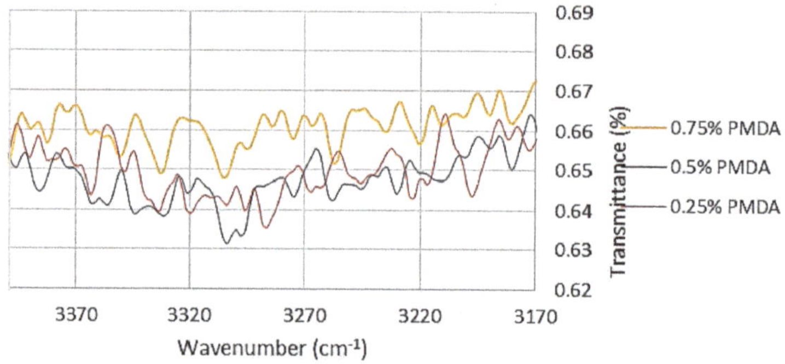

Figure 4. Carboxyl end-group region for modified R-PET.

3.4. Manufacturing a Lab-Made 3D-Printing Filament

Samples were ground and dried prior to extrusion for filament making. The first step in the experiment was to establish a steady flow of warm water into/out of the cooling water bath. The temperature of the cooling water in the bath was in the range of 40~47 °C throughout the experiment. Then, extrusion was started at 280 °C and 35 rpm. The melt was pulled manually, then immersed in the water bath and fed to the spooler. The spooling speed was adjusted to have the filament at the desirable diameter and also to maintain tension on the melt being pulled from the extruder's die. The filament diameter was measured by a digital dial caliper tool, and the spooler's rotational speed was adjusted accordingly.

In this study, a few short segments of modified R-PET that have on-spec filament dimensions could be produced when PMDA concentration was around 0.75 wt%. Figure 5 shows a magnified image of the filament's cross-sectional area that validates the good ovality that was achieved. Additionally, two other attempts were made to pull a filament of R-PET modified with PMDA concentrations of 0.5 wt% and 0.25%, but these were not successful due to melt breakage.

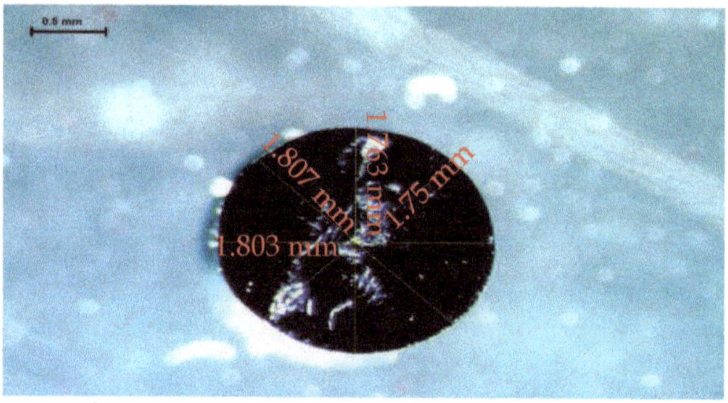

Figure 5. Magnified image of cross-sectional area of lab-made filament from modified R-PET showing good ovality (magnification: X12.6).

3.5. D-Printing with Lab-Made R-PET Filament

Finally, using the lab-made filament, with an extrusion temperature of 275 °C and a bed temperature of 120 °C, a first layer of a small object was successfully 3D-printed, as shown in Figure 6. As shown in Figure 6, the green object was made with ABS material, while the black object was made by using the modified R-PET obtained in our study. This clearly shows that modified R-PET can be successfully used for 3D printing by adjusting its melt flow characteristics with the proposed PMDA additive.

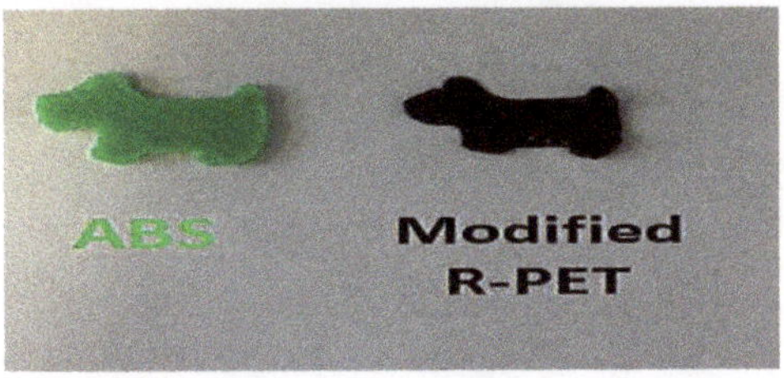

Figure 6. Small dog shape 3D-printed with modified R-PET (right).

4. Conclusions

Modification of R-PET was performed by reactive extrusion with the chain extender PMDA in order to enhance its melt flow, which is desirable for FDM 3D-printing applications. The MFI results showed that PMDA successfully increased the viscosity of the used polymer when used as chain extender. A decrease in MFI from 90 to 1.2 (g/10 min) was recorded when PMDA was added at 0.75 wt%. Furthermore, FT-IR results showed a mild indication of lower carboxyl end-group content for the sample that had PMDA at 0.75 wt% concentration. Furthermore, a 3D-printing filament was made from the modified R-PET via by a mimicked filament-making process which included an extrusion stage, a water bath cooling stage and a spooling stage. R-PET was processed with 0.75% PMDA, and segments of on-spec filaments were obtained. Finally, an attempt was made to print with the produced filament, and the printing temperature was set at 275 °C. This attempt yielded the successful printing of a small object.

Author Contributions: Conceptualization, M.A., H.A., L.S., K.Y., C.M.R.M. and A.E.; data curation, M.A., H.A., K.Y. and A.E.; formal analysis, M.A., H.A., L.S., K.Y. and C.M.R.M.; investigation, M.A., H.A., L.S., K.Y. and A.E.; methodology, M.A., H.A., L.S., K.Y., C.M.R.M. and A.E.; project administration, A.E.; resources, M.A.; software, M.A. and H.A.; supervision, A.E.; validation, H.A., L.S., K.Y. and A.E.; visualization, L.S.; writing—original draft, H.A.; writing—review and editing, C.M.R.M. and A.E. All authors have read and agreed to the published version of the manuscript.

Funding: This research has been financially supported by the Ministry of Education, Saudi Arabia.

Institutional Review Board Statement: Not applicable.

Informed Consent Statement: Not applicable.

Data Availability Statement: Not applicable.

Conflicts of Interest: The authors declare that there are no conflicts of interest including any financial, personal, or other relationships with other people or organizations.

References

1. Plastic Atlas Asia Edition. Facts and Figures About the World of Synthetic Polymers. 2021. Available online: https://hk.boell.org/en/2021/04/22/plastic-atlas-asia-edition (accessed on 18 December 2021).
2. Bremer, M.; Janoschek, L.; Kaschta, D.; Schneider, N.; Wahl, M. Influence of plastic recycling—A feasibility study for additive manufacturing using glycol modified polyethylene terephthalate (PETG). *SN Appl. Sci.* **2022**, *4*, 156. [CrossRef]
3. Pan, Y.; Zhang, Y.; Zhang, D.; Song, Y. 3D printing in construction: State of the art and applications. *Int. J. Adv. Manuf. Technol.* **2021**, *115*, 1329–1348. [CrossRef]
4. Camacho, D.D.; Clayton, P.; O'Brien, W.J.; Seepersad, C.; Juenger, M.; Ferron, R.; Salamone, S. Applications of additive manufacturing in the construction industry—A forward-looking review. *Autom. Constr.* **2018**, *89*, 110–119. [CrossRef]
5. Craveiro, F.; Duarte, J.P.; Bartolo, H.; Bartolo, P.J. Additive manufacturing as an enabling technology for digital construction: A perspective on Construction 4.0. *Autom. Constr.* **2019**, *103*, 251–267. [CrossRef]
6. Lee, E. Additive Manufacturing Accelerating Change in Auto Industry. *WardsAuto*, 19 May 2022. Available online: https://www.wardsauto.com/industry-news/additive-manufacturing-accelerating-change-auto-industry (accessed on 28 July 2022).
7. Liu, G.; Zhang, X.; Chen, X.; He, Y.; Cheng, L.; Huo, M.; Yin, J.; Hao, F.; Chen, S.; Wang, P.; et al. Additive manufacturing of structural materials. *Mater. Sci. Eng. R Rep.* **2021**, *145*, 100596. [CrossRef]

8. BMW Group Press Club. Industrial-Scale 3D Printing Continues to Advance at BMW Group. 10 December 2020. Available online: https://www.press.bmwgroup.com/global/article/detail/T0322259EN/industrial-scale-3d-printing-continues-to-advance-at-bmw-group?language=en (accessed on 19 March 2021).
9. Xometry. 3D Printing in the Food Industry: History, Benefits, Materials, and Examples. 8 August 2022. Available online: https://www.xometry.com/resources/3d-printing/3d-food-printing/#:~:text=3D%20printing%20in%20the%20food%20industry%20refers%20to%20the%20process,layer%20by%20layer%20until%20complete (accessed on 20 September 2022).
10. Blakey-Milner, B.; Gradl, P.; Snedden, G.; Brooks, M.; Pitot, J.; Lopez, E.; Leary, M.; Berto, F.; du Plessis, A. Metal additive manufacturing in aerospace: A review. *Mater. Des.* **2021**, *209*, 110008. [CrossRef]
11. Campbell, R.I.; Telma, F. Additive Manufacturing and Design Strategies for Customized Jewellery Production. In Proceedings of the International Conference on Competitive Manufacturing COMA'13, Stellenbosch, South Africa, 21 August 2013; Available online: https://www.academia.edu/26604201/Additive_Manufacturing_and_Design_Strategies_for_Customized_Jewellery_Production (accessed on 20 March 2021).
12. Salmi, M. Additive Manufacturing Processes in Medical Applications. *Materials* **2021**, *14*, 191. [CrossRef]
13. Culmone, C.; Smit, G.; Breedveld, P. Additive manufacturing of medical instruments: A state-of-the-art review. *Addit. Manuf.* **2019**, *27*, 461–473. [CrossRef]
14. Patel, P.; Gohil, P. Role of additive manufacturing in medical application COVID-19 scenario: India case study. *J. Manuf. Syst.* **2020**, *60*, 811–822. [CrossRef]
15. Soleyman, E.; Aberoumand, M.; Soltanmohammadi, K.; Rahmatabadi, D.; Ghasemi, I.; Baniassadi, M.; Abrinia, K.; Baghani, M. 4D printing of PET-G via FDM including tailormade excess third shape. *Manuf. Lett.* **2022**, *33*, 1–4. [CrossRef]
16. Soleyman, E.; Aberoumand, M.; Rahmatabadi, D.; Soltanmohammadi, K.; Ghasemi, I.; Baniassadi, M.; Abrinia, K.; Baghani, M. Assessment of controllable shape transformation, potential applications, and tensile shape memory properties of 3D printed PETG. *J. Mater. Res. Technol.* **2022**, *18*, 4201–4215. [CrossRef]
17. Palermo, E. Fused Deposition Modeling: Most Common 3D Printing Method. Available online: http://www.livescience.com/39810-fused-deposition-modeling.html (accessed on 3 December 2022).
18. CPIA (Canadian Plastics Industry Association). *2014 Postconsumer Plastics Recycling in Canada (Rep.)*; CPIA: Mississauga, ON, Canada, 2016.
19. Nisticò, R. Polyethylene terephthalate (PET) in the packaging industry. *Polym. Test.* **2020**, *90*, 106707. [CrossRef]
20. Makkam, S.; Harnnarongchai, W. Rheological and Mechanical Properties of Recycled PET Modified by Reactive Extrusion. *Energy Procedia* **2014**, *56*, 547–553. [CrossRef]
21. Venkatachalam, S.; Nayak, S.G.; Labde, J.V.; Gharal, P.R.; Rao, K.; Kelkar, A. Degradation and Recyclability of Poly (Ethylene Terephthalate). In *Polyester*; El-Din, M.S., Ed.; Intech: Rijeka, Croatia, 2012; pp. 75–98. [CrossRef]
22. Assadi, R.; Colin, X.; Verdu, J. Irreversible structural changes during PET recycling by extrusion. *Polymer* **2004**, *45*, 4403–4412. [CrossRef]
23. Incarnato, L.; Scarfato, P.; Di Maio, L.; Acierno, D. Structure and rheology of recycled PET modified by reactive extrusion. *Polymer* **2000**, *41*, 6825–6831. [CrossRef]
24. Cardi, N.; Po, R.; Giannotta, G.; Occhiello, E.; Garbassi, F.; Messina, G. Chain extension of recycled poly(ethylene terephthalate) with 2,2′-Bis(2-oxazoline). *J. Appl. Polym. Sci.* **1993**, *50*, 1501–1509. [CrossRef]
25. Japon, S.; Boogh, L.; Leterrier, Y.; Månson, J.-A. Reactive processing of poly(ethylene terephthalate) modified with multifunctional epoxy-based additives. *Polymer* **2000**, *41*, 5809–5818. [CrossRef]
26. Karayannidis, G.P.; Psalida, E.A. Chain extension of recycled poly(ethylene terephthalate) with 2,2′-(1,4-phenylene)bis(2-oxazoline). *J. Appl. Polym. Sci.* **2000**, *77*, 2206–2211. [CrossRef]
27. Tzoganakis, C. Reactive extrusion of polymers: A review. *Adv. Polym. Technol.* **1989**, *9*, 321–330. [CrossRef]
28. Qin, D.; Wang, C.; Wang, H.; Jiang, X. Chain Extension and Thermal Behavior of Recycled Poly(Ethylene Terephthalate) Modified by Reactive Extrusion with Triphenyl Phosphite. In Proceedings of the MATEC Web of Conferences, Cape Town, South Africa, 1–3 February 2016; Volume 67, p. 06052.
29. Yang, Z.; Xin, C.; Mughal, W.; Li, X.; He, Y. High-melt-elasticity poly(ethylene terephthalate) produced by reactive extrusion with a multi-functional epoxide for foaming. *J. Appl. Polym. Sci.* **2017**, *135*, 45805. [CrossRef]
30. Xia, T.; Xi, Z.; Liu, T.; Pan, X.; Fan, C.; Zhao, L. Melt foamability of reactive extrusion-modified poly(ethylene terephthalate) with pyromellitic dianhydride using supercritical carbon dioxide as blowing agent. *Polym. Eng. Sci.* **2014**, *55*, 1528–1535. [CrossRef]
31. Awaja, F.; Daver, F.; Kosior, E. Recycled poly(ethylene terephthalate) chain extension by a reactive extrusion process. *Polym. Eng. Sci.* **2004**, *44*, 1579–1587. [CrossRef]
32. Giles, H.F., Jr.; Wagner, J.R., Jr.; Mount, E.M., III. *Extrusion: The Definitive Processing Guide and Handbook*; Andrew, Elsevier: Oxford, UK, 2005; pp. 83–89.
33. Chalmers, J.M.; Meier, R.J. *Molecular Characterization and Analysis of Polymers*; Elsevier: Amsterdam, The Netherlands, 2008.

34. Al-AbdulRazzak, S.; A Lofgren, E.; A Jabarin, S. End-group determination in poly(ethylene terephthalate) by infrared spectroscopy. *Polym. Int.* **2002**, *51*, 174–182. [CrossRef]
35. Hase, Y.; Kawai, K.; Sala, O. The infrared and Raman spectra of pyromellitic dianhydride. *J. Mol. Struct.* **1975**, *26*, 297–302. [CrossRef]

sustainability

MDPI

Article

A Superstructure Mixed-Integer Nonlinear Programming Optimization for the Optimal Processing Pathway Selection of Sludge-to-Energy Technologies

Omar Morsy [1], Farzad Hourfar [1], Qinqin Zhu [1], Ali Almansoori [2] and Ali Elkamel [1,2,*]

[1] Department of Chemical Engineering, University of Waterloo, Waterloo, ON N2L 3G1, Canada
[2] Department of Chemical Engineering, Khalifa University, Abu Dhabi P.O. Box 127788, United Arab Emirates
* Correspondence: aelkamel@uwaterloo.ca

Abstract: The perception of sewage sludge has increasingly changed from being a waste, which is a burden to the environment and society, to a useful resource of materials and renewable energy. There are several available technologies at different stages of maturity that aim to convert sludge to energy in the form of electricity and/or fuels. In this paper, a decision-making support tool is proposed to help in choosing the optimal pathway for the sludge-to-energy conversion from a techno-economic perspective. The conversion technologies under study are: (1) anaerobic digestion, (2) pyrolysis, (3) gasification, (4) incineration, (5) supercritical water oxidation, (6) supercritical water gasification, as well as the corresponding dewatering and drying methods for each technology. Different synergies between the available technologies are compared by the formulation of a superstructure optimization problem expressed in a mixed-integer non-linear program (MINLP) model. The applicability of the proposed model is explored via a case study for a hypothetical sludge treatment plant with a capacity of 100 tons of dry solids (tDS) per day. The model is solved via the BARON solver using GAMS software within a reasonable processing time. According to the obtained results, the fast pyrolysis technology, coupled with filter press dewatering and thermal drying as pre-treatment steps, show the most promising outcomes with the minimum treatment cost of USD 180/tDS. Fast pyrolysis converts the sludge to bio-oil, which can be used as an alternative fuel after further refining, and biochar, which can be used for soil amendment or adsorption purposes. The model parameters are subject to uncertainty that is addressed in the sensitivity analysis section of this paper. Moreover, the pyrolysis pathway shows a high degree of robustness in most of the sensitivity analysis scenarios. Meanwhile, anaerobic digestion coupled with fast pyrolysis demonstrates the best energy recovery performance upon increasing electricity prices.

Keywords: sludge-to-energy; mixed-integer nonlinear programming; wastewater treatment; pyrolysis; optimization

check for updates

Citation: Morsy, O.; Hourfar, F.; Zhu, Q.; Almansoori, A.; Elkamel, A. A Superstructure Mixed-Integer Nonlinear Programming Optimization for the Optimal Processing Pathway Selection of Sludge-to-Energy Technologies. *Sustainability* **2023**, *15*, 4023. https://doi.org/10.3390/su15054023

Academic Editor: José Carlos Magalhães Pires

Received: 17 January 2023
Revised: 15 February 2023
Accepted: 17 February 2023
Published: 22 February 2023

1. Introduction

1.1. Motivation

Wastewater treatment plants (WWTPs) have been a crucial element of maintaining the health and environment of modern societies. However, these facilities require a significant amount of energy and operational costs. It has been estimated that WWTPs account for 3% of the total electricity consumption in the United States [1]. The treatment and handling of sewage sludge, which is the solid byproduct of WWTPs, accounts for approximately 30% of this electricity consumption [2] and 50% of the annual operating costs of a WWTP [3]. In addition, 73% of the treated sludge is eventually either landfilled or sent for land applications [4], which impose regularly updating stringent disposal requirements. Thus, the need for cost-effective, energy-efficient, and sustainable methods of sludge handling has recently become an active research field.

In the past decade and coinciding with the efforts to combat global warming and climate change, there has been a paradigm shift taking place towards sludge. It has shifted from being perceived only as waste and burden to society and the environment, to being looked at as a useful resource of materials and renewable energy. Several studies in the literature [5–9] reviewed available and potential technologies for energy recovery from sewage sludge in the form of electricity, heat, and/or fuels. These energy products can help in offsetting the energy consumption of the wastewater treatment facilities and thus reducing their carbon footprints as well as generating a revenue stream from products that can be sold in the market. Yet, there have been few efforts put into developing frameworks that quantitatively compare those sludge-to-energy alternatives from an economic perspective.

On the other hand, for relatively similar feedstock materials such as biomass, microalgae, and municipal solid wastes, superstructure optimization approaches have been successfully used [10–16]. Therefore, the purpose of this paper is to first provide a brief overview of a set of the most promising sludge-to-energy conversion technologies. Afterward, a mathematical model is developed, using a superstructure optimization-based approach as a decision-making support tool. Clearly, the proposed solution is useful for both researchers as well as stakeholders in municipalities looking forward to implementing master plans and strategies for biosolid handling for a sustainable future.

1.2. Background Information

1.2.1. Sludge Characterization

Sewage sludge is composed of a complex series of microorganisms, organic and inorganic solid compounds (total solids) that coexist in water heterogeneously. The organic compounds, commonly called volatile solids (VS), originate from several sources such as fecal material, plants, paper, and oils. They contain a variety of complex molecular structures from polysaccharides, lipids, proteins, and peptides to plant macromolecules, and micropollutant organic compounds such as dibenzofurans and polycyclic aromatic hydrocarbons (PAHs) [17]. The energy recovery potential in the sewage sludge is highly dependent on the amount of VS present in the sludge (i.e., the higher the percentage of volatile solids, the higher the energy content of the sludge) [18]. The inorganic compounds, also referred to as ash, are mainly composed of minerals such as silica (quartz), calcites, or microclines. Trace amounts of heavy metals are also present in sewage sludge, and examples are chromium, copper, nickel, zinc, mercury, cadmium, and lead [19]. Finally, nutrients in the form of nitrogen, potassium (potash), and phosphorus are found in the sludge and are one of the main criteria upon which the suitability of the treated sludge for usage as a fertilizer or soil conditioner depends.

1.2.2. Anaerobic Digestion

Anaerobic digestion is the most common process to stabilize sewage sludge in today's market [20]. In this process, a portion of the biodegradable organic compounds in the sludge is decomposed in an oxygen-free environment to a methane-rich gaseous mixture called "biogas" [17]. The unconverted portion of the organic compounds in the digester together with the inorganic compounds and moisture exit the process and are named "digested sludge" or "digestate".

The digestion process takes place in a series of complex biochemical reactions. Hydrolysis converts insoluble and high molecular weight organic compounds such as polysaccharides, proteins, and lipids into soluble amino and fatty acids. Those soluble compounds from hydrolysis are additionally split to form volatile fatty acids in the acidogenesis step. Acetogenesis is the step in which the organic acids and alcohols generated in acidogenesis are converted to acetic acid together with hydrogen and carbon dioxide. Finally, the methanogenesis step is where methane gas is predominantly produced by two different methanogenic groups of bacteria: one of them decomposes acetate to CH_4 and CO_2 and the other group utilizes H_2 as an electron donor and CO_2 as an acceptor to produce CH_4 [21]. The hydrolysis step is generally deemed as the rate-limiting one.

1.2.3. Incineration

Incineration is a process in which waste combustion takes place in a controlled manner producing flue gas, ash, and heat that can be recovered. Incineration and combustion of sewage sludge are sometimes used interchangeably; however, it needs to be noted that there is a subtle difference between both terms. Combustion is a more general term that refers to a thermochemical exothermic reaction between excess oxygen and organic material of a fuel that is completely oxidized to CO_2 and H_2O at high temperatures. Incineration on the other hand is a special case of combustion where the combustible material originates from a waste that needs to be disposed of. The main purpose of combustion is the energy recovery from the fuel in the form of heat that can then be used in steam generation which in turn can produce electricity upon passing through steam turbines, while incineration's main purpose is the destruction of the harmful material in the waste and reducing its volume upon disposal [22]. For the purpose of our work, where energy recovery is the main interest, incineration and combustion of sewage sludge will refer to the same concept.

1.2.4. Gasification

Gasification is another thermochemical conversion process in which the organic components of sludge are transformed in a net reducing environment to a combustible gas called syngas, while the remaining sludge constituents are converted to ash [23]. There are lots of similarities between gasification and combustion, but they mainly differ in the lower requirement of sludge moisture content fed to the gasifier (below 15 wt%) and that oxidants are present in amounts below the stoichiometric quantities required for complete combustion or oxidation [24]. Syngas or synthesis gas is a mixture that consists mainly of hydrogen (8.89–11.17 vol%), carbon monoxide (6.28–10.77 vol%), lower percentages of methane (1.26–2.09 vol%), and C2s (0.75–1.2 vol%), along with CO_2 and the gasification medium [25]. The gasification medium, also called the gasifying agent, is the fluid which reacts with the sludge carbonaceous components to partially oxidize them to syngas. Typically, air with oxygen amounts of 20–40% less than that required for complete combustion is used as a gasifying agent. Nevertheless, the following media have been also studied and used in sludge gasification: pure oxygen, steam, steam–air mixture, steam–O_2, steam–CO_2, and pure CO_2 as reported in [22]. The gasification medium has a significant impact on the composition and accordingly the heating value of the produced syngas ranges from 4 to 12 MJ/Nm^3, where the highest values are obtained from gasification with pure oxygen [24]. Steam gasification increases the yield of H_2 in the syngas mixture compared to CO which can be attributed to both the reforming of methane and the water–gas shift reaction promoted by steam. Higher H_2/CO ratios correspond to higher syngas calorific values as well [26].

1.2.5. Pyrolysis

Pyrolysis is a thermochemical process in which the organic components of the sludge are destructed at temperatures between 300 °C and 700 °C in an oxygen-free environment [22]. Unlike combustion, which is an exothermic process, pyrolysis requires a significant amount of heat (in the range of 100 MJ/tDS) for its reactions to occur [27]. It also has a much lower moisture content tolerance to the sludge that enters the reactor (<10 wt%) and thus requires a higher drying energy [24]. The first step of the process takes place when the sludge is heated to temperatures in the range of 100–200 °C, where the remaining moisture associated with the sludge is evaporated and volatile gaseous products start to form, leaving a solid residue with non-volatiles referred to as char. These products are the result of several bond-breaking and forming reactions and are called primary pyrolysis. This is the same initial step in other thermochemical processes discussed as combustion and gasification [28]. With further heating, the next step, called secondary pyrolysis, takes place at temperatures close to 600 °C where the volatile gaseous products undergo further decomposition into simpler low molecular weight gases and stable aromatic compounds.

The vapor product is then sent for cooling and is separated into a liquid product called bio-oil and non-condensable gases (syngas).

1.2.6. Supercritical Water Treatment Methods

The thermochemical sludge treatment methods discussed so far, i.e., incineration, gasification, and pyrolysis, all require a drying step before the main sludge processing. The fact that raw and/or digested sludges have a significantly high moisture content, makes those processing routes rather more capital and energy intensive. An innovative way to stabilize sludge while eliminating the need for a pre-drying step is to treat it in the supercritical water (SCW) phase [29]. Supercritical water is a phase that takes place when critical temperature and pressure values of water exceed 374 °C and 22.1 Bar, respectively [29]. At such a state, one cannot distinguish between water in its liquid and vapor phase (steam) and water has unique properties. In this section, two SCW treatment methods are briefly discussed, namely supercritical water oxidation (SCWO) and supercritical water gasification (SCWG).

Supercritical Water Oxidation (SCWO)

SCWO occurs at high temperatures and pressures (around 600 °C and 25 Bar), conditions that are well suited for the disintegration of sewage sludge [17]. Much higher oxidation rates are observed in supercritical conditions compared to subcritical ones, which can aid in the complete destruction of organic constituents of the sludge [30]. Organic compounds are mainly composed of carbon, hydrogen, nitrogen, sulfur, and phosphorus, which are oxidized to CO_2. H_2O, N_2, SO_4^{2-}, and PO_4^{3-}, respectively, while heavy metals are oxidized to their respective oxides [17]. Most of the oxidation reactions occur at a conversion rate of 99.9% and reaction times of 30 s or less at a temperature of 600°C, which results in relatively small reactor dimensions [31]. Another advantage of SCWO compared to incineration is the simple treatment required for the off-gas released, which is a major cost in incineration plants. Since SCWO is an exothermic reaction, energy recovery can be achieved either from heat exchange with the reactor vessel directly, or with its effluent product to produce steam [17].

Supercritical Water Gasification (SCWG)

Similar to conventional gasification, SCWG decomposes the organic constituents of the sewage sludge into a gaseous mixture called syngas. However, the composition of the syngas from SCWG is much richer in hydrogen, which makes this technology especially attractive. Supercritical water gasification (SCWG) of sewage sludge has been studied in several research works for the purpose of hydrogen production. This technology has not been implemented yet at full scale but shows great potential for future adoption. Some of the main advantages of SCWG of biomass in general, which apply to sewage sludge as well, are as follows [32]:

- No need for prior drying of the feedstock to the SCWG reactor. Conversely, the moisture content of the feed is necessary for the reaction;
- Higher yield of H_2 compared to CO in the syngas product whereas in dry gasification processes CO is the main constituent of syngas and an extra water–gas shift process is required to achieve such high H_2:CO ratios;
- Lower amounts of coke and tar formation;
- Salts remain in the aqueous solution which avoids corrosion problems during the treatment of the produced gas.

Depending on the production scale, the hydrogen product from SCWG can be sold in the market as fuel for H_2 fuel cells, used in refineries, or other industrial uses (ammonia, methanol, etc.) [33].

1.2.7. Dewatering and Drying

The water content removal is an essential step in any sludge treatment plant to achieve a volume reduction in the stabilized product for further disposal or treatment. Such a reduction has a significant effect on the transportation and/or energy costs. There are four different categories of water/moisture present in sewage sludge: free water, adsorbed water, capillary water, and cellular water. Free water is the easiest of them to remove and is achieved by simple flotation or gravitation methods. Adsorbed and capillary waters on the other hand require much higher forces compared to free water. These higher forces can be accomplished mechanically by dewatering equipment such as centrifuges or filter presses, or chemically by the employment of flocculants. A final product called "cake" or "dewatered sludge" with a concentration greater than 30% dry solids (DS) can be achieved. This product has a semi-solid appearance and compatibility with a belt conveyor transfer or manipulation of spades. The removal of the three categories of water discussed so far can result in a volume reduction in the range of 90–95% to an effluent originally at 2% DS. The last category, cellular water, is the hardest to remove and requires even higher forces that can only be achieved thermally. Thermal dryers can produce a granular product with up to 95% DS in an efficient manner [34].

The water removal steps which lie within the scope of this study are dewatering and thermal drying. Prior to sludge dewatering, an important pre-treatment is required referred to as sludge conditioning. This step is crucial in impacting the efficiency and ease of sludge dewatering and can be achieved via different methods: thermal pre-treatment, or the use of organic and/or inorganic chemicals. The most popular chemical conditioners are inorganic lime and ferric chloride and organic polymers. Chemical type and dosage rates depend on the sludge characteristics and dewatering method/equipment type. The most common dewatering methods are belt presses, centrifuges, vacuum filters, plate, or diaphragm filter presses, and exclusively for digested sludges, sludge lagoons and drying beds [35].

On the other hand, thermal drying can be achieved either by direct or indirect methods, where the difference lies in whether the heating medium is in direct contact with the sludge or not. Direct drying methods are more commonly used. Examples of direct dryer technologies are rotary dryers, fluidized bed dryers, and belt dryers. One of the advantages of thermal drying is that it acts as both a further stabilization and volume reduction method of the sludge. The end product can be sold as Class A biosolids (pathogen-free), which are used in agricultural applications such as fertilizers. However, the high operating costs associated with drying are usually not offset by the revenues generated from selling the dried product [36]. Moreover, another problem associated with sludge drying is the potential production of odors and volatile organic compounds (VOCs) [37].

1.3. Sludge-to-Energy Fundamental Concepts

1.3.1. Sludge-to-Energy Decision-Making Frameworks

There are significant efforts being made in relation to the development of decision-making support frameworks or tools that help in ranking different sludge-to-energy alternatives. Multi-criteria decision-making (MCDM) methodologies have been applied to the problem of sludge management in [38,39]. The former study is based on traditional grey relational analysis (GRA) modified to allow for linguistic inputs, while the latter study is based on Dempster–Shafer theory and fuzzy best-worst method. Both studies consider environmental, technological, social, and economic criteria. Tang et al. proposed another MCDM framework for prioritizing different sludge technologies using four different methodologies combined with triangular fuzzy numbers to deal with hybrid-data types [40]. This work also contains a recent review of other related studies in the area of decision-making for sustainable sewage sludge management. Although MCDM tools can be useful, they are not flexible in assessing and synthesizing innovative combinations of various technologies at different capacities to maximize economic or environmental benefits. In addition, many of these tools rely on "experts' opinion", which might lead to more subjective or biased results. A more suitable approach to address those limitations would be to formulate an optimization of mathematical models for superstructure mapping of the different alternatives. Typically, these optimization problems are modelled and solved by mixed-integer linear programming (MILP) or mixed-integer nonlinear programming (MINLP) models.

1.3.2. Sludge Management Optimization Models

A few studies are available in the literature which utilized MILP in solving a sludge management-related problem. A case study in [41] compared alternatives for the thermal treatment of digested sludge in the region of Zurich in Switzerland. A multi-objective MILP was developed to find the optimal environmental performance of the following technologies: sludge mono-incineration, co-incineration with municipal solid waste (MSW), and co-processing for cement manufacturing. This study did not cover any energy recovery method other than incineration, and it also did not consider the economic performance and costs associated with the potential pathways. Moreover, in [42], a stochastic multi-objective MILP model was utilized to compare different sludge utilization pathways, namely, anaerobic digestion with thermal hydrolysis, lime stabilization, incineration, land application, and selling of Class A biosolids in the market as a fertilizer. Moreover, several utilization paths for the produced biogas were considered, such as electricity production, and upgrading to compressed natural gas (CNG). The economic performance in terms of capital and operating expenditure as well as revenue from valuable products was studied. In addition, environmental performance in terms of CO_2 emissions and energy costs were investigated. However, similar to the case study in [41], only a few energy recovery technologies were included in the model.

The work completed in [43] focused on the whole sludge supply chain in a certain region in north-western Europe considering the synergies between 241 WWTPs. A generic decision framework called OPTIMASS, originally created for optimizing biomass supply chains [44], was customized to fit the specific application of sewage sludge. However, only a limited number of energy recovery alternatives were included in the model with the following processing equipment/routes: thickening, dewatering, MAD, thermal drying, mono/co-incineration, and utilization in the cement industry. Another shortcoming of that study was the unavailability of the parameters used in the model due to privacy agreements. Finally, in [45] anaerobic digestion, hydrothermal liquefaction, and catalytic hydrothermal gasification pathways were compared using a multi-objective superstructure optimization methodology. The developed MILP model considered both economic and environmental aspects, while CAPEX and OPEX were assumed to have linear relations. Although more technologies were assessed in that study in comparison to the former ones, the study did not consider some of the most studied sludge-to-energy technologies such as incineration, gasification, and pyrolysis.

1.3.3. Waste-to-Energy Optimization Models

Aside from sewage sludge, there is some available literature on the application of superstructure optimization or mixed-integer programming methodologies to find the optimal processing pathway for energy recovery from other types of wastes. The majority of those studies are related to the different types of MSW such as plastics, metals, glass, and various other organic wastes (paper, textile, food waste, etc.). For example, a fuzzy multi-objective superstructure optimization methodology with the aim of cost-minimizing while maximizing waste reduction and electricity generation was introduced in [46]. LP and MINLP (linearized to MILP) superstructure optimization models were proposed in [14,15], respectively, with a single objective function of maximizing net profit for the selected technology pathway. The presented work in [47] was not limited to only optimal technology selection, it also considered the complete supply chain of MSW including transportation between different cities. The objective function to be optimized in the MILP formulation of that study aimed at maximizing the economic benefit while considering the incurred environmental cost because of CO_2 emissions. Another study in [48] looked at supply chain optimization together with technology selection via a multi-objective MILP model. The multiple objectives were: (1) minimizing economic and environmental costs, and (2) minimizing the associated risks with the chosen pathway. The latter study also included a comprehensive list of many of the works related to MSW optimization modelling frameworks.

Poultry litter is another type of waste that decision-making tools based on optimization mathematical models have been applied to. The recent work of [16,49] studied the comparison of thermochemical valorization pathways, developing mixed-integer (non)linear fractional programming models. A parametric algorithm was proposed for linearizing the optimization models to a series of MILP problems to obtain solutions in a relatively less computationally intensive way. The first study aimed at just technology selection while maximizing the return on investment (ROI). This objective function was the source of the fractional nonlinearity of the model due to the presence of a ratio of two linear equations. The second study focused on the comparison of two pyrolysis pathways, slow and fast pyrolysis, for the valorization of poultry waste, considering multiple objectives, the first being maximizing annualized profit per unit waste and the second being minimizing the equivalent CO_2 emissions from the chosen pathway. This study also considered optimizing the whole supply chain including the selection of the optimal location of pyrolysis facilities in relation to the waste sources taking into account transportation costs. The proposed methodology was applied to a case study for the poultry waste supply chain in the state of Georgia in the United States.

2. Methodology

2.1. Overview

The first step in the presented methodology was to identify candidate technologies that had the ability to convert sewage sludge to energy products. This was completed according to the outcomes of a comprehensive literature review process in which the strengths and drawbacks of each technology were extracted. The second step was to develop a superstructure mapping for those various alternative technologies. Subsequently, a mathematical model formulation for the optimization problem was developed in order to aid in the selection of an optimal pathway. After that, a case study was developed to test the applicability of the model by defining all the economic and technical parameters and solving for the decision variables. Finally, a sensitivity analysis was conducted over the parameters defined in the case study to assess the impacts of inherent uncertainty on the optimal solutions. The schematic of these steps is illustrated in Figure 1. In the following subsections, each of these steps is further elaborated on in detail.

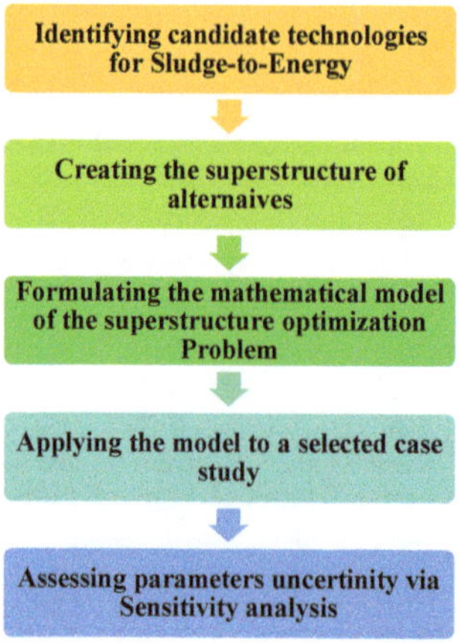

Figure 1. Decision-making framework for sludge-to-energy process synthesis [50].

2.2. Superstructure Development

The superstructure of alternatives in this work refers to a graphical representation of a network that shows the connections and relationships between the feed stream(s) being processed, potential processing technologies, and intermediate and final products. In the considered problem, there is a single feed stream crossing the boundary limit of the superstructure which is thickened sewage sludge. The processing units are categorized into biochemical processes, thermochemical processes, and intermediate processes. As shown in the schematic in Figure 2, the biochemical processes covered in this superstructure are MAD, and MAD + THP. The thermochemical processes include Incineration, Gasification, Pyrolysis, SCWO, and SCWG. The intermediate processes comprise mechanical processes such as sludge dewatering and thermal processes such as sludge thermal drying. Intermediate processes are duplicated to differentiate between those processing digested sludge and those processing undigested sludge. This is because depending on whether a biochemical technology is selected or not, intermediate processes can have varying capacities and are part of different pathways. Three different dewatering technology options were modelled, namely, belt filter dewatering, filter press dewatering, and low-speed centrifuge dewatering.

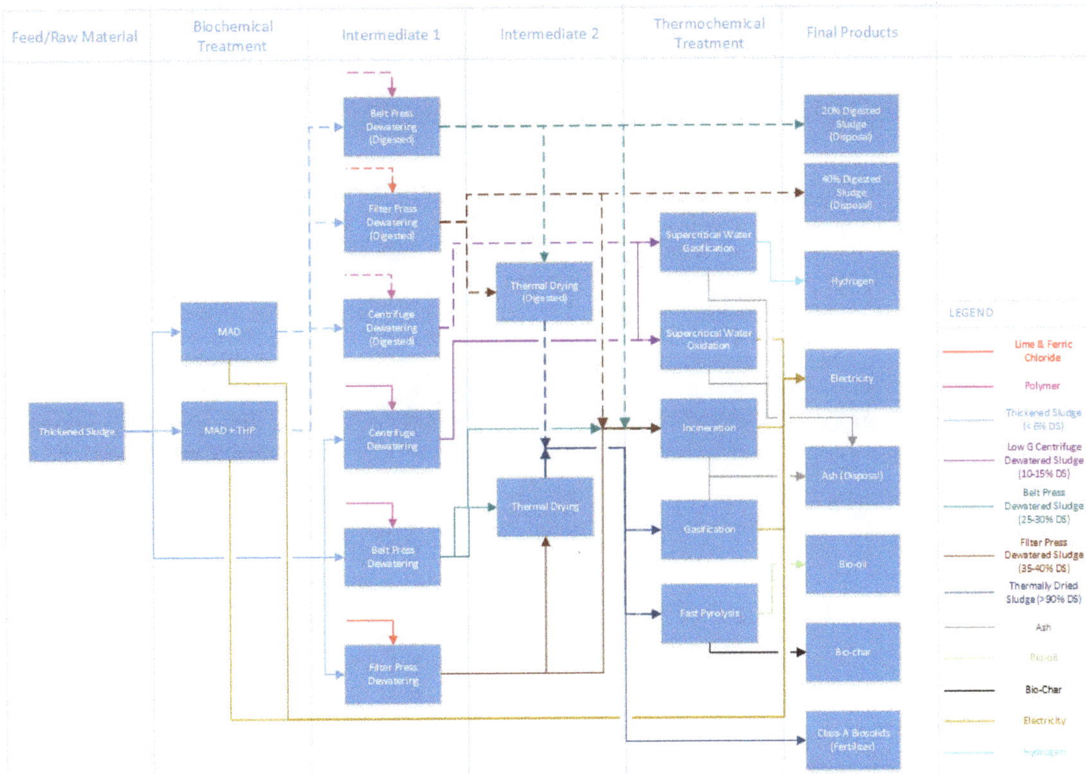

Figure 2. Superstructure representation of Sludge-to-Energy alternatives.

Each of the biochemical and thermochemical units includes an energy recovery facility that produces energy in the form of electricity or fuels. The final products shown in the superstructure are either value-added products or residual/waste products. Value-added products are those that can be sold in the market, such as electricity, Class A biosolids, bio-oil, biochar, and hydrogen. Residual products, such as dewatered sludge and ash, are cost-incurring ones that can be disposed into landfills or sent for beneficial use (i.e., use in cement industry for ash, land application for dewatered sludge). For ease of presentation, each material stream is given a distinct color as explained in the legend of the superstructure diagram. In addition, digested sludge products are differentiated graphically by using dashed lines compared to solid lines for undigested sludge streams. This superstructure illustration represents the foundation for the logical relationships of the building blocks of the mathematical model formulation as demonstrated in the next subsection.

2.3. *Mathematical Model Formulation*

2.3.1. General

Any optimization problem involves the minimization or maximization of a certain function, called the objective function, which is subject to a set of equality and inequality constraints. Superstructure optimization problems formulations follow the same concept and can be mathematically expressed as follows [51]:

$$\begin{gathered} \min_{x,z}\ C = c^T z + p(x) \\ \text{s.t.}\quad r(x) = 0 \\ s(x) + Bz\ \leq 0 \\ x \in R^n, \qquad z \in \{0,\ 1\}^l \end{gathered} \tag{1}$$

where the objective cost function C consists of: (a) costs related to a discrete decision integer variables vector z which is multiplied by a matrix of relevant cost coefficients c (this matrix usually consists of capital cost parameters), and (b) costs related to continuous variables vector x represented in functions $p(x)$ and those are typically costs related to operation and maintenance, or revenues from product sales. The objective function is constrained by the physical performance of the process or technology efficiency, which is modelled using an equality functions vector $r(x)$, and the logical relationships are dictated by inequality functions $s(x)$ that relate to the discrete integer decision variables vector via a coefficient matrix B. Depending on whether functions $p(x)$, $r(x)$, and $s(x)$ are all linear or any of them are non-linear, the problem becomes a mixed-integer linear program (MILP) or mixed-integer non-linear program (MINLP), respectively, where each type has its applicable algorithms to be solved.

Equation (1) represents the generalized high-level architecture of such problems. However, in this paper, a detailed model following the same general approach, but customized to suit the specific needs of our problem, is formulated. At the beginning, the relationships between elements of the superstructure are described in this section. The proposed framework consists of a group of sets, parameters, variables, and equations. The sets are expressed by a number of bold roman letters (example: **I**), parameters use light italic roman letters (example: *I*), and variables are expressed by italic bold letters (example ***X***). Model element identifiers (subscripts and superscripts) express process blocks (italic letters), streams, and their components (normal roman letters). Generic identifiers are light formatted, while if a specific identifier is used, it is **bolded**. The sets can be grouped into two main groups: sets that define the main model elements (i.e., Feed sources, technologies, process streams, components, and final products), and sets that define the relationships between those elements. The identifiers that are used to describe individual model elements that belong to a corresponding set(s) are listed in Table 1:

Table 1. Model element identifiers (subscripts and superscripts).

Superstructure Element	Identifier	Description
General	*i,j,k*	Aliases of subscripts identifiers for feed, process, and product blocks.
	s	Generic identifier of a process stream
	c	Generic identifier of a component in a stream
Feed Source	*TH*	Thickened Sludge
Technologies/Processes	*MAD*	Mesophilic Anaerobic Digestion
	MADT	MAD + Thermal Hydrolysis Pretreatment
	CD	Centrifuge dewatering for digested sludge
	CU	Centrifuge dewatering for undigested sludge
	BPD	Belt press dewatering for digested sludge
	BPU	Belt press dewatering for undigested sludge
	FPD	Filter press dewatering for digested sludge
	FPU	Filter press dewatering for undigested sludge
	TD	Thermal Drying
	INC	Incineration
	GN	Gasification
	PY	Fast Pyrolysis
	SCO	Supercritical Water Oxidation
	SCG	Supercritical Water Gasification
Final Products	*DS20*	20% dewatered digested sludge
	DS40	40% dewatered digested sludge
	ASH	Ash
	E	Electricity
	FERT	Class A Biosolids (Fertilizer)
	BO	Bio-oil from pyrolysis
	BC	Biochar from pyrolysis
	H2	Hydrogen
Process Streams	THS	Thickened Sludge
	ADS	Anaerobically Digested Sludge
	E	Electricity
	P	Polymer for chemical conditioning
	L	Lime for chemical conditioning
	FC	Ferric chloride for chemical conditioning
	DWS	Dewatered Sludge
	TDS	Thermally dried sludge
	ASH	Ash
	BO	Bio-oil
	BC	Biochar
	H2	Hydrogen
Components in process streams	VS	Total volatile solids
	ASH	Ash
	DS	Total dry solids (VS + Ash)
	W	Water or moisture in the sludge/biosolids
	E	Electricity
	BO	Bio-oil
	BC	Biochar
	H2	Hydrogen

Moreover, the sets describing the model elements and their relationships are described in Table 2.

Table 2. Sets of model elements and their relationships.

Set	Description
I	Combined set of feed, process, and final product blocks
FEED	Subset of feed blocks, **FEED** ⊂ **I**
PROCESS	Subset of processing technologies, **PROCESS** ⊂ **I**
PRODUCT	Subset of final products, **PRODUCT** ⊂ **I**
STR	Set of process streams
CHEM	Subset of chemicals streams used for conditioning **CHEM** ⊂ **STR**
COMP	Set of components of process streams
$\mathbf{S}_i$	Set of descendant block(s) from block $i \in$ **FEED** ∪ **PROCESS**. Where $\mathbf{S}_i \subset$ **PROCESS** ∪ **PRODUCT**
$\mathbf{P}_i$	Set of precedent block(s) of block $i \in$ **PROCESS** ∪ **PRODUCT**. Where $\mathbf{P}_i \subset$ **FEED** ∪ **PROCESS**
$\mathbf{STF}_i$	Set of inlet stream(s) applicable with process $i \in$ **PROCESS**. Where $\mathbf{STF}_i \subset$ **STR**
$\mathbf{STPR}_i$	Set of outlet stream(s) applicable with process $i \in$ **PROCESS**. Where $\mathbf{STPR}_i \subset$ **STR**
$\mathbf{SCOMP}_s$	Set of component(s) applicable to stream $s \in$ **STR**. Where $\mathbf{SCOMP}_s \subset$ **COMP**
$\mathbf{FPCO}_i$	Set of component(s) used for specifying the revenue/disposal cost of a final product $i \in$ **PRODUCT**. Where $\mathbf{FPCO}_i \subset$ **COMP**

After specifying the sets, defining model elements, and their relationships, a group of performance and economic parameters applicable to all the processing technologies are defined (Table 3).

Table 3. Parameters applicable to all processing technologies.

Parameter	Description
CAP_i	Maximum processing capacity of a certain process $i \in$ **PROCESS** in tDS/day.
BCC_i	Base (reference) capital cost of process $i \in$ **PROCESS** in **USD** (USD 2019)
BQ_i	Base (reference) processing capacity of process $i \in$ **PROCESS** used in capital cost calculation.
α_i	Economies of scale exponent of process $i \in$ **PROCESS**.
POC_i	Operating cost parameter for a certain process $i \in$ **PROCESS**.
DPY	Days of operation per year

The next component to be defined for the model formulation is the decision variables. The variables can be grouped in several ways: process variables versus economic variables, continuous variables versus integer and/or binary variables, and dependent variables versus independent variables. In terms of the mathematical model formulation, what matters the most is the distinction between continuous and integer/binary variables, because this will play a key role in determining the type of the optimization problem and its solution. Table 4 lists the different variables that are part of the general model formulation.

Table 4. Variables for general model formulation.

Variable	Type	Description
$FI_i^{s,c}$	Process, continuous, dependent	Total inlet flowrate of a component c ∈ **COMP** within a process stream s ∈ **STR** into process i ∈ **PROCESS**.
$FO_i^{s,c}$	Process, continuous, dependent	Total outlet flowrate of a component c ∈ **COMP** within a process stream s ∈ **STR** out of process i ∈ **PROCESS**.
$X_{i,j}^{s,c}$	Process, continuous, dependent	Flowrate of a component c ∈ **COMP** within a process stream s ∈ **STR** going from any block i ∈ **I** to another block j ∈ **I**.
$SF_{i,j}^{s}$	Process, continuous, independent	Split factor of a process stream s ∈ **STR** going from any block i ∈ **I** to another block j ∈ **I**.
z_i	Process, binary, independent	Binary variable that dictates whether a certain process i ∈ **PROCESS** exists or not. $z_i \in \{0,1\}$
CC_i	Economic, continuous, dependent	Capital cost of a certain process i ∈ **PROCESS** in USD (USD 2019)
OC_i	Economic, continuous, dependent	Operating cost of a certain process i ∈ **PROCESS** in USD/yr (USD 2019).

The relationship between the different process variables is represented graphically in Figure 3. For a given process, the flowrate of each component c in an inlet stream s is calculated by summing all the individual flowrates of the same component and stream from the preceding blocks of that process. On the other hand, the individual flowrate of a certain component in a stream going from a certain block to a subsequent one is dictated by a split factor $SF_{i,j}^{s}$ ranging from 0 to 1 which is specific to each stream, origin process, and destination block. These concepts are mathematically represented in Equations (2)–(7). Equation (8) forces the total split factors originating from a certain process, which are equal to zero in case the process is not chosen. Similarly, Equation (9) forces the total sludge dry solids inlet flowrate to a certain process equal to zero in case the process is not chosen. If the process is selected, this equation ensures the flowrate does not exceed the maximum capacity. Equation (10) forces a minimum flow of 10% of the maximum capacity to enter a certain process if it is selected. The relationship between total inlet flows and outlet flows of relevant components and streams of a certain process is discussed for each block in the next sections.

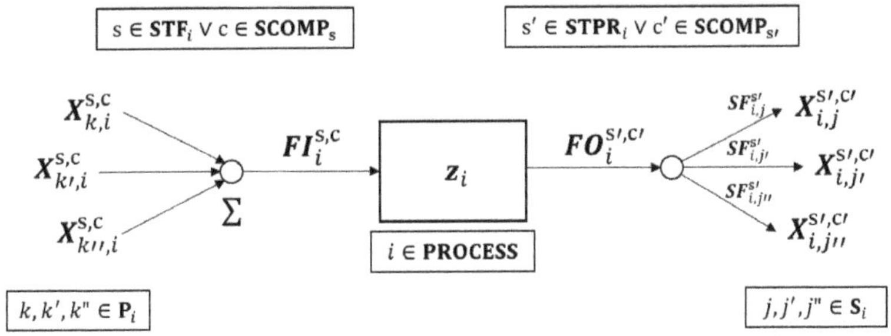

Figure 3. Graphical representation of relationships between the model's process variables.

$$FI_i^{s,c} = \sum_{k \in \mathbf{P}_i} X_{k,i}^{s,c}, \ \forall \, (i \in \mathbf{PROCESS} \wedge s \in \mathbf{STF}_i \wedge c \in \mathbf{SCOMP}_s) \quad (2)$$

$$FI_i^{s,c} = 0, \ \forall \, (i \in \mathbf{PROCESS} \wedge (s \notin \mathbf{STF}_i \vee c \notin \mathbf{SCOMP}_s)) \quad (3)$$

$$FO_i^{s,c} = 0, \ \forall \, (i \in \mathbf{PROCESS} \wedge (s \notin \mathbf{STPR}_i \vee c \notin \mathbf{SCOMP}_s)) \quad (4)$$

$$X_{i,j}^{s,c} = SF_{i,j}^{s} * FO_i^{s,c}, \ \forall \, (i \in \mathbf{PROCESS} \wedge j \in \mathbf{S}_i \wedge s \in \mathbf{STPR}_i \wedge c \in \mathbf{SCOMP}_s) \quad (5)$$

$$0 \leq SF_{i,j}^{s} \leq 1, \ \forall (i \in \mathbf{PROCESS} \wedge j \in \mathbf{S}_i \wedge s \in \mathbf{STPR}_i) \quad (6)$$

$$SF^{s}_{i,j} = 0,\ \forall\ (i \in \mathbf{PROCESS} \wedge (s \notin \mathbf{STF}_j,\ j \in \mathbf{S}_i)\ ) \tag{7}$$

$$\sum_{j \in \mathbf{S}_i} SF^{s}_{i,j} = z_i,\ \forall\ (i \in \mathbf{PROCESS} \wedge s \in \mathbf{STPR}_i) \tag{8}$$

$$FI^{s,DS}_{i} \le z_i * CAP_i,\ \forall\ (i \in \mathbf{PROCESS} \wedge s \in \mathbf{STF}_i) \tag{9}$$

$$FI^{s,DS}_{i} \ge 0.1 * z_i * CAP_i,\ \forall\ (i \in \mathbf{PROCESS} \wedge s \in \mathbf{STF}_i) \tag{10}$$

It should be noted that the sets, parameters, variables, and equations stated above are not conclusive of all the mathematical model formulations. So, more sets, parameters, variables, and equations specific to each block are defined in the next sections.

2.3.2. Thickened Sludge Block

The thickened sludge block represents the feed stream that is distributed among the different subsequent alternatives. Table 5 lists all the parameters that are exclusively relevant to this block.

Table 5. Model elements applicable to the thickened sludge block.

Symbol	Type	Description	Units/Set Elements
$FTHS$	Parameter	Flowrate of thickened sludge to be processed in tons of dry solids per day	tDS/day
$FPVS$	Parameter	Feed volatile solids mass percentage of total dry solids flowrate	%
$FPASH$	Parameter	Ash mass percentage of dry solids	%
PDS_{THS}	Parameter	Dry solids mass percentage of total sludge flowrate	%

Equations (11)–(13) define the total flowrates of the various components in the thickened sludge stream. Equations (14)–(18) define the individual flowrates of those components going to any of the applicable descendant blocks.

$$FO^{THS,VS}_{TH} = FPVS * FTHS \tag{11}$$

$$FO^{THS,ASH}_{TH} = FPASH * FTHS \tag{12}$$

$$FO^{THS,DS}_{TH} = FTHS \tag{13}$$

$$X^{THS,\ VS}_{TH,j} = SF^{THS}_{TH,j} * FO^{THS,VS}_{TH},\ \forall\ j \in \mathbf{S}_{TH} \tag{14}$$

$$X^{THS,\ ASH}_{TH,j} = SF^{THS}_{TH,j} * FO^{THS,ASH}_{TH},\ \forall\ j \in \mathbf{S}_{TH} \tag{15}$$

$$X^{THS,\ DS}_{TH,j} = X^{THS,\ VS}_{TH,j} + X^{THS,\ ASH}_{TH,j},\ \forall\ j \in \mathbf{S}_{TH} \tag{16}$$

$$X^{THS,W}_{TH,j} = X^{THS,\ DS}_{TH,j} * \frac{1 - PDS_{THS}}{PDS_{THS}},\ \forall\ j \in \mathbf{S}_{TH} \tag{17}$$

$$\sum_{j \in \mathbf{S}_{TH}} SF^{THS}_{TH,j} = 1 \tag{18}$$

2.3.3. Anaerobic Digestion Blocks

The anaerobic digestion blocks convert the thickened sludge stream into electricity, which is exported to the grid or used onsite. Meanwhile, the digested sludge is sent to any of the available dewatering options. Table 6 lists all the model elements that are exclusively relevant to this block.

Table 6. Model elements applicable to anaerobic digestion blocks.

Symbol	Type	Description	Units/Set Elements
AD	Set	Subset of anaerobic digestion blocks **DW** ⊂ **PROCESS**	$\{MAD, MADT\}$
$VSD_i,\ i \in \mathbf{AD}$	Parameter	Volatile solids destruction percentage	%
$Y_i^E,\ i \in \mathbf{AD}$	Parameter	Yield of net electricity per ton of dry volatile solids destructed	kWh/tVSD

Equations (19)–(21) define the yield of each component in the outlet product stream of digested sludge, while Equation (22) defines the second outlet product stream of electricity generated from biogas utilization.

$$\boldsymbol{FO}_i^{\mathbf{ADS,VS}} = \boldsymbol{FI}_i^{\mathbf{THS,VS}} * (1 - VSD_i),\ \ \forall i \in \mathbf{AD} \tag{19}$$

$$\boldsymbol{FO}_i^{\mathbf{ADS,ASH}} = \boldsymbol{FI}_i^{\mathbf{THS,ASH}},\ \ \forall i \in \mathbf{AD} \tag{20}$$

$$\boldsymbol{FO}_i^{\mathbf{ADS,DS}} = \boldsymbol{FO}_i^{\mathbf{ADS,VS}} + \boldsymbol{FO}_i^{\mathbf{ADS,ASH}},\ \ \forall i \in \mathbf{AD} \tag{21}$$

$$\boldsymbol{FO}_i^{\mathbf{E,\ E}} = \boldsymbol{FI}_i^{\mathbf{THS,VS}} * VSD_i * Y_i^E,\ \ \forall i \in \mathbf{AD} \tag{22}$$

Moreover, Equations (23) and (24) define the capital and operating costs of any of the anaerobic digestion blocks, respectively.

$$\boldsymbol{CC}_i = BCC_i * \left(\frac{\boldsymbol{FI}_i^{\mathbf{THS,DS}}}{BQ_i}\right)^{\alpha_i},\ \ \forall i \in \mathbf{AD} \tag{23}$$

$$\boldsymbol{OC}_i = POC_i * \boldsymbol{FI}_i^{\mathbf{THS,DS}} * DPY,\ \ \forall i \in \mathbf{AD} \tag{24}$$

2.3.4. Dewatering Blocks

The function of dewatering blocks is to reduce the moisture content of the influent sludge after being conditioned with a certain chemical that enhances its dewaterability. Three dewatering methods are available in the superstructure, namely centrifuge, belt press, and filter press, each of which is capable of achieving a different degree of cake dryness. For each dewatering method, a distinct block is modelled depending on the type of feed sludge entering, undigested thickened sludge, or anaerobically digested sludge. The subsequent processing step/destination differs depending on the dewatering method and its feed. Table 7 lists all the sets, variables, and parameters that are relevant to this block.

Table 7. Model elements applicable to dewatering blocks.

Symbol	Type	Description	Units/Set Elements
DW	Set	Subset of dewatering processes **DW** ⊂ **PROCESS**.	$\{CU, CD, BPU, BPD, FPU, FPD\}$
CH$_i$	Set	Set of matching a certain chemical conditioning stream s ∈ **CHEM** to a corresponding dewatering process i ∈ **DW**.	{P} for $i = CU, CD, BPU,$ *and* BPD {L, FC} for $i = FPU,$ *and* FPD
DR^s	Parameter	Dosage rate of conditioning chemical stream s ∈ **CHEM**.	ton/tDS
PDS_i	Parameter	Percentage of total dry solids in dewatering process i ∈ **DW**.	%
$\boldsymbol{CH}_i^s$	Variable	Flowrate of conditioning chemical s ∈ **CHEM** to a certain dewatering technology i ∈ **DW**	ton/day

In Equation (25), the flowrate of the relevant conditioning chemical to a certain dewatering process is defined as a function of the sludge dry solids flowrate multiplied by the dosage rate parameter. Equations (26)–(29) define the yield of each component in the outlet product stream of the dewatered sludge.

$$CH_i^s = \sum_{s' \in \mathbf{STF}_i} FI_i^{s',\mathbf{DS}} * DR^s, \ \forall (i \in \mathbf{DW} \wedge s \in \mathbf{CH}_i) \quad (25)$$

$$FO_i^{\mathbf{DWS,VS}} = \sum_{s' \in \mathbf{STF}_i} FI_i^{s',\mathbf{VS}}, \ \forall i \in \mathbf{DW} \quad (26)$$

$$FO_i^{\mathbf{DWS,ASH}} = \sum_{s' \in \mathbf{STF}_i} FI_i^{s',\mathbf{ASH}} + \sum_{s \in \mathbf{CH}_i} CH_i^s, \ \forall i \in \mathbf{DW} \quad (27)$$

$$FO_i^{\mathbf{DWS,DS}} = FO_i^{\mathbf{DWS,VS}} + FO_i^{\mathbf{DWS,ASH}}, \ \forall i \in \mathbf{DW} \quad (28)$$

$$FO_i^{\mathbf{DWS,W}} = FO_i^{\mathbf{DWS,DS}} * \frac{1 - PDS_i}{PDS_i}, \ \forall i \in \mathbf{DW} \quad (29)$$

Equations (30) and (31) define the capital and operating costs of the various sludge dewatering blocks, respectively.

$$CC_i = BCC_i * \left(\frac{FI_i^{s,\mathbf{DS}}}{BQ_i} \right)^{\alpha_{DW}}, \ \forall (i \in \mathbf{DW} \wedge s \in \mathbf{STF}_i) \quad (30)$$

$$OC_i = POC_{DW} * FI_i^{s,\mathbf{DS}} * DPY, \ \forall (i \in \mathbf{DW} \wedge s \in \mathbf{STF}_i) \quad (31)$$

2.3.5. Thermal Drying Block

The thermal drying block further reduces the moisture content in the sludge using heat. A single block is modelled to receive sludge from the existing dewatering blocks. The dried sludge is routed to the possible subsequent options, namely, pyrolysis and/or being sold as Class A biosolids fertilizer. The model elements applicable to the thermal drying block are identified in Table 8.

Table 8. Model elements for the thermal drying block.

Symbol	Type	Description	Units
PDS_{TD}	Parameter	Percentage of total dry solids from thermal drying	%
FWE_{TD}	Variable	Total flowrate of water evaporated in the thermal dryer	tonH_2O/day

Equations (32)–(35) define the yield of each component in the outlet product stream of the thermally dried sludge. Equation (36) defines the amount of water/moisture evaporated in the dryer, which is a key parameter in sizing the dryer for cost estimating.

$$FO_{TD}^{\mathbf{TDS,VS}} = FI_{TD}^{\mathbf{DWS,VS}} \quad (32)$$

$$FO_{TD}^{\mathbf{TDS,ASH}} = FI_{TD}^{\mathbf{DWS,ASH}} \quad (33)$$

$$FO_{TD}^{\mathbf{TDS,DS}} = FI_{TD}^{\mathbf{DWS,DS}} \quad (34)$$

$$FO_{TD}^{\mathbf{TDS,W}} = FO_{TD}^{\mathbf{TDS,DS}} * \frac{1 - PDS_{TD}}{PDS_{TD}} \quad (35)$$

$$FWE_{TD} = FI_{TD}^{\mathbf{DWS,W}} - FO_{TD}^{\mathbf{TDS,W}} \quad (36)$$

Equations (37) and (38) define the capital and operating costs of the thermal drying block, respectively.

$$CC_{TD} = BCC_{TD} * \left(\frac{FWE_{TD}}{BQ_{TD}} \right)^{\alpha_{TD}} \quad (37)$$

$$OC_{TD} = POC_{TD} * FWE_{TD} * DPY \quad (38)$$

Table 6. Model elements applicable to anaerobic digestion blocks.

Symbol	Type	Description	Units/Set Elements
AD	Set	Subset of anaerobic digestion blocks **DW** ⊂ **PROCESS**	{*MAD, MADT*}
VSD_i, $i \in \mathbf{AD}$	Parameter	Volatile solids destruction percentage	%
$Y_i^{\mathbf{E}}$, $i \in \mathbf{AD}$	Parameter	Yield of net electricity per ton of dry volatile solids destructed	kWh/tVSD

Equations (19)–(21) define the yield of each component in the outlet product stream of digested sludge, while Equation (22) defines the second outlet product stream of electricity generated from biogas utilization.

$$FO_i^{\mathbf{ADS,VS}} = FI_i^{\mathbf{THS,VS}} * (1 - VSD_i), \ \forall\, i \in \mathbf{AD} \tag{19}$$

$$FO_i^{\mathbf{ADS,ASH}} = FI_i^{\mathbf{THS,ASH}}, \ \forall\, i \in \mathbf{AD} \tag{20}$$

$$FO_i^{\mathbf{ADS,DS}} = FO_i^{\mathbf{ADS,VS}} + FO_i^{\mathbf{ADS,ASH}}, \ \forall\, i \in \mathbf{AD} \tag{21}$$

$$FO_i^{\mathbf{E,\,E}} = FI_i^{\mathbf{THS,VS}} * VSD_i * Y_i^E, \ \forall\, i \in \mathbf{AD} \tag{22}$$

Moreover, Equations (23) and (24) define the capital and operating costs of any of the anaerobic digestion blocks, respectively.

$$CC_i = BCC_i * \left(\frac{FI_i^{\mathbf{THS,DS}}}{BQ_i}\right)^{\alpha_i}, \ \forall\, i \in \mathbf{AD} \tag{23}$$

$$OC_i = POC_i * FI_i^{\mathbf{THS,DS}} * DPY, \ \forall\, i \in \mathbf{AD} \tag{24}$$

2.3.4. Dewatering Blocks

The function of dewatering blocks is to reduce the moisture content of the influent sludge after being conditioned with a certain chemical that enhances its dewaterability. Three dewatering methods are available in the superstructure, namely centrifuge, belt press, and filter press, each of which is capable of achieving a different degree of cake dryness. For each dewatering method, a distinct block is modelled depending on the type of feed sludge entering, undigested thickened sludge, or anaerobically digested sludge. The subsequent processing step/destination differs depending on the dewatering method and its feed. Table 7 lists all the sets, variables, and parameters that are relevant to this block.

Table 7. Model elements applicable to dewatering blocks.

Symbol	Type	Description	Units/Set Elements
DW	Set	Subset of dewatering processes **DW** ⊂ **PROCESS**.	{*CU, CD, BPU, BPD, FPU, FPD*}
$\mathbf{CH}_i$	Set	Set of matching a certain chemical conditioning stream s ∈ **CHEM** to a corresponding dewatering process $i \in \mathbf{DW}$.	{P} for $i = CU, CD, BPU, and\ BPD$ {L, FC} for $i = FPU, and\ FPD$
DR^s	Parameter	Dosage rate of conditioning chemical stream s ∈ **CHEM**.	ton/tDS
PDS_i	Parameter	Percentage of total dry solids in dewatering process $i \in \mathbf{DW}$.	%
CH_i^s	Variable	Flowrate of conditioning chemical s ∈ **CHEM** to a certain dewatering technology $i \in \mathbf{DW}$	ton/day

In Equation (25), the flowrate of the relevant conditioning chemical to a certain dewatering process is defined as a function of the sludge dry solids flowrate multiplied by the dosage rate parameter. Equations (26)–(29) define the yield of each component in the outlet product stream of the dewatered sludge.

$$CH_i^s = \sum_{s' \in \mathbf{STF}_i} FI_i^{s',DS} * DR^s, \ \forall \left(i \in \mathbf{DW} \wedge \ s \in \mathbf{CH}_i\right) \tag{25}$$

$$FO_i^{DWS,VS} = \sum_{s' \in \mathbf{STF}_i} FI_i^{s',VS}, \ \forall i \in \mathbf{DW} \tag{26}$$

$$FO_i^{DWS,ASH} = \sum_{s' \in \mathbf{STF}_i} FI_i^{s',ASH} + \sum_{s \in \mathbf{CH}_i} CH_i^s, \ \forall i \in \mathbf{DW} \tag{27}$$

$$FO_i^{DWS,DS} = FO_i^{DWS,VS} + FO_i^{DWS,ASH}, \ \forall i \in \mathbf{DW} \tag{28}$$

$$FO_i^{DWS,W} = FO_i^{DWS,DS} * \frac{1 - PDS_i}{PDS_i}, \ \forall i \in \mathbf{DW} \tag{29}$$

Equations (30) and (31) define the capital and operating costs of the various sludge dewatering blocks, respectively.

$$CC_i = BCC_i * \left(\frac{FI_i^{s,DS}}{BQ_i}\right)^{\alpha_{DW}}, \ \forall \left(i \in \mathbf{DW} \wedge \ s \in \mathbf{STF}_i\right) \tag{30}$$

$$OC_i = POC_{DW} * FI_i^{s,DS} * DPY, \ \forall \left(i \in \mathbf{DW} \wedge \ s \in \mathbf{STF}_i\right) \tag{31}$$

2.3.5. Thermal Drying Block

The thermal drying block further reduces the moisture content in the sludge using heat. A single block is modelled to receive sludge from the existing dewatering blocks. The dried sludge is routed to the possible subsequent options, namely, pyrolysis and/or being sold as Class A biosolids fertilizer. The model elements applicable to the thermal drying block are identified in Table 8.

Table 8. Model elements for the thermal drying block.

Symbol	Type	Description	Units
PDS_{TD}	Parameter	Percentage of total dry solids from thermal drying	%
FWE_{TD}	Variable	Total flowrate of water evaporated in the thermal dryer	$tonH_2O/day$

Equations (32)–(35) define the yield of each component in the outlet product stream of the thermally dried sludge. Equation (36) defines the amount of water/moisture evaporated in the dryer, which is a key parameter in sizing the dryer for cost estimating.

$$FO_{TD}^{TDS,VS} = FI_{TD}^{DWS,VS} \tag{32}$$

$$FO_{TD}^{TDS,ASH} = FI_{TD}^{DWS,ASH} \tag{33}$$

$$FO_{TD}^{TDS,DS} = FI_{TD}^{DWS,DS} \tag{34}$$

$$FO_{TD}^{TDS,W} = FO_{TD}^{TDS,DS} * \frac{1 - PDS_{TD}}{PDS_{TD}} \tag{35}$$

$$FWE_{TD} = FI_{TD}^{DWS,W} - FO_{TD}^{TDS,W} \tag{36}$$

Equations (37) and (38) define the capital and operating costs of the thermal drying block, respectively.

$$CC_{TD} = BCC_{TD} * \left(\frac{FWE_{TD}}{BQ_{TD}}\right)^{\alpha_{TD}} \tag{37}$$

$$OC_{TD} = POC_{TD} * FWE_{TD} * \ DPY \tag{38}$$

2.3.6. Incineration Block

The incineration block is modelled to have a single input which is the sludge (either digested or not), and two outputs which are net electricity generated and the residual ash. The net electricity generated is calculated in two steps. First, the heat losses from the incinerator and the heat required for moisture evaporation are both subtracted from the lower heating value of the sludge. This difference resembles the recovered heat in the waste heat boiler which generates steam. The second step is to multiply the calculated steam enthalpy by the efficiency of the Rankine cycle to obtain the produced net electricity. Model elements relevant to the incineration block are listed in Table 9. It should be noted that the LHV of the sludge can be impacted by the addition of lime as a conditioner for the filter press dewatering step. Lime stabilizes/inhibits some of the volatile solids in the sludge; the exact value of the reduction is uncertain and will be subject to sensitivity analysis.

Table 9. Model elements applicable to the incineration block.

Symbol	Type	Description	Units
LHV_{VS}	Parameter	Lower heating value parameter (coefficient) for sludge	MJ/tVDS
λ_W	Parameter	Latent heat of vaporization of water	MJ/ton
HLF	Parameter	Heat Loss Factor in the incinerator	Dimensionless
η_R	Parameter	Efficiency of the Rankine cycle	%
CF_{MJ2kWh}	Parameter	Conversion factor of MJ to kWh	Dimensionless
H_{INC}^{VS}	Variable	Heat flow of volatile solids entering the incineration block	kWh(th)/day
H_{INC}^{W}	Variable	Heat required to evaporate moisture in sludge entering the incineration block	kWh(th)/day
H_{INC}	Variable	Net heat recovered from incineration	kWh(th)/day

Equation (39) specifies that the yield of the ash produced out of incineration is equal to that fed from the incoming dewatered sludge. Equation (10) defines the heat content of the sludge based on its volatile solids content. Equation (41) calculates the amount of heat required to evaporate the moisture content of the sludge. Equation (43) defines the net amount of electricity which can be recovered via a Rankine cycle using the heat input calculated in Equation (42).

$$\boldsymbol{FO_{INC}^{ASH,ASH}} = \boldsymbol{FI_{INC}^{DWS,ASH}} \tag{39}$$

$$\boldsymbol{H_{INC}^{VS}} = LHV_{VS} * \boldsymbol{FI_{INC}^{DWS,VS}} * CF_{MJ2kWh} \tag{40}$$

$$\boldsymbol{H_{INC}^{W}} = \lambda_W * \boldsymbol{FI_{INC}^{DWS,W}} * CF_{MJ2kWh} \tag{41}$$

$$\boldsymbol{H_{INC}} = \left(\boldsymbol{H_{INC}^{VS}} - \boldsymbol{H_{INC}^{W}}\right) * (1 - HLF) \tag{42}$$

$$\boldsymbol{FO_{INC}^{E,E}} = \boldsymbol{H_{INC}} * \eta_R \tag{43}$$

Equations (44) and (45) define the capital and operating costs of the incineration block, respectively.

$$CC_{INC} = BCC_{INC} * \left(\frac{\boldsymbol{FI_{INC}^{DWS,DS}}}{BQ_{INC}}\right)^{\alpha_{INC}} + 1147\left(\boldsymbol{FO_{INC}^{E,E}}\right)^{0.695} \tag{44}$$

$$OC_{INC} = (POC_{INC} * \boldsymbol{FI_{INC}^{DWS,DS}} + POC_{ST} * \boldsymbol{FO_{INC}^{E,E}}) * DPY \tag{45}$$

where ST refers to a steam turbine unit for electricity generation.

2.3.7. Gasification Block

For modelling the gasification block, the thermal drying unit is included inside its boundaries. The reason behind this assumption is the recycling of heat from syngas combustion is utilized to both dry the sludge and provide the necessary heat for the gasifier (overall endothermic reaction). Modelling the blocks separately with recycle streams will be challenging. So, in order to simplify the problem, the units are combined together since there are available data in the literature about net electricity generated from such a configuration.

The effect of moisture in the dewatered sludge entering the gasification block on the produced net electricity is negligible at moisture contents below 80% [52]. Therefore, the model will be insensitive on whether the sludge to gasification is from belt press, or from filter press dewatering units.

The yield of net electricity obtained in [52] was for certain conditions, i.e., sludge composition, temperature, pressure, and sludge drying level. For our modelling purposes, all the conditions are assumed to remain the same for the optimized design, except for the sludge composition (%volatile solids). Accordingly, the net power produced in [52] will be divided by the amount of volatile solids entering in this study and is assumed to increase linearly with %VS. The only additional parameter to be defined that is exclusively applicable to gasification is YF_{GN}^{E}, which resembles the yield of net electricity in kWh per ton dry volatile solids fed to the gasifier.

Equations (46) and (47) define the yields of gasification products as net electricity and ash, respectively.

$$\boldsymbol{FO}_{GN}^{\mathbf{E,E}} = \boldsymbol{FI}_{GN}^{\mathbf{DWS,VS}} * YF_{GN}^{E} \tag{46}$$

$$\boldsymbol{FO}_{GN}^{\mathbf{ASH,ASH}} = \boldsymbol{FI}_{GN}^{\mathbf{DWS,ASH}} \tag{47}$$

Equations (48) and (49) define the capital and operating costs of the gasification block.

$$CC_{GN} = BCC_{GN} * \left(\frac{\boldsymbol{FI}_{GN}^{\mathbf{DWS,DS}}}{BQ_{GN}}\right)^{\alpha_{GN}} \tag{48}$$

$$OC_{GN} = POC_{GN} * \boldsymbol{FI}_{GN}^{\mathbf{DWS,DS}} * DPY \tag{49}$$

2.3.8. Pyrolysis Block

For modelling purposes, linear empirical equations were found in the literature that predict the yield of both bio-oil and biochar [20]. The yield parameters are a function of the percentage of volatile solids and total dry solids entering the pyrolysis reactor. The yield of syngas is usually not accounted for since it is of negligible heating value. Accordingly, the syngas stream is excluded from our model. The correction factors CF_{PY}^{BO} and CF_{PY}^{BC} are added to Equations (46) and (47), respectively, to account for the uncertainty in the coefficients of the empirical equation upon doing the sensitivity analysis for model parameters. Pyrolysis is an endothermic reaction, so the need for an auxiliary fuel exists to reach the required operating conditions. The heat duty required is calculated as the summation of the heat of drying any residual moisture, the sensible heat to reach reaction temperature, and the heat of reaction. From the calculated heat duty, the amount of required natural gas is obtained to satisfy the energy balance and also to estimate the operating cost parameters for the unit.

Equations (50) and (51) define the yields of fast pyrolysis products, bio-oil and biochar, in their respective order.

$$FO_{PY}^{\mathbf{BO,BO}} = \left(63.68\% * FI_{PY}^{\mathbf{TDS,VS}} - 11.34\% * FI_{PY}^{\mathbf{TDS,DS}}\right) * CF_{PY}^{BO} \tag{50}$$

$$FO_{PY}^{\mathbf{BC,BC}} = -78.95\% * FI_{PY}^{\mathbf{TDS,VS}} + 98.79\% * FI_{PY}^{\mathbf{TDS,DS}} * CF_{PY}^{BC} \tag{51}$$

Equations (52) and (53) define the capital and operating costs of the pyrolysis block, respectively.

$$CC_{PY} = BCC_{PY} * \left(\frac{FI_{PY}^{\mathbf{DWS,DS}}}{BQ_{PY}}\right)^{\alpha_{PY}} \tag{52}$$

$$OC_{PY} = POC_{PY} * FI_{PY}^{\mathbf{DWS,DS}} * DPY \tag{53}$$

2.3.9. SCWO and SCWG blocks

The SCWO and SCWG blocks are modelled in a way to simply convert the volatile solids portion of the sludge fed into the block to the energy product of each unit. Table 10 lists the yield parameters defined for each of these two processes in the model.

Table 10. Model parameters applicable to the SCWO and SCWG blocks.

Symbol	Type	Description	Units
YF_{SCO}^{E}	Parameter	Yield of net electricity per ton dry volatile solids fed to the SCWO block	kWh/tVS
YF_{SCG}^{H2}	Parameter	Yield of hydrogen per ton dry volatile solids fed to the SCWG block	kgH2/tVS

Equation (54) defines the electricity product yield from SCWO while Equation (56) defines that of hydrogen from SCWG. Equations (55) and (57) calculate the ash product yield from SCWO and SCWG, respectively.

$$FO_{SCO}^{\mathbf{E,E}} = FI_{SCO}^{\mathbf{DWS,VS}} * Y_{SCO}^{E} \tag{54}$$

$$FO_{SCO}^{\mathbf{ASH,ASH}} = FI_{SCO}^{\mathbf{DWS,ASH}} \tag{55}$$

$$FO_{SCG}^{\mathbf{H2,H2}} = FI_{SCG}^{\mathbf{DWS,VS}} * Y_{SCG}^{H2} \tag{56}$$

$$FO_{SCG}^{\mathbf{ASH,ASH}} = FI_{SCG}^{\mathbf{DWS,ASH}} \tag{57}$$

The economic variables of capital and operating costs of SCWO and SCWG are defined in Equations (58) and (59).

$$CC_{SCO} = BCC_{SCO} * \left(\frac{FI_{SCO}^{\mathbf{DWS,DS}}}{BQ_{SCO}}\right)^{\alpha_{SCO}} \tag{58}$$

$$OC_{SCO} = POC_{SCO} * FI_{SCO}^{\mathbf{DWS,VS}} * DPY \tag{59}$$

$$CC_{SCG} = BCC_{SCG} * \left(\frac{FI_{SCG}^{\mathbf{DWS,DS}}}{BQ_{SCG}}\right)^{\alpha_{SCG}} \tag{60}$$

$$OC_{SCG} = POC_{SCG} * FI_{SCG}^{\mathbf{DWS,DS}} * DPY \tag{61}$$

2.3.10. Objective Function

The objective function to be minimized in the optimization problem is the net annual cost defined in Equation (62), which is the difference between annual costs and annual revenues. The annual costs comprise of the total annualized capital costs (defined by Equations (63) and (64)), total annual operating costs of the optimal pathway technologies chosen by the model, as in Equation (65), and the total disposal costs of the produced byproducts (Equation (66)). The total annual revenue from the sales of final products is specified in Equation (67). The total flow of each final product or byproduct is defined in Equation (68) and it is used in the calculation of revenue and disposal costs variables. The model elements for defining the equations related to the objective function are listed in Table 11.

Table 11. Model elements applicable to the objective function definition.

Symbol	Type	Description	Units/Set Elements
REVP	Set	Subset of revenue-generating products **REVP** ⊂ **PRODUCT**	{*E, Fert, BO, BC, H2*}
DISP	Set	Subset of cost-incurring products to be disposed **DISP** ⊂ **PRODUCT**	{*DS20, DS40, ASH*}
NETCOST	Variable	Objective function variable to be minimized representing the net production cost of the chosen pathway	USD/yr (USD 2019).
TACC	Variable	Total annualized capital costs of the chosen processes in the optimal pathway.	USD/yr (USD 2019).
TOC	Variable	Total annualized operating costs of the chosen processes in the optimal pathway.	USD/yr (USD 2019).
TADC	Variable	Total annual disposal costs from the disposal of final byproducts.	USD/yr (USD 2019).
TREV	Variable	Total revenues from selling of final products.	USD/yr (USD 2019).
FPI_i	Variable	Total flowrate of a final product $i \in$ **PRODUCT**	unit product/day
AF	Parameter	Annualized capital charge ratio	dimensionless
d	Parameter	Interest/discount rate	%
n	Parameter	Number of years of the project life	yr
SP_i	Parameter	Price of selling of a final product $i \in$ **REVP**.	USD/unit product
DC_i	Parameter	Disposal cost of a final product $i \in$ **DISP**.	USD/unit product

$$NETCOST = TACC + TOC + TADC - TREV \quad (62)$$

$$TACC = AF * \sum_{i \in \mathbf{PROCESS}} z_i * CC_i \quad (63)$$

$$AF = \frac{d * (1 + d)^n}{(1 + d)^n - 1} \quad (64)$$

$$TOC = \sum_{i \in \mathbf{PROCESS}} z_i * OC_i \quad (65)$$

$$TADC = \sum_{i \in \mathbf{DISP}} FPI_i * DC_i * DPY \quad (66)$$

$$TREV = \sum_{i \in \mathbf{REVP}} FPI_i * SP_i * DPY \quad (67)$$

$$FPI_i = \sum_{s \in \mathbf{STF}_i} \sum_{c \in \mathbf{FPCO}_i} \sum_{k \in \mathbf{P}_i} X_{k,i}^{s,c}, \ \forall i \in \mathbf{PRODUCT} \quad (68)$$

3. Case Study

3.1. Case Study Parameters

The parameters in the mathematical model formulation discussed above were given appropriate values for the purpose of performing a case study for a sludge treatment plant. Those values were either reasonably assumed or extracted from various sources to help illustrate the applications of the developed optimization model. Table 12 lists the feed property-related parameters, while Table 13 lists the capital and operating costs for each technology. Moreover, Table 14 presents the selling prices and disposal costs of the final products, and finally, Table 15 lists the process performance-related parameters. As the capital and operating costs were gathered from different sources in the literature that vary in currency and year of study, those values were adjusted for inflation, using the Chemical Engineering Plant Cost Index (CEPCI) [53], for the year 2019, and were then converted to US dollars (USD/$) for consistency. The economies of scale exponent α_i was assumed to be 0.6 for all the technologies. The case study was evaluated for a project lifetime (n) of 20 years at an interest/discount rate (d) of 7.5% [15]. Continuous chemical processing plants typically operate 8000 h per year; hence, the DPY parameter was assumed to be 333 days/year.

Table 12. Feed Property Parameters.

Parameter	Value	Units
$FTHS$	100	tDS/day
$FPVS$	70	%
$FPASH$	30	%
PDS_{THS}	5	%

Table 13. Capital and Operating Costs of Technologies.

Technology	BCC_i (MMUSD)	BQ_i (tDS/day)	POC_i (USD/tDS)	Ref.
MAD	31.86	100	52	[54]
MADT	33.26	100	62	[54]
CD	2.16	50	58	[35]
CU	2.16	50	58	[35]
BPD	6.6	50	69	[35]
BPU	6.6	50	69	[35]
FPD	8.2	50	134	[35]
FPU	8.2	50	134	[35]
TD	12.59	480 *	26 **	[55]
INC	34.62	130	95	[56,57]
GN	2.09	5	154	[58]
PY	8.26	50	100	[57]
SCO	9	14	113 ***	Correspondence with SCFI [59]
SCG	18.44	24	175	[32]

* tH_2O(evaporated)/day; ** USD/tH_2O (evaporated) ; *** USD/tVS.

Table 14. Final Product Disposal Costs and Selling Prices.

Final Product	DC_i (USD/ton)	SP_i (USD/ton)	Ref.
DS20	250	N/A	[56]
DS40	125	N/A	[56]
ASH	77	N/A	[56]
E	N/A	0.08	[60]
FERT	N/A	30	[61]
BO	N/A	285 *	[62]
BC	N/A	200	[63]
H2	N/A	2 **	[64]

* Assuming a price equivalent to 70% of crude oil of price ≈ 60 USD/bbl. ** USD/kg.

Table 15. Technology Performance-Related Parameters.

Parameter	Value	Units	Ref.
VSD_{MAD}	50	%	[54]
Y^{E}_{MAD}	2390	kWh/tVSD	[54]
VSD_{MADT}	60	%	[54]
Y^{E}_{MADT}	2390	kWh/tVSD	[54]
DR^{P}	0.004	ton/tDS	[34]
DR^{L}	0.1	ton/tDS	[34]
DR^{FC}	0.07	ton/tDS	[34]
PDS_{CU}, PDS_{CD}	10	%	[35]
PDS_{BPU}, PDS_{BPD}	20	%	[35]
PDS_{FPU}, PDS_{FPD}	40	%	[35]
PDS_{TD}	90	%	Typical
LHV_{VS}	21,000	MJ/tVDS	[65]
λ_{W}	2260	MJ/tonne	Steam Table
HLF	0.05	Dimensionless	assumed
η_{R}	25	%	[57]
CF_{MJ2kWh}	0.27778	Dimensionless	
YF^{E}_{GN}	1368	kWh/tVS	[58]
CF^{BO}_{PY}	1	Dimensionless	
CF^{BC}_{PY}	1	Dimensionless	
YF^{E}_{SCO}	825	kWh/tVS	Correspondence with SCFI [59]
YF^{H2}_{SCG}	112	kgH2/tVS	[32]

3.2. Sensitivity Analysis

The model economic parameters used in this case study were subject to several sources of uncertainty rooting from: (1) the inconsistent basis for factors used in capital cost calculations, (2) the assumption that the operating costs varied linearly with the processing capacity, (3) volatility of products selling prices and the market demand, and (4) uncertainty in possible government incentives for each technology. The model technical performance-related parameters were also prone to a level of uncertainty due to: (1) the infancy of some of the new developed technologies, (2) different sludge characteristics that the original sources relied on, (3) scalability issues, etc. Consequently, the need to assess the model sensitivity to each of the relevant uncertain parameters was a necessity. The capital and operating costs of each technology were usually estimated from preliminary techno-economic studies for feasibility purposes. This type of study corresponded to a Class 4 cost estimate as defined by the Association for the Advancement of Cost Engineering (AACE) and could have an accuracy between +50% and −30% [66]. This range was used for the sensitivity analysis on capital and operating costs parameters.

In this study, selling prices were assessed for the following ranges: Electricity price from 6 to 30 cents per kWh, fertilizer price from 20 to 100 USD/ton, bio-oil from 100 to 500 USD/ton (15 to 70 USD/bbl of bio-oil), biochar from 100 to 500 USD/ton, and H_2 from 1 to 5 USD/kg. Disposal costs of dewatered sludge and ash varied from 25 to 75 USD/wet ton of solids and from 40 to 100 USD/ton of ash, respectively. The discount rate also varied between 5% and 10%. As far as performance-related parameters were concerned, yield parameters and sludge LHV varied by ±30%. The percentage of dry solids produced from belt press and filter press dewatering varied between 12–37% and 27–46%, respectively; these ranges covered the whole spectrum of dewatering efficiencies for those two technologies to account for extreme cases of sludge composition variations. Finally, the feed characteristics were examined as follows: inlet flowrates from 50 tDS/day to 150 tDS/day and composition of sludge VS% from 50% to 80% with a corresponding Ash% of 50% to 20%. In this study, the results of the sensitivity analysis are reported to understand how a change in objective function value was accompanied by a change in the optimal processing pathway.

3.3. *Results and Discussion*

3.3.1. Base Case Results

The proposed optimization model formulation together with the case study parameters values were entered into GAMS software. The BARON solver [67] was used to solve the MINLP model guaranteeing global optimality within a reasonable runtime. The model and solver characteristics are summarized in Table 16.

Table 16. Model and Solver Characteristics.

Model Statistics		Solver Statistics	
Single Equations	635	Solver	BARON
Single Variables	418	Optimality Tolerance	10^{-6}
Non-linear matrix entries	371	Branch-and-reduce iterations	41
Discrete Variables	14	Max. no. of nodes in memory	21
Non-zero elements	1700	CPU Time (s)	70.72

The optimal processing pathway was determined by looking at the results of both the discrete variable z_i and the continuous variable $\boldsymbol{SF^s_{i,j}}$, where the former states the choice of a certain technology, and the latter foresees whether a certain technology product stream is split between more than one destination. The non-zero z variables obtained in the solution were for process identifiers ***FPU, TD, PY*** with the final products being bio-oil and biochar as per the split factor results. Figure 4 shows a schematic for the optimal processing route with stream flowrates.

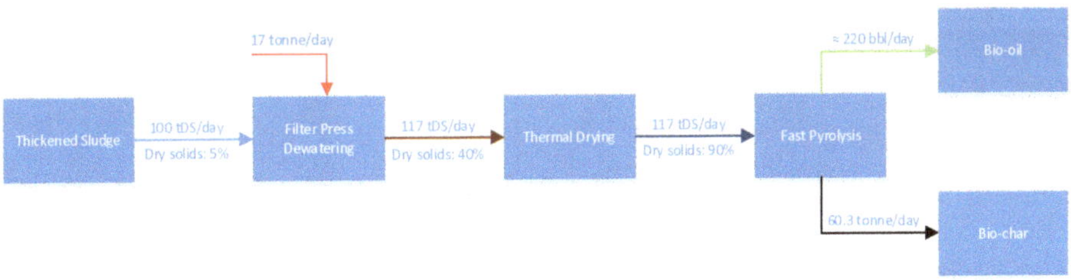

Figure 4. Optimal Solution Pathway with product flowrates.

The annual net cost for this pathway was approximately 6 million USD/year for a daily load of 100 tDS of sludge, while the specific cost of sludge treatment came to 180 USD/tDS. This cost of treatment per ton of dry sludge was in the same order of magnitude with reported ranges between 100 and 800 USD/tDS of various conventional sludge handling methods (i.e., landfilling, land application, and incineration) [68]. This implies that the parameters used for the case study were reliable enough for demonstrating the applicability of the proposed optimization model. The annual costs were close to 13 million USD/year where 75% of that cost was attributed to the operating costs of the different technologies and the remaining 25% was related to the annualized capital cost payments. The annual revenues were 7 million USD/year with bio-oil sales contributing to 43% of the total revenue, and a biochar share of 57%. Figure 5a shows the revenues from product sales of the selected pathway in comparison to the total costs. Figure 5b illustrates the breakdown of costs between the different technologies in the selected pathway. It is worth mentioning that the operating cost of the filter press dewatering process accounted for the highest portion of the total costs with 34%, followed by pyrolysis operating costs with 30%. The capital cost of the dewatering step was also comparable to that of the pyrolysis. This shows how significant and important the dewatering step is in the whole processing route.

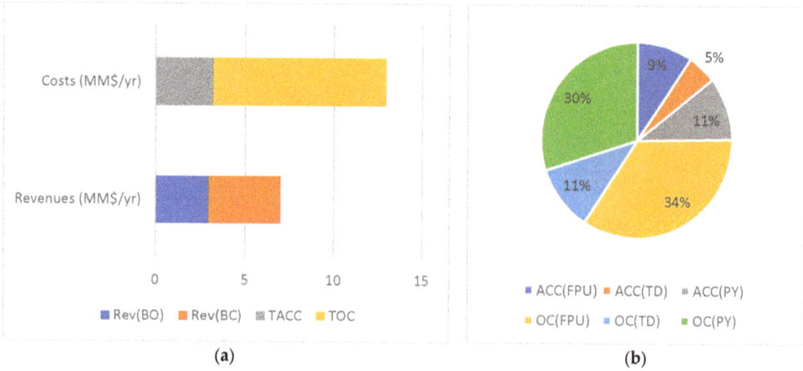

Figure 5. (**a**) Optimal Pathway Costs and Revenues (**b**) Optimal Pathway Costs Breakdown.

3.3.2. Sensitivity Analysis Results

Feed Characteristics

- Feed Flowrate

The changes of feed flowrate in the studied ranges, from 50 to 150 tDS/day, did not have any impact on the optimal processing route which was selected in the base case scenario. Nevertheless, there were obvious changes in the value of the objective function or the net cost value as presented in Figure 6. To see the effect of economies of scale, the percentage of net cost increase per a 10 tDS/day rise in feed flowrate was added in the same plot. As expected, the percentage of additional net annual costs decreased with increasing the capacity from 15% per each extra 10 tDS/day at 60 tDS/day capacity, to 5.7% at 150 tDS/day.

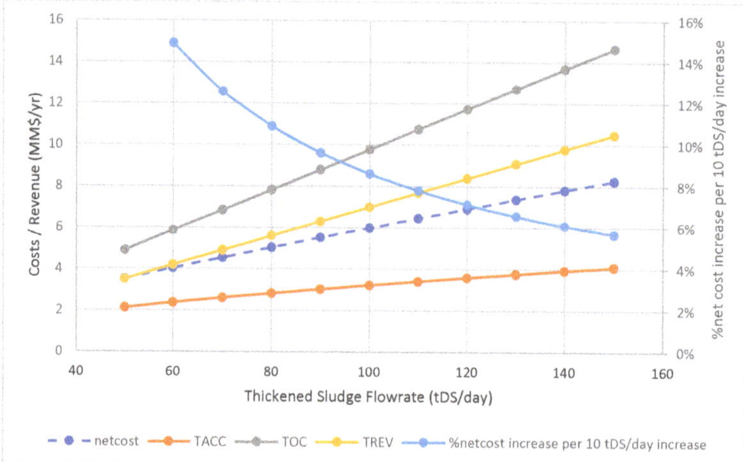

Figure 6. Sensitivity Analysis of Economic Variables (MMUSD/yr) with Feed Flowrate.

However, it is not yet clear from the above results which of the net cost components (i.e., ***TACC***, ***TOC***, and ***TREV***) had the most underlying impact on the net cost reduction. Therefore, another method to examine the capacity effects is presented, which aimed to conduct the comparison against the various economic variables, but per unit ton of dry sludge treated. The specific costs/revenue variables are suffixed by the asterisk symbol (*) and were plotted against different feed flowrate values as shown in Figure 7. It was observed that the specific operating costs and revenues in USD/tDS were constant across the whole spectrum of capacities studied at values of 293 USD/tDS and 210 USD/tDS, respectively. Thus, these two components did not play a significant role in the economies of scale. On the other hand, the annualized capital costs per unit of sludge exponentially reduced with increasing capacities. Hence, they were the sole drivers behind the changes in the specific net cost results. The rate of change in specific annual capital costs, and accordingly, that of specific net costs, decelerated with increasing capacity from a 4% reduction per each extra 10 tDS/day of feed flowrate at an initial capacity of 60 tDS/day to 1.5% at 150 tDS/day. This indicates that the effects of scaling economies are minimal at capacities higher than 150–200 tDS/day and that the specific net costs will asymptote at values close to 155–160 USD/tDS.

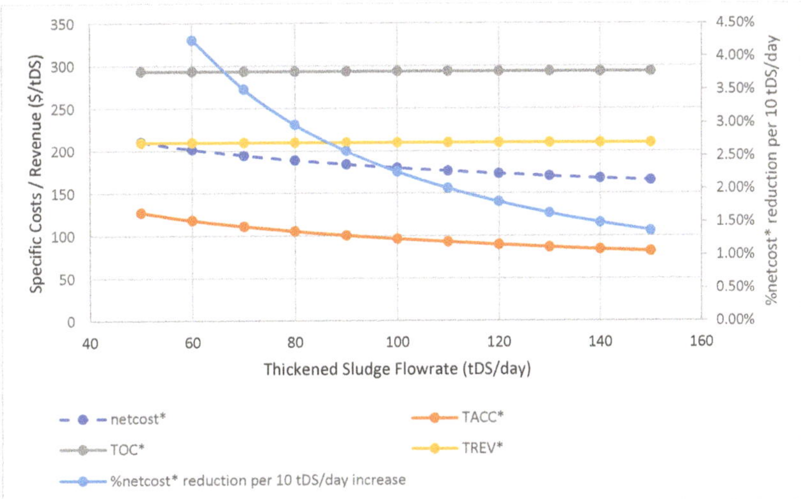

Figure 7. Sensitivity Analysis of Specific Economic Variables (USD/tDS) with Feed Flowrate.

- Feed Composition

The impact of changing the composition of the sewage sludge on the objective function was seen to be minimal. As shown in Figure 8, a 5% increase in the sludge volatile solids percentage only led to a reduction of approximately 0.65% in the net cost primarily due to the revenue increases from higher bio-oil yields at the expense of biochar. Thus, the maximum variation expected in the net cost for a composition change of 30% (i.e., from 50% to 80%) is close to 3.8% which is not trivial if compounded annually; however, it does not undermine the feasibility of the processing route if a certain wastewater treatment facility is generating sludge with lower organic contents. The optimal processing pathway did not change with varying the composition and this is visible also in the figure from the constant capital and operating costs that were dependent mainly on the total amount of dry solids, regardless of their components analysis for the chosen processing route.

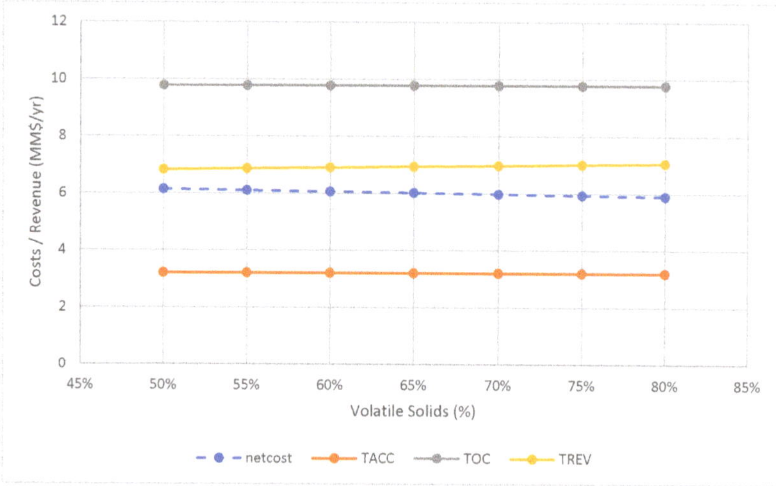

Figure 8. Sensitivity Analysis of Economic Variables (MMUSD/yr) with Feed Composition.

Economic Parameters

- Capital and Operating Costs

The objective function value is sensitive to any capital and/or operating cost variations of all the technologies present in the optimal pathway of the base case, namely: ***FPU***, ***TD***, and ***PY***. The percentage of change in capital cost and operating cost parameters of those technologies were plotted against the % change in the objective function (compared to base case results) and demonstrated in Figures 9 and 10, respectively. The objective function value plateaued after an increase greater than 20% for the pyrolysis operating cost; this was because at such value the optimization model decides to discard pyrolysis technology from the optimal pathway and the thermally dried biosolids are chosen to be sold as fertilizers instead of being further processed. Similarly, with an increase in ***FPU*** operating costs higher than 10%, the net cost stagnates and the optimal pathway changes to ***BPU***, ***TD***, ***PY***. Reductions in ***FPU*** operating costs had the most significant impacts on net cost where at −30%, the corresponding decrease in objective function value was 22%.

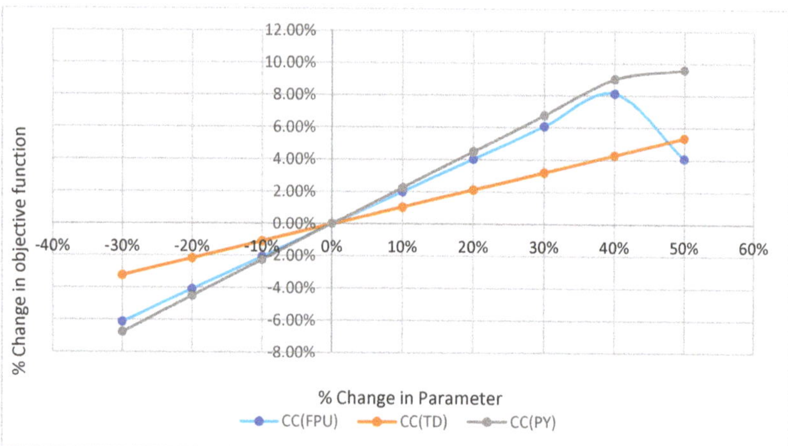

Figure 9. Sensitivity Analysis of Optimal Pathway Technologies' Capital Cost Parameters.

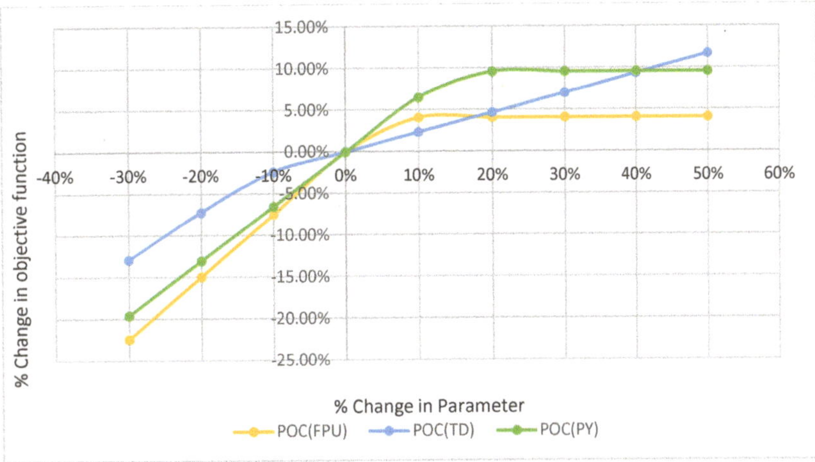

Figure 10. Sensitivity Analysis of Optimal Pathway Technologies' Operating Cost Parameters.

MADT was the only technology outside the base case optimal pathway that the capital cost parameter of which had an impact on the net cost (objective function). This impact appeared only at a 30% reduction in ***MADT*** capital cost, which led to a 4% decrease in the net cost, and a different optimal pathway, as shown in Figure 11 where the thickened sludge underwent thermal hydrolysis pre-treatment and anaerobic digestion before eventually being sent to the pyrolysis pathway of the base case. This adds an additional energy product in the form of electricity recovered from MAD biogas.

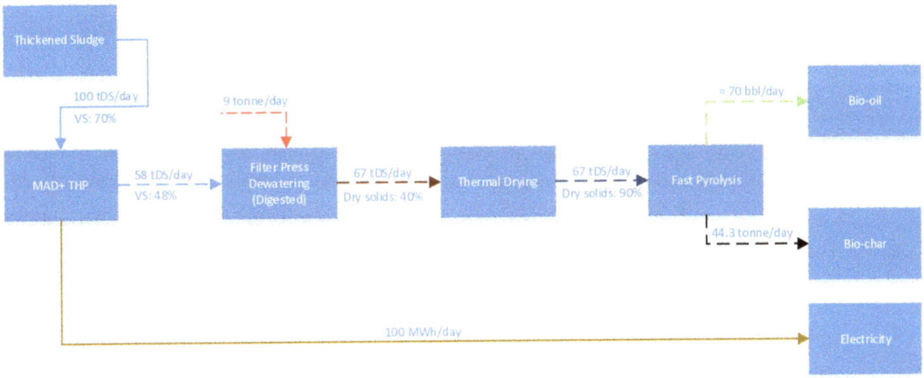

Figure 11. Optimal Pathway at 30% Reduction in MADT Capital Cost.

As far as the remaining technologies' operating cost parameters are concerned, at −30% for ***FPD***, a slight reduction in the net cost of approximately 1% can be attributed to a corresponding optimal processing pathway similar to that presented in Figure 11. Furthermore, reductions in operating costs of ***BPU*** by 20% and 30% corresponded to the objective function value dropping by 7% and 3.5%, respectively. The resulting optimal pathway was similar to that of the base case, except that ***FPU*** was replaced by ***BPU***, which produced a sludge of 20% solids in comparison to the 40% solids produced by ***FPU***. The remaining operating cost parameters for the other technologies were found to be insensitive to both the objective function as well as the optimal pathway choice.

- Product Selling and Disposal Prices

All the final products' selling prices had an impact on the objective function value when compared to the base case. Electricity prices appeared to be the most sensitive parameter compared with the remaining final products. The highest price studied for electricity (30 cents/kWh) brought the net cost down by −130% compared to the base case results, which actually led to having a net profit at such a rate (Figure 12). The optimal pathway chosen at this electricity price was ***MADT, BPD, GN***. Such a price was much higher than the average prices for industrial use, thus accounting for more optimistic scenarios of government incentives to electricity from waste such as feed-in-tariff (FIT) and/or tax credits policies. Followed by electricity, the selling prices of biochar and hydrogen were the second most sensitive to changes. At their higher limits (500 USD/ton of biochar and 5 USD/kg H_2), they caused a reduction close to 100% of the objective function value compared to the base case scenario. The optimal pathway for biochar's highest price stayed the same as the base case, while for H_2 it changed to ***CU*** + ***SCG*** even at prices as low as 3 USD/kgH_2. This shows the promising potential of that technology, especially with the future higher demands of a sustainable hydrogen economy. The objective function value was least sensitive to the prices of bio-oil and Class A biosolids fertilizer relative to the remaining products. However, significant reductions of approximately 40% to the net costs were achievable at the higher end of the studied price range.

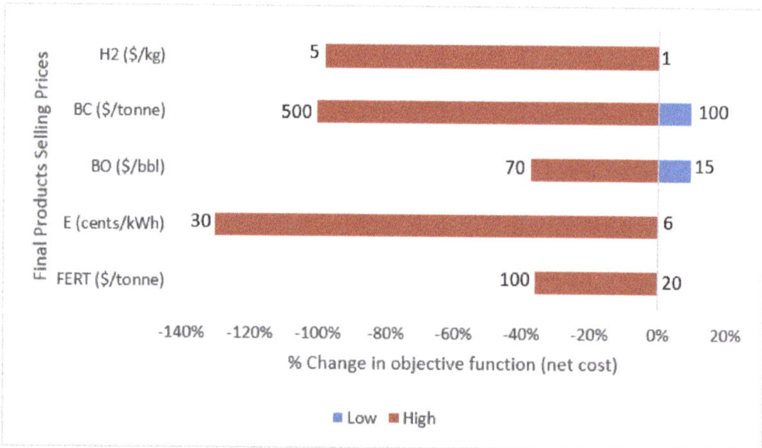

Figure 12. Sensitivity Analysis of Final Product Selling Prices.

On the other hand, changes in the disposal costs of the by-products (i.e., dewatered sludge, ash) did not impact on either the optimal pathway selected by the model nor the objective function value.

- Discount Rate

The objective function value was sensitive to variations of the discount rate *d*; however, the optimal pathway stayed the same. As shown in Figure 13, an incremental change of ±0.5% in *d*, led to a ±2% change in the objective function, which is a significant change. This suggests that at higher rates of inflation trends, the investment in such projects can be less attractive without further incentives.

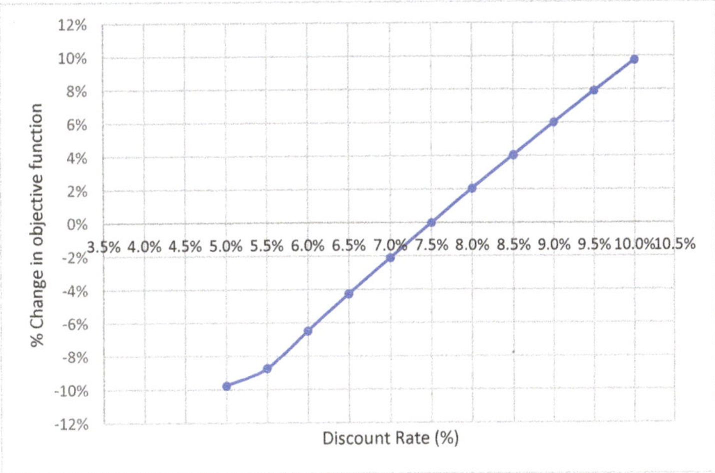

Figure 13. Sensitivity Analysis of the Discount Rate.

Performance-Related Parameters

Changing the yield parameters of products of ***MAD***, ***MADT***, ***GN***, ***INC***, ***SCO***, and ***SCG*** by ±30% had no effect on both the objective function and optimal processing pathway. This indicates the robustness of the pyrolysis pathway against a wide range of process efficiencies of the competing technologies. As far as pyrolysis product yields are concerned, an inverse proportion relationship existed between them and the objective function value. As shown in Figure 14, a 10% increase in bio-oil yield caused a 5% decrease in net cost and vice versa. However, at bio-oil yield reductions below 20%, the optimal pathway changed to only filter press dewatering followed by thermal drying, where the dried sludge could be sold as fertilizer causing no more additional reduction in net costs. A similar relationship existed between biochar yield and the objective function value. However, a 10% increase in the yield caused a 6.7% reduction in the net cost. Biochar yield reduction increased the net cost also by 6.7% until the objective function reduction value stagnated at any yield reduction below 15% as shown in Figure 15. This indicates that the objective function value was more sensitive to changes in the biochar yield than bio-oil which was expected based on the sensitivity analysis results of products' selling prices as it was discussed earlier.

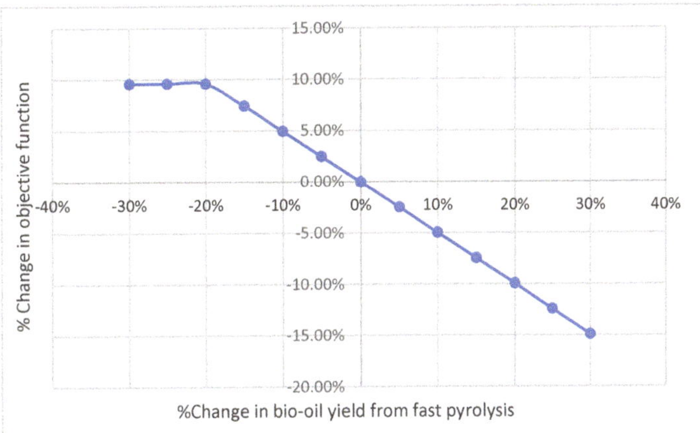

Figure 14. Sensitivity Analysis of the Bio-oil Yield Parameter.

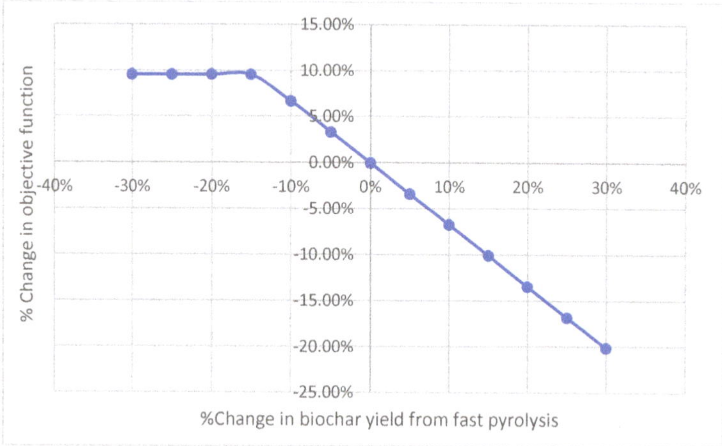

Figure 15. Sensitivity Analysis of the Biochar Yield Parameter.

The efficiency of the dewatering processes was found to have an impact on the results. For belt dewatering, the assessed values were between 12% and 37% with a 5% increment. There was no change in results for dry solids% up to 22%. Starting from 27%, the optimal pathway favored a pathway of ***BPU*** followed by ***TD*** with the final product being Class A biosolids sold as fertilizers. Results for the latter three scenarios are listed in Table 17.

Table 17. Sensitivity analysis results for belt press dewatering efficiency runs.

PDS_{BP} (Cases)	22% (and Lower)	27%	32%	37%
Optimal Pathway	*FPU* + *TD* + *PY*	*BPU* + *TD*	*BPU* + *TD*	*BPU* + *TD*
***TACC* (MMUSD/yr)**	3.21	1.84	1.72	1.62
***TOC* (MMUSD/yr)**	9.77	4.54	4.03	3.67
***TREV* (MMUSD/yr)**	6.99	1.00	1.00	1.00
***NETCOST* (MMUSD/yr)**	5.99	5.37	4.75	4.28

The percentage of dry solids produced from filter press dewatering varied between 27% and 48% with increments of 4%. The objective function value changed for all the assessed values. Optimal processing pathway swapped ***FPU*** in the base case results with ***BPU*** at outlet dry solids values between 27% and 35%. From 39% to 48%, the same pathway as that of the base case remained unchanged. Table 18 shows the summary of economic parameter values at those different values.

Table 18. Sensitivity analysis results for filter press dewatering efficiency runs.

PDS_{FP} (Cases)	35% (and Lower)	39%	43%	48%
Optimal Pathway	BPU + TD + PY	FPU + TD + PY	FPU + TD + PY	FPU + TD + PY
***TACC* (MMUSD/yr)**	3.30	3.23	3.16	3.10
***TOC* (MMUSD/yr)**	9.01	9.83	9.59	9.39
***TREV* (MMUSD/yr)**	6.08	6.99	6.99	6.99
***NETCOST* (MMUSD/yr)**	6.24	6.07	5.76	5.50

It can be concluded from the results of both dewatering processes that more efficient belt press dewatering can yield higher cost savings compared to filter press dewatering assuming the same capital and operating costs with the higher efficiency.

The last studied performance-related parameter was the LHV value of the sludge, which could potentially have an impact on favoring the incineration technology at higher values. Nevertheless, even at an increase of 30% of LHV value, neither the objective function value nor the chosen optimal pathway changed.

3.3.3. Complementary Analysis

It can be observed from the above results that the following technologies were not selected in any of the studied scenarios under the sensitivity analysis: ***MAD, INC,*** and ***SCO***. Hence, additional runs were performed on GAMS software for further investigation. In each run, the binary variable z_i of one of these technologies was forced to equal 1, while all other conversion technologies z_i values were set to 0. Table 19 lists the processing pathway resulting from each run as well as the change in the different components of the objective function compared to the base case scenario results. It is clearly evident from the results that the main driver for the lower objective function value of the base case scenario in comparison to the other three pathways was the higher annual revenues from selling biochar and bio-oil compared to electricity. In addition, there were no additional disposal costs required for pyrolysis products compared to the dewatered sludge from ***MAD*** and ash from ***INC*** and ***SCO*** that required transportation and disposal expenses. The operating and maintenance cost of all three pathways were significantly lower when compared to the base case; however, this offset the remaining objective function components.

Table 19. Additional Runs Results Summary.

Processing Pathway	Δ *TACC*	Δ *TOC*	Δ *TADC*	Δ *TREV*	Δ *NETCOST*
MAD + ***FPD***	+0.80 MMUSD/yr [+25%]	−5.13 MMUSD/yr [−52%]	+3.17 MMUSD/yr [N/A]	−4.76 MMUSD/yr [−68%]	+3.65 MMUSD/yr [+61%]

4. Conclusions

The main objective of this study was to propose a decision-making support tool for choosing the most economic pathway of sludge-to-energy technologies via superstructure optimization techniques. A literature review was conducted to assess the state of research in the addressed problem and the gaps were identified in utilizing mixed-integer optimization for sludge-to-energy decision-making frameworks. A mathematical model customized for the problem was formulated and its applicability was tested via a case study for a hypothetical treatment facility with a capacity of 100 tDS/day.

One of the main outcomes from the case study was that although the proposed model is a MINLP formulation, which is usually difficult to solve, global optimal solutions were found efficiently within a reasonable computing time. The base case results showed that a combination of filter press dewatering followed by thermal drying and fast pyrolysis was deemed to be the most economical pathway among the available alternatives. The products of such a pathway were bio-oil, which can be used as an alternative fuel upon refining, and biochar, which has a variety of useful applications such as an adsorption material or in agriculture. The estimated net specific cost for processing a ton of dry sludge using this pathway was USD 180, which is in the same order of magnitude of current ranges of conventional sludge handling methods, but with added environmental and social benefits.

The parameters used in the case study were extracted from various sources in the literature and vendor documents. So, the model parameters defined in the case study were subjected to a high degree of uncertainty related to the reliability and availability of high-quality data in the literature, as well as uncertainties related to the market volatility of the final product prices and government subsidies or incentives that can be provided to such products. Therefore, a sensitivity analysis for wide expected ranges of each of those uncertain parameters was conducted to determine the impact of each individual parameter variation on the objective function value represented by the annual net costs and/or changes in the optimal pathway that is selected by the model. In this regard, the following results were achieved:

- The technology selection route was sensitive to the capital cost parameter of ***MADT***, and operating costs of ***FPD*** and ***BPU***. Changes in the remainder of the technologies' capital and operating cost parameters did not impact the model outputs;
- Variations in the final products' prices also had a significant effect on the selected optimal pathway and the net costs of the selected plant. Electricity price was the most sensitive parameter followed by hydrogen and biochar prices, while bio-oil and Class A biosolids (fertilizer) prices were found to have the least relative effect on the objective function values;
- The objective function values were also sensitive to the value of discount rates; however, the technology selection did not change with reasonable interest variations;
- Changing the yield parameters of technologies other than fast pyrolysis had no influence on the solution. This indicates the robustness of the pyrolysis pathway against a wide range of process efficiencies of the competing technologies;
- The objective function was highly sensitive to all the parameters related to the technologies in the base case optimal pathway, which proved the applicability of the proposed model and provided sensible results;
- The feed characteristics affected the optimal cost value, which was explained by the economies of scale. The inverse relationship between net cost and process capacity effects started to diminish at capacities above 200 tDS/day. There were slight impacts from changing the composition of the sewage sludge where cost reductions were observed at a higher %VS due to an increased yield of energy products correlated with an increase in the organic contents of the sludge.

Finally, it should be noted that although the optimization framework proposed in this study can be used as an early screening tool for decision-makers to assess different sludge-to-energy pathways, it can be further extended to account for different feedstocks and also to consider the effects of environmental constraints (CO_2 emissions).

Author Contributions: Methodology, O.M.; Software, O.M.; Supervision, Q.Z., A.A. and A.E.; Writing—original draft, O.M. and F.H.; Writing—review and editing, F.H., Q.Z. and A.E. All authors have read and agreed to the published version of the manuscript.

Funding: This research received no external funding.

Data Availability Statement: The data presented in this study are available on request from the corresponding author.

Conflicts of Interest: The authors declare no conflict of interest.

References

1. U.S. Environmental Protection Agency. *Wastewater Management Fact Sheet—Energy Conservation*; U.S. Environmental Protection Agency: Washington, DC, USA, 2006.
2. EPRI; WRF. *Electricity Use and Management in the Municipal Water Supply and Wastewater Industries*; EPRI: Palo Alto, CA, USA, 2013.
3. Spinosa, L.; Ayol, A.; Baudez, J.-C.; Canziani, R.; Jenicek, P.; Leonard, A.; Rulkens, W.; Xu, G.; Van Dijk, L. Sustainable and Innovative Solutions for Sewage Sludge Management. *Water* **2011**, *3*, 702–717. [CrossRef]
4. Basic Information about Biosolids | US EPA. Available online: https://www.epa.gov/biosolids/basic-information-about-biosolids (accessed on 26 February 2022).
5. Rulkens, W.H.; Bien, J.D. Recovery of energy from sludge—Comparison of the various options. *Water Sci. Technol.* **2004**, *50*, 213–221. [CrossRef]
6. Pérez-Elvira, S.I.; Nieto Diez, P.; Fdz-Polanco, F. Sludge minimisation technologies. *Rev. Environ. Sci. Biotechnol.* **2006**, *5*, 375–398. [CrossRef]
7. Zhang, L.; Xu, C.; Champagne, P.; Mabee, W. Overview of current biological and thermo-chemical treatment technologies for sustainable sludge management. *Waste Manag. Res.* **2014**, *32*, 586–600. [CrossRef]
8. Raheem, A.; Sikarwar, V.S.; He, J.; Dastyar, W.; Dionysiou, D.D.; Wang, W.; Zhao, M. Opportunities and challenges in sustainable treatment and resource reuse of sewage sludge: A review. *Chem. Eng. J.* **2018**, *337*, 616–641. [CrossRef]

9. Adar, E.; Karatop, B.; Ince, M.; Bilgili, M.S. Comparison of methods for sustainable energy management with sewage sludge in Turkey based on SWOT-FAHP analysis. *Renew. Sustain. Energy Rev.* **2016**, *62*, 429–440. [CrossRef]
10. Santibañez-Aguilar, J.E.; González-Campos, J.B.; Ponce-Ortega, J.M.; Serna-González, M.; El-Halwagi, M.M. Optimal Planning of a Biomass Conversion System Considering Economic and Environmental Aspects. *Ind. Eng. Chem. Res.* **2011**, *50*, 8558–8570. [CrossRef]
11. Kim, J.; Sen, S.M.; Maravelias, C.T. An optimization-based assessment framework for biomass-to-fuel conversion strategies. *Energy Environ. Sci.* **2013**, *6*, 1093–1104. [CrossRef]
12. Rizwan, M.; Lee, J.H.; Gani, R. Optimal processing pathway for the production of biodiesel from microalgal biomass: A superstructure based approach. *Comput. Chem. Eng.* **2013**, *58*, 305–314. [CrossRef]
13. Rizwan, M.; Lee, J.H.; Gani, R. Optimal design of microalgae-based biorefinery: Economics, opportunities and challenges. *Appl. Energy* **2015**, *150*, 69–79. [CrossRef]
14. Satchatippavarn, S.; Martinez-Hernandez, E.; Leung Pah Hang, M.Y.; Leach, M.; Yang, A. Urban biorefinery for waste processing. *Chem. Eng. Res. Des.* **2016**, *107*, 81–90. [CrossRef]
15. Rizwan, M.; Saif, Y.; Almansoori, A.; Elkamel, A. Optimal processing route for the utilization and conversion of municipal solid waste into energy and valuable products. *J. Clean. Prod.* **2018**, *174*, 857–867. [CrossRef]
16. Ma, J.; You, F. Superstructure optimization of thermal conversion based poultry litter valorization process. *J. Clean. Prod.* **2019**, *228*, 1111–1121. [CrossRef]
17. Tyagi, V.K.; Lo, S.L. Sludge: A waste or renewable source for energy and resources recovery? *Renew. Sustain. Energy Rev.* **2013**, *25*, 708–728. [CrossRef]
18. Stone, L.; Kuchenrither, R.; Quintanilla, A.; Torres, E.; Groome, M.; Pfeifer, T.; Dominak, R.; Taylor, D. *Renewable Energy Resources: Banking on Biosolids*; NACWA: Washington, DC, USA, 2010.
19. Fonts, I.; Azuara, M.; Gea, G.; Murillo, M.B. Study of the pyrolysis liquids obtained from different sewage sludge. *J. Anal. Appl. Pyrolysis* **2009**, *85*, 184–191. [CrossRef]
20. Cao, Y.; Pawłowski, A. Sewage sludge-to-energy approaches based on anaerobic digestion and pyrolysis: Brief overview and energy efficiency assessment. *Renew. Sustain. Energy Rev.* **2012**, *16*, 1657–1665. [CrossRef]
21. Appels, L.; Baeyens, J.; Degrève, J.; Dewil, R. Principles and potential of the anaerobic digestion of waste-activated sludge. *Prog. Energy Combust. Sci.* **2008**, *34*, 755–781. [CrossRef]
22. Syed-Hassan, S.S.A.; Wang, Y.; Hu, S.; Su, S.; Xiang, J. Thermochemical processing of sewage sludge to energy and fuel: Fundamentals, challenges and considerations. *Renew. Sustain. Energy Rev.* **2017**, *80*, 888–913. [CrossRef]
23. Manara, P.; Zabaniotou, A. Towards sewage sludge based biofuels via thermochemical conversion—A review. *Renew. Sustain. Energy Rev.* **2012**, *16*, 2566–2582. [CrossRef]
24. Oladejo, J.; Shi, K.; Luo, X.; Yang, G.; Wu, T. A review of sludge-to-energy recovery methods. *Energies* **2019**, *12*, 60. [CrossRef]
25. Fytili, D.; Zabaniotou, A. Utilization of sewage sludge in EU application of old and new methods—A review. *Renew. Sustain. Energy Rev.* **2008**, *12*, 116–140. [CrossRef]
26. Samolada, M.C.; Zabaniotou, A.A. Comparative assessment of municipal sewage sludge incineration, gasification and pyrolysis for a sustainable sludge-to-energy management in Greece. *Waste Manag.* **2014**, *34*, 411–420. [CrossRef]
27. Khiari, B.; Marias, F.; Zagrouba, F.; Vaxelaire, J. Analytical study of the pyrolysis process in a wastewater treatment pilot station. *Desalination* **2004**, *167*, 39–47. [CrossRef]
28. Alvarez, J.; Lopez, G.; Amutio, M.; Artetxe, M.; Barbarias, I.; Arregi, A.; Bilbao, J.; Olazar, M. Characterization of the bio-oil obtained by fast pyrolysis of sewage sludge in a conical spouted bed reactor. *Fuel Process. Technol.* **2016**, *149*, 169–175. [CrossRef]
29. Zhang, L.; Xu, C.; Champagne, P. Energy recovery from secondary pulp/paper-mill sludge and sewage sludge with supercritical water treatment. *Bioresour. Technol.* **2010**, *101*, 2713–2721. [CrossRef]
30. Paul, E.; Liu, Y. *Biological Sludge Minimization and Biomaterials/Bioenergy Recovery Technologies*; John Wiley and Sons: Hoboken, NJ, USA, 2012.
31. Svanström, M.; Modell, M.; Tester, J. Direct energy recovery from primary and secondary sludges by supercritical water oxidation. *Water Sci. Technol.* **2004**, *49*, 201–208. [CrossRef]
32. Gasafi, E.; Reinecke, M.Y.; Kruse, A.; Schebek, L. Economic analysis of sewage sludge gasification in supercritical water for hydrogen production. *Biomass Bioenergy* **2008**, *32*, 1085–1096. [CrossRef]
33. Binder, M.; Kraussler, M.; Kuba, M.; Luisser, M.; Rauch, R. *Hydrogen from Biomass Gasification. IEA Bioenergy*; IEA: Paris, France, 2018.
34. Vitorio, C.; Andreoli, M.; von Sperling, F.; Fernandes, F. *Sludge Treatment and Disposal*; IWA Publishing: London, UK, 2007; Volume 6.
35. U.S. Environmental Protection Agency. *Handbook: Estimating Sludge Management Costs*; Water Engineering Research Laboratory: Cincinnati, OH, USA, 1985.
36. Metcalf & Eddy Inc. *Wastewater Engineering: Treatment and Resource Recovery*, 5th ed.; McGraw-Hill Education: New York, NY, USA, 2013.
37. Ding, W.; Li, L.; Liu, J. Investigation of the effects of temperature and sludge characteristics on odors and VOC emissions during the drying process of sewage sludge. *Water Sci. Technol.* **2015**, *72*, 543–552. [CrossRef]

38. Ren, J.; Liang, H.; Dong, L.; Gao, Z.; He, C.; Pan, M.; Sun, L. Sustainable development of sewage sludge-to-energy in China: Barriers identification and technologies prioritization. *Renew. Sustain. Energy Rev.* **2017**, *67*, 384–396. [CrossRef]
39. Liu, Y.; Ren, J.; Man, Y.; Lin, R.; Lee, C.K.M.; Ji, P. Prioritization of sludge-to-energy technologies under multi-data condition based on multi-criteria decision-making analysis. *J. Clean. Prod.* **2020**, *273*, 123082. [CrossRef]
40. Tang, C.; Xu, D.; Chen, N. Sustainability prioritization of sewage sludge to energy scenarios with hybrid-data consideration: A fuzzy decision-making framework based on full consistency method and fusion ranking model. *Environ. Sci. Pollut. Res.* **2020**, *285*, 5548–5565. [CrossRef]
41. Vadenbo, C.; Guillén-Gosálbez, G.; Saner, D.; Hellweg, S. Multi-objective optimization of waste and resource management in industrial networks—Part II: Model application to the treatment of sewage sludge. *Resour. Conserv. Recycl.* **2014**, *89*, 41–51. [CrossRef]
42. U-Tapao, C.; Gabriel, S.A.; Peot, C.; Ramirez, M. Stochastic, Multiobjective, Mixed-Integer Optimization Model for Wastewater-Derived Energy. *J. Energy Eng.* **2015**, *141*, B5014001. [CrossRef]
43. De Meyer, A.; Cattrysse, D.; Ostermeyer, P.; Van Orshoven, J. Implementation of OPTIMASS to optimise municipal wastewater sludge processing chains: Proof of concept. *Resour. Conserv. Recycl.* **2016**, *114*, 168–178. [CrossRef]
44. De Meyer, A.; Cattrysse, D.; Van Orshoven, J. A generic mathematical model to optimise strategic and tactical decisions in biomass-based supply chains (OPTIMASS). *Eur. J. Oper. Res.* **2015**, *245*, 247–264. [CrossRef]
45. Amoedo, R.C.; Damartzis, T.; Granacher, J.; Marechal, F.M.A. System Design and Performance Evaluation of Wastewater Treatment Plants Coupled With Hydrothermal Liquefaction and Gasification. *Front. Energy Res.* **2020**, *8*, 233. [CrossRef]
46. Ng, W.P.Q.; Lam, H.L.; Varbanov, P.S.; Klemeš, J.J. Waste-to-Energy (WTE) network synthesis for Municipal Solid Waste (MSW). *Energy Convers. Manag.* **2014**, *85*, 866–874. [CrossRef]
47. Mohammadi, M.; Harjunkoski, I. Performance analysis of waste-to-energy technologies for sustainable energy generation in integrated supply chains. *Comput. Chem. Eng.* **2020**, *140*, 106905. [CrossRef]
48. Yousefloo, A.; Babazadeh, R. Designing an integrated municipal solid waste management network: A case study. *J. Clean. Prod.* **2020**, *244*, 118824. [CrossRef]
49. Zhao, N.; Lehmann, J.; You, F. Poultry Waste Valorization via Pyrolysis Technologies: Economic and Environmental Life Cycle Optimization for Sustainable Bioenergy Systems. *ACS Sustain. Chem. Eng.* **2020**, *8*, 4633–4646. [CrossRef]
50. Morsy, O. Optimal Processing Pathway for Sludge-to-Energy Technologies: A Superstructure Optimization-Based MINLP Model. Master's Thesis, University of Waterloo, Waterloo, ON, Canada, 2022. Available online: http://hdl.handle.net/10012/18184 (accessed on 1 November 2022).
51. Bertran, M.O.; Frauzem, R.; Sanchez-Arcilla, A.S.; Zhang, L.; Woodley, J.M.; Gani, R. A generic methodology for processing route synthesis and design based on superstructure optimization. *Comput. Chem. Eng.* **2017**, *106*, 892–910. [CrossRef]
52. Lumley, N.P.G. Techno-Economic Analysis of Wastewater Sludge Gasification: A Decentralized Urban Perspective. Master's Thesis, Colorado School of Mines, Golden, CO, USA, 2013. Available online: https://repository.mines.edu/bitstream/handle/11124/12062/Lumley_mines_0052N_10344.pdf?sequence=1&isAllowed=y (accessed on 1 February 2022).
53. The Chemical Engineering Plant Cost Index—Chemical Engineering. Available online: https://www.chemengonline.com/pci-home (accessed on 19 January 2022).
54. Mills, N. Unlocking the Full Energy Potential of Sewage Sludge. Ph.D. Thesis, University of Surrey, Guildford, UK, 2015.
55. Haque, N.; Somerville, M. Techno-economic and environmental evaluation of biomass dryer. *Procedia Eng.* **2013**, *56*, 650–655. [CrossRef]
56. 2011 Sewage Sludge Incineration (SSI) New Source Performance Standards (NSPS) and Emission Guidelines (EG) Source Data Files | US EPA. Available online: https://www.epa.gov/stationary-sources-air-pollution/2011-sewage-sludge-incineration-ssi-new-source-performance (accessed on 19 January 2022).
57. Bridgwater, A.V.; Toft, A.J.; Brammer, J.G. A techno-economic comparison of power production by biomass fast pyrolysis with gasification and combustion. *Renew. Sustain. Energy Rev.* **2002**, *6*, 181–246. [CrossRef]
58. Lumley, N.P.G.; Ramey, D.F.; Prieto, A.L.; Braun, R.J.; Cath, T.Y.; Porter, J.M. Techno-economic analysis of wastewater sludge gasification: A decentralized urban perspective. *Bioresour. Technol.* **2014**, *161*, 385–394. [CrossRef]
59. Hydrothermal Oxidation Technology | AquaCritox | Scfi. Available online: https://www.scfi.eu/water/the-technology/what-is-aquacritox/ (accessed on 3 October 2021).
60. Electric Power Monthly—U.S. Energy Information Administration (EIA). Available online: https://www.eia.gov/electricity/monthly/epm_table_grapher.php?t=epmt_5_6_a (accessed on 20 January 2022).
61. Darvodelsky, P. Biosolids Snapshot. 2012. Available online: https://www.awe.gov.au/sites/default/files/documents/biosolids-snapshot.pdf (accessed on 20 January 2022).
62. Xin, C.; Zhao, J.; Ruan, R.; Addy, M.M.; Liu, S.; Mu, D. Economical feasibility of bio-oil production from sewage sludge through Pyrolysis. *Therm. Sci.* **2018**, *22*, 459–467. [CrossRef]
63. Patel, S.; Kundu, S.; Paz-Ferreiro, J.; Surapaneni, A.; Fouche, L.; Halder, P.; Setiawan, A.; Shah, K. Transformation of biosolids to biochar: A case study. *Environ. Prog. Sustain. Energy* **2019**, *38*, 13113. [CrossRef]
64. Hydrogen Production Costs by Production Source, 2018—Charts—Data & Statistics—IEA. Available online: https://www.iea.org/data-and-statistics/charts/hydrogen-production-costs-by-production-source-2018 (accessed on 20 January 2022).

65. Schaum, C.; Lensch, D.; Cornel, P. Evaluation of the energetic potential of sewage sludge by characterization of its organic composition. *Water Sci. Technol.* **2016**, *73*, 3072–3079. [CrossRef]
66. AACE International. *Cost Estimate Classification System—As Applied in Engineering, Procurement, and Construction for the Process Industries*; AACE International Recommended Practice No. 18R-97 (Rev. 1 March 2016); AACE International: Morgantown, WV, USA, 2016.
67. Tawarmalani, M.; Sahinidis, N.V. A polyhedral branch-and-cut approach to global optimization. *Math. Program.* **2005**, *103*, 225–249. [CrossRef]
68. Peccia, J.; Westerhoff, P. We Should Expect More out of Our Sewage Sludge. *Environ. Sci. Technol.* **2015**, *49*, 8271–8276. [CrossRef]

MDPI AG
Grosspeteranlage 5
4052 Basel
Switzerland
Tel.: +41 61 683 77 34

Sustainability Editorial Office
E-mail: sustainability@mdpi.com
www.mdpi.com/journal/sustainability

www.ingramcontent.com/pod-product-compliance
Lightning Source LLC
Chambersburg PA
CBHW040838250926
51850CB00007BA/1347
* 9 7 8 3 7 2 5 8 1 5 4 2 5 *